CORROSION

FOR STUDENTS OF
SCIENCE AND ENGINEERING

DANGER:Government Wealth WARNING:
CORROSION CAN
SERIOUSLY DAMAGE YOUR WEALTH

CORROSION

FOR STUDENTS OF SCIENCE AND ENGINEERING

Kenneth R Trethewey BSc PhD CChem MRSC MICorrST

Senior Lecturer in Engineering Materials
Royal Naval Engineering College
Manadon, Plymouth, UK

John Chamberlain CEng MIM

Principal Scientific Officer
Admiralty Research Establishment
Portland, Dorset, UK

Longman
Scientific &
Technical

Copublished in the United States with
John Wiley & Sons, Inc., New York

Longman Scientific & Technical,
Longman Group UK Limited,
Longman House, Burnt Mill, Harlow,
Essex CM20 2JE, England
and Associated Companies throughout the world.

Copublished in the United States with
John Wiley & Sons, Inc., 605 Third Avenue, New York, NY 10158

First published 1988
Third impression 1992

British Library Cataloguing in Publication Data
Trethewey, Kenneth R.
  Corrosion: for students of science and engineering.
  1. Corrosion and anti-corrosives
  I. Title   II. Chamberlain, John, *1934–*
  620.1'1223   TA462

ISBN 0-582-45089-6

Library of Congress Cataloging in Publication Data
Trethewey, Kenneth R. (Kenneth Richard), 1950–
  Corrosion for students of science and engineering.

  Bibliography: p.
  Includes index.
  1. Corrosion and anti-corrosives. I. Chamberlain,
  John, 1934–   II. Title.
  TA462.T696   1988   620.1'1223   86-34417
  ISBN 0-470-20794-9 (USA only)

Set in 10/12pt Linotron 202 Times Romen

Produced by Longman Singapore Publishers Pte Ltd
Printed in Singapore

CONTENTS

Contents

PREFACE

This book is an attempt to present in simple form a subject which affects every one of us. Without the use of metals, the advance of society would not have occurred, yet we allow our most valuable natural resources to be wasted through corrosion. In many instances, corrosion is inevitable, but it can be controlled. Engineers have a major contribution to make in reducing the unnecessary levels of corrosion but it is the responsibility of everyone to ensure that society uses its metals to the best advantage. An understanding of corrosion and its control is important for all of us, if only for the selfish reason that it saves us money. Studies of the effects of corrosion in society have shown that we seem unable to learn from our mistakes. This problem can be properly addressed only through education and it is therefore the aim of this book to bring to the attention of as many people as possible the reasons for the occurrence of corrosion and the methods of controlling it.

The subject material is a slight expansion of a course of 25 lectures given to engineering undergraduates. Much of the additional material has been added to enable the book to reach a wider readership. In particular, Chapter 2 contains some elementary material drawn from chemistry, physics, metallurgy and electrical science in an attempt to draw together students from various disciplines who may have a weakness in one or other of the subject areas. Many undergraduate students should be confident enough to omit most of this chapter.

Corrosion is a very practical subject and emphasis is placed on experiments, case histories, questions and answers. The experiments in Chapter 3 are simple and can be performed with the minimum of materials and equipment, but are important to the development of an understanding of corrosion. The authors regard them as an essential part of any teaching course. They have been carefully selected and written in such a way that they can be performed with no previous knowledge of the subject. Students should find them both entertaining and instructive. Alternatively, the teacher may prefer to use them as class demonstrations, in which case they may be easily refined with more dedicated equipment. Other, rather more specific but equally important experiments are placed at intervals throughout the early chapters.

The authors make no apology for the simplifications to the corrosion theory. Our experience has shown that the theory given in many other texts defeats all but the brightest students. We have therefore attempted to reduce what is a very complicated subject to a limited number of logical concepts, presented in a more relaxed style with the aid of simplified equations and figures. The question and answer approach furthers this aim by trying to anticipate students' questions at various points of the discussion.

The case histories which are reported have been collected from many sources, but a large proportion are from the museum collection at the Royal Naval Engineering College. These, together with the authors' own experience, often relate to marine corrosion, but it is considered that any textual bias towards the marine environment should not affect the student's appreciation of corrosion in other environments. A good number of the case histories have been reproduced with the kind permission of Dr Oliver Seibert from his two excellent papers, 'Classic blunders in corrosion protection.' The authors found these papers quite inspirational and are sincerely grateful to Dr Seibert for his assistance.

Section 18.1 contains a selection of worked examples, but solutions to the questions in section 18.2 have not been provided. In most of these cases, the solutions may be lengthy, as well as subjective, and it is expected that these will be discussed with the course teacher.

Some readers may consider that the space allocated to high-temperature corrosion is disproportionately small. In the authors' opinion this area of study is more specialised and, for a text which is intended as a first level of study, the treatment given is deemed to be appropriate.

The authors wish to express their sincere gratitude to all those who have helped in the preparation of this book. Indeed, it might never have been written without the first suggestion by Clive Holly to whom the authors are indebted. In particular, acknowledgement is made to the Captain, Royal Naval Engineering College, Manadon, for permission to use facilities and materials. Support and encouragement by the Dean of RNEC, Captain G C George BSc MSc CEng MIM ndc Royal Navy, are gratefully acknowledged. Dr D J Tighe-Ford BSc PhD CBiol MIBiol kindly allowed us to use the caption to the frontispiece, as well as providing the focus for many lively discussions. Mr D A Sargeant CEng DipEE MIERE AMIEE carefully checked the electrical science section, Dr D L Bartlett BSc MSc PhD assisted with numerous theoretical aspects of the work, and Lieutenant-Commander S J Bates BSc PhD Royal Navy advised on the contents of Chapter 9. Professor R N Parkins BSc PhD DSc FIM of the University of Newcastle upon Tyne kindly gave constructive criticism of Chapter 10, while Dr Peter Scott of the Materials Development Division, AERE, Harwell read the section on corrosion fatigue and kindly gave permission for the reproduction of some of his work. Dr Geoff Backhouse of Subspection Ltd provided useful material concerning cathodic protection, while Lieutenant-

Commander J M Harrison BSc MPhil Royal Navy offered much help with the section on hot corrosion.

Dr Michael Rodgers was sufficiently daring to accept the original idea and then proceeded to offer every possible assistance in the preparation of the book, including psychological support during moments of authors' desperation! The authors are very grateful to him and his team at Longman. Thanks are due also to an anonymous British referee for checking the whole text so carefully, and to Professor M Marek of the Georgia Institute of Technology who read the first draft and made some very helpful suggestions for improvement. The greatest debt of gratitude is owed to Captain J N McGrath MSc PhD MIM Royal Navy, without whose constructive criticism and constant encouragement the book would have been much the poorer.

Finally, we thank our wives for their patience and support, and our friends and colleagues in the Engineering Materials Section at RNEC for forming such a pleasant social and professional environment over many years.

1 CORROSION AND SOCIETY

> Lay not up for yourselves treasures upon earth, where moth and rust doth corrupt, and where thieves break through and steal.
>
> (Matthew 6: 19)

The lessons of history

In 1763 a report fell upon a polished oak desktop in an office of their Lordships of the Admiralty in London.[1] Its concise and detailed script, though of only modest scientific significance at the time, was one of the earliest examples of a solution to a practical engineering problem caused by corrosion. In common with many similar studies carried out since then, however, the essence of its conclusions has been frequently disregarded by engineers throughout the world.

Two years before the appearance of the report the 32-gun frigate, HMS *Alarm* (Fig. 1.1), had had its hull completely covered with a thin copper sheathing. The purpose of the sheathing was twofold. Firstly it was intended to reduce the considerable damage caused by the teredo woodworm, and secondly the well-established toxic property of copper was expected to lessen the speed-killing barnacle growth which always occurred on ships' hulls. After a two-year deployment to the West Indies HMS *Alarm* was beached in order to examine the effects of the experiment. It was soon discovered that the sheathing had become detached from the hull in many places because the iron nails which had been used to fasten the copper to the timbers had been 'much rotted'. Closer inspection revealed that some nails, which were less corroded, were insulated from the copper by brown paper which was trapped under the nail head. The copper had been delivered to the dockyard wrapped in the paper which was not removed before the sheets were nailed to the hull. The obvious conclusion therefore, and the one which was contained in that report of 1763, was that iron should not be allowed direct contact with copper in a sea water environment if severe corrosion of the iron was to be avoided. This type of corrosion by two dissimilar metals in contact was to become known as galvanic corrosion, though it is more precisely called bimetallic, or dissimilar metal, corrosion.

It would seem that the first occasion in which this advice went unheeded was in 1769 when Commodore the Hon. John Byron began a circumnavigation of the globe in his coppered ship, *Dolphin*.[2] In addition to the fright-

1

Fig. 1.1 HMS *Alarm*, the Royal Navy frigate which in 1763 was the subject of the first recorded study of bimetallic corrosion.

ening prospect of an uncharted Coral Sea reef scything through the hull, 'Foul-Weather Jack' lost much sleep because of a thump-thump-thump below the stern windows of his cabin. He recorded in his log that he feared that the *Dolphin's* very loose rudder would drop off at any time for its iron pintles, which were in contact with the copper sheathing, were corroded away to needle thinness. With no prospect of a repair in such a dire event, this was a quite unnecessary worry in addition to the multitude of others with which he had to contend.

Since those times, the phenomenon of bimetallic corrosion, in common with the many other forms of corrosion described in this book, has continued to cause service failures, despite its apparently well-publicised effects. In 1962 a report fell upon a plastic desktop in the British Ministry of Defence. This study bore a striking resemblance to its bicentennial ante-cedent for it revealed that the copper alloy end-plate had fallen off a sea-water evaporator on board a submarine because the steel bolts with which it was secured had effectively dissolved through galvanic action. The remains of the bolts are shown in Fig. 1.2.

In 1982 the nose wheels failed on two Royal Navy Sea Harriers which had returned from the Falklands conflict (Fig. 1.3). Studies showed that the same galvanic action which had been so clear to the scientists who had studied HMS *Alarm* 219 years earlier had occurred between the magnesium alloy wheel hub and its stainless-steel bearing.

These are examples from one industry alone: despite these three cases, the Royal Navy is by no means atypical in its proportion of corrosion fail-

Fig. 1.2 The remains of steel bolts which in 1962 had been used to hold a copper alloy end-plate on to an evaporator on board a Royal Navy submarine.

Fig. 1.3 A Royal Navy Sea Harrier which suffered nose wheel collapse because of bimetallic corrosion between the bearing and the wheel.

ures. Examples abound throughout the engineering world which illustrate quite clearly that, as in all other sections of society, we do not always learn the lessons of past experience.

1.1   RUST

To the great majority of people, *corrosion* means *rust*, an almost universal object of hatred. 'Rust' is, of course, the name which has more recently been specifically reserved for the corrosion of iron, while 'corrosion' is the destructive phenomenon which affects almost all metals. Although iron was not the first metal used by man, it has certainly been the most used, and must have been one of the first with which serious corrosion problems were obtained. It is not, therefore, surprising that the terms corrosion and rust are almost synonymous.

The great Roman philosopher, Pliny, AD23–79, wrote at length about *ferrum corrumpitur*, or spoiled iron,[3] for by his time the Roman Empire had been established as the world's foremost civilisation, a distinction due partly to the extensive use of iron for weaponry and other artefacts. To the fighters of old, rust was something of a mixed blessing. In the eleventh century, a Norman knight, William de Lacey, lost his way during a hunting expedition into the thickly wooded and swampy Vale of Ewas, in Wales.[4] He came across the remains of St David's hermitage whereupon, overcome by an urge to mend his sinful ways, he decided to dedicate the remainder of his days to religious contemplation and rebuilding the chapel. Legend has it that he never for the rest of his life removed his armour. One explanation for this strange behaviour was that it was a self-imposed penance. More likely, however, is that he was prevented from doing so because of corrosion brought about by the dank atmosphere of the valley.

Corrosion of arms and armour has also been advantageous. The techniques of blueing and gilding were frequently used to protect steel objects, for, it was found that the application of a variety of heat treatments created highly protective films of oxide (rust). These, with skill, could turn functional weaponry into beautiful works of art.

The Romans must have been vexed by the susceptibility of iron to rust, for, in true scientific fashion, Pliny asked himself the question: why should iron corrode more easily than other metals? Lacking the ability to investigate the problem experimentally, he arrived at the metaphysical solution that it is because iron is both the best and the worst of man's servants. Although very useful domestically, it is also the metal of war, slaughter and brigandage. Writing about iron arrows, Pliny said,

> It is a great evil that to enable death to reach human beings more quickly we have taught iron how to fly.*

He believed that the nuisance of corrosion compensates for the advantages of the metal because,

---

* Copyright. Reprinted by permission of William Heinemann Ltd and the Loeb Classical Library.

the same benevolence of nature has limited the power of iron by inflicting on it the penalty of rust, and the same foresight has made nothing in the world more mortal than that which is most hostile to mortality.

A remarkable consequence of his philosophy was the technique of corrosion control by religious ceremony. He recorded that this had been used to protect the chains of a suspension bridge built for Alexander the Great, but Pliny was sceptical because the method had failed on previous occasions.

Had Pliny been transported to the twentieth century, he may well have concluded that:

Corrosion Costs Society

It is unfortunate that modern usage of the word **cost** implies only financial penalty. Corrosion **costs** society in three ways:

(a) *It is extremely expensive financially.*

Fact: In 1980 in the US, the Battelle Institute estimated that $70 billion was being lost annually to the American economy because of corrosion.[5]

(b) *It is extremely wasteful of natural resources.*

Fact: It has been calculated that in the UK, 1 tonne of steel is converted completely into rust every 90 seconds. Apart from the waste of metal, the energy required to produce a tonne of steel from iron ore is sufficient to provide an average family home with energy for three months.

(c) *It causes considerable inconvenience to human beings, and sometimes even loss of life.*

Fact: In 1985, the roof of a 13-year-old swimming pool in Switzerland collapsed, killing 12 and injuring others. It is thought that the failure resulted from stress-corrosion cracking of the exposed stainless steel hangers which supported the 200-tonne suspended reinforced concrete roof. The corrosion was probably caused by attack on the stainless steel by chlorine in the atmosphere.[6]

We shall first consider category (a) by describing the monetary considerations of corrosion. Categories (b) and (c) will be combined in section 1.3 under the heading, 'Social implications'.

1.2   MONETARY CONSIDERATIONS

The most recent major study of the cost of corrosion in the UK was carried out by the Government Committee on Corrosion and Protection[7] for its report of 1971. A summary of its findings is reproduced in Table 1.1.

Table 1.1 The UK national cost of corrosion and protection in 1971 (Reproduced by permission of the Controller, HMSO)

Industry	Estimated cost (£m. p.a.)	Estimated potential saving (£m. p.a.)
Building	250	50
Food	40	4
General engineering	110	35
Government departments	55	20
Marine	280	55
Metal refining	15	2
Oil and chemical	180	15
Power	60	25
Transport	350	100
Water	25	4
Total	1365	310

The Committee concluded that the total cost to the national economy was a staggering £1365 million (1971 prices) and about 3.5 per cent of the Gross National Product. Of this, about one-quarter could be saved by better and wider use of well-established corrosion protection techniques. Even more surprisingly, the survey did not include the agricultural industry. This was covered later in a report by the University of Manchester Institute of Science and Technology[8] which reported in 1981 that corrosion was costing the agricultural industry about £600 m. per year and that about half of this could be saved by existing corrosion control technology.

The enormity of corrosion costs such as those in Table 1.1 is at first surprising. How can the cost of replacing corroded components be so great? The answer to this question lies in the popular misconception that the cost of corrosion is only that of replacement. Costs almost always arise in addition to the actual cost of replacement. These are known as *indirect* costs and may occur as a result of any of the following:

(a) *Lost production during a shutdown or as a result of a failure.*

Fact: It costs £5000 per day (1977) to take a typical 400 kV transmission line out of service to deal with corrosion damage.[6]

The replacement of a small corroded pipe in a chemical plant may be effected at a cost of just a few pounds, but the loss of product during the time that the process is stopped can run into thousands of pounds an hour. In the worst possible case the whole plant may be taken out and loss of life may occur, as at Flixborough in 1974 when a bypass assembly introduced into a train of six cyclohexane reactors failed and a catastrophic explosion ensued. Twenty-nine people died.

(b) *High maintenance costs.*

Fact: The cost of repainting the steelwork on the Gravelly Hill,

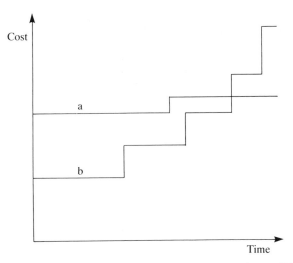

Fig. 1.4 The economics of correct materials selection. a. High initial cost, low maintenance. b. Low initial cost, high maintenance.

Birmingham, motorway interchange over an 18-month period is £1.2 m.[6]

The selection of the material or protection system which optimises corrosion resistance for a given component lifetime greatly reduces the need for costly maintenance during that lifetime. Though the initial cost may be higher, the overall cost is usually much less. This principle of corrosion control economics is illustrated in Fig. 1.4.

The essential factor is the choice of the correct design lifetime. Problems arise when a time scale chosen at the design stage to fit the various cost parameters is extended later in the life of the component. Alternatively, even though maintenance schedules are laid down, they are not adhered to. Both can have disastrous consequences.

Fact: Four men died and five were seriously injured at the site of a power station in the UK, when the single suspension rope of the hoist cage in which they were travelling broke at a point weakened by corrosion and lack of lubricant. The safety gear failed to operate, also because of corrosion, and the cage fell more than 30 m to the bottom of a 60 m shaft. At the inquiry it was stated that the required six-monthly inspection was overdue.[9]

Problems have frequently arisen in the consumer industry when the manufacturer and the consumer have differed in their opinions as to the desirable lifetime of the product. Manufacturers often deliberately select inferior materials for their products using the philosophy that:

- manufacturing costs are kept to a minimum
- the product is therefore competitively priced
- the product lifetime is short

- turnover is maintained at a high level
- large profits ensue.

In the event, the consumer always pays the price. This applies, in particular, to the motor industry where for several decades corrosion resistance was consistently poor. The Hoar Committee[7] noted that automobile manufacturers had been the subject of more incidental comment during the survey than any other branch of industry. Evidence in support of a poor track record is clearly visible in Table 1.1. In their defence, the car and bus manufacturers blamed the increased use of road salt and the reluctance of the public to pay for increased protection as the major causes of corrosion. However, the Committee highlighted the case of exhaust systems as an important example. Here, the replacement of the unprotected mild steel exhaust by, say, an aluminised steel system, would lead to an increase of life from the then average of two years to an expected six years. For very little extra cost the national annual saving was estimated at £55 million on this single item alone.

There is, however, a *non sequitur* in the manufacturers' philosophy above, for consumers who find the performance of a product to be unsatisfactory will, at the very least, buy the competitor's product. Recently it has become common for consumers to form associations with the power to demand quality improvements from offending companies, as well as compensation for the inconvenience of their members. Hence the indirect cost of:

(c) *Warranty claims on corroded consumer durables and the consequent loss of customer confidence and sales.*

Fact: The Ontario Rusty Ford Owners Association obtained substantial compensation from the courts for the corrosion to their cars.[5]

The reputations acquired by certain motor manufacturers for producing 'rust buckets' during the 1960s and early 1970s has led to a remarkable change in manufacturing policy and a much greater awareness on the part of both consumer and producer that corrosion resistance must be improved.

In other areas of industry, consumer demands for better quality control have increased overall production costs because of the need to overcome the risk of:

(d) *Loss of product quality in a plant owing to contamination from corrosion of the materials used to make the production line.*

In general, the heavy-chemical, oil and petrochemical industries were found[7] to be far more corrosion conscious than the pharmaceutical industry. This was suggested to be because of the former industries' experience in the use and storage of highly corrosive substances, and a consequent tendency to overdesign. In less than 10 per cent of the pharmaceutical companies questioned was a corrosion specialist employed, and in the majority of cases problems were dealt with by a maintenance engineer. The industry was found to be very conscious of product quality, though suffering higher costs from corrosion than was necessary.

In the food industry it was reported[7] that companies were suffering from a wide variety of corrosion problems, which had led to high maintenance costs, but that there was a lack of experience to tackle the problems. The food industry was said to be conservative and reluctant to change processes and equipment that had been proved to give satisfactory products.

Another possible source of indirect cost is:

(e) *High fuel and energy costs as a result of steam, fuel, water or compressed air leakage from corroded pipes.*

Serious problems have arisen recently in the US where in 1981 the report of the Nuclear Regulatory Commission to Congress in Washington[10] stated that the 'vast majority' of steam-generating pressurised water reactors are suffering failures of stainless steel cooling tubes. While stressing that there was no danger to the public it was estimated[11] that the maintenance bill would top $6 billion.

A final source of indirect cost is:

(f) *Extra working capital because of increased labour requirements and larger stocks of spare parts.*

A medium-sized engineering company in the UK, believing that it had no problems with corrosion, decided to look deeper into the question. It discovered that corrosion was in fact costing £43,000 a year. By improving materials handling and paying greater attention to stock control and records, the company was able to save over £10,000 of this sum.[6]

Consideration of all the above factors illustrates quite clearly that:

> **If management is fundamentally concerned with maximising profit, no manager can afford to ignore corrosion.**

1.3 SOCIAL IMPLICATIONS

Not enough members of society are aware of the enormous amount of damage being caused by corrosion. Although most people would admit to a dislike of corrosion, their experience of it is usually limited to consumer durables. It is probably because their motor cars corrode that most people dislike rust so much, and a study of the corrosion of motor cars alone could fill a large volume. This is because almost everyone owns a motor car and, because it is such an expensive item, is therefore concerned to keep it in good condition. If individuals were as aware of corrosion in other sections of the economy as they are of motor cars there is a good chance that government and industry would be under greater pressure to reduce the overall effects of corrosion highlighted in this chapter.

Fig. 1.5 The consumer's problem. **Note:** It is not intended to imply that this make of vehicle is worse than any other.

We shall begin this section with some examples from the motor industry because cars are familiar to most of us. The Production Manager of British Leyland Body and Assembly Division was quoted as saying:[12]

> No consumer product takes as much abuse from the elements as a motor car. It is exposed to high and low temperatures, rain, sleet, ice, snow, and road salt. It is expected to be able to withstand the grit blasting of stones thrown up by other vehicles and also its own wheels, atmospheric fallout from industrial, domestic and agricultural sources, as well as being used as an armoured battering ram. Despite this, the car is still expected to maintain its showroom appearance over many years of service.

In 1981, passenger cars and commercial vehicles were being scrapped in Europe at the rate of 3 million a year, attributable to one of three causes: accidents, obsolescence and corrosion. Since the amount of corrosion often determines whether a car can be economically repaired, the number scrapped as a result of accidents is affected by the amount of corrosion. McArthur[13] published evidence that there might be a correlation between the number of serious injuries suffered in road accidents and the age (i.e. amount of corrosion) of the vehicle. He argued that it was reasonable to expect that a corroded vehicle was less able to absorb the energy of an impact than was a new vehicle.

McArthur also reported the results of a survey which found that the average life of a motor car was increasing, both in total miles run and lifetime (50 per cent survival of initial registrations). The lifetime of the average car was 9.9 years, while for Volvo the figure was 17.9 years. In the US, Chandler[14] reported that in 1974 the 'median life' of a car was 4.9 years, which in 1984 had risen to 6.0 years, evidence in support of what many people already believed: that the motor industry had at last reacted to the public outcry for more corrosion protection.

Meanwhile, many car owners, in their rightful attempts to own vehicles which remained rust-free, had resorted to one of the many rustproofing treatments which could be applied to new vehicles. In 1982 a report by the office of the Attorney General in New York State[10] claimed that consumers were being defrauded of at least $11 m. annually because of the poor quality of treatments. A remarkable 83 per cent of all cars inspected failed to meet acceptable standards. The most damning comments were that:

> most of the manufacturers' rustproofing warranties contain limitations or conditions that rendered them essentially worthless . . . The sale of dealer-applied rustproofing by New York City dealers is permeated by gross negligence.

Interestingly, it was planned[15] that the now infamous De Lorean sports car would adopt an 'ethical sports car' image by having a life of at least 15 years. The few cars which reached the US showrooms were sold for $25,000, a not inexpensive price-tag, and there is no doubt that the 'built-in' corrosion resistance contributed considerably to the cost. It is obviously much easier to prolong the life of motor cars when economic considerations are less important, perhaps the ultimate example being the Rolls Royce, a car which is particularly long lasting.

Steel has been used as a civil engineering material, both for all-steel structures and for reinforced concrete, since Bessemer's process was devised in 1856. Many examples can be found of steel bridges which, with the benefit of nineteenth-century overdesign philosophy, have remained functional after only modest maintenance. The following examples all concern bridges in a marine, salt-laden environment.

Figure 1.6 shows the Royal Albert Bridge at Saltash, built in 1859 by Brunel to carry a main railway line across the River Tamar between Devon and Cornwall. The 450 m-long bridge has remained in good condition with only the usual amount of maintenance and a recent survey gave it a clean bill of health for at least another 50 years.[16] Brunel's other famous bridge at Bristol, the Clifton suspension bridge, has been maintained in safe condition by regular attention, while Telford's Menai suspension bridge in North Wales has also stood the test of time. First opened in 1826, it was reconstructed in 1938–40 to meet the traffic requirements current at the time, and was the first large bridge to be metal sprayed in the UK. It provides an excellent example of the intelligent use of protective coatings in a hostile environment, for, after four decades, the original protection was recently shown to be undamaged.[11]

In cases such as these, the engineers and designers of the past applied common-sense principles born out of sound engineering practices. The results were often long-lasting structures which showed good corrosion resistance and were comfortably within the margins of safety to allow normal operation if something unforeseen should occur. In today's society,

Fig.1.6 The Royal Albert Bridge at Saltash, Cornwall. Built largely of wrought iron by I K Brunel in 1859, it has spent its lifetime in a salt-laden marine atmoshpere, yet still provides the main arterial rail link into Cornwall.

the burden of impossible financial restrictions and highly competitive tenders has enforced the use of the most advanced and often untried engineering techniques in order to save money. Operating margins have been trimmed to such a fine extent that when a problem arises, as it must in any novel situation, severe penalties ensue. Thus, the supposed advantages of modern engineering design seem to have created a plethora of problems which merely serve to underline the conclusions of the Hoar Report. The following examples serve to illustrate this.

In the UK, the innovative design of the suspension bridge over the River Severn was heralded with great optimism when the bridge was opened in 1966. Problems were first noticed in October 1978 when broken wires within the raked steel hangers were revealed by cracking in the paintwork. Within a short time it was found that cracking of the steel deck box-welds had occurred, and by 1983 grave concern was expressed over the future of the main arterial link between England and South Wales. The Department of Transport insisted that while there may have been some fatigue failure of wires in the suspension/hanger system, the problem had been caused by unexpectedly high traffic loading, aggravated by corrosion, as a result of inadequate protection against the salt-laden atmosphere of the Severn estuary. Estimates of the cost of repair spiralled upwards and in 1983 reached £30 m, compared with a bill of £100 m for a new bridge.[17] Meanwhile, as the engineers wrangled with the accountants, the traffic had been reduced to a single line in each direction, with frequent delays . . .

Problems were experienced with the 350 m-long Pelham bridge in

12

Lincoln. A dual carriageway supported on 23 rows of columns, it had been constructed in 1957, at a time when it was impossible to foresee the great increase in the use of de-icing salt on roads which was to occur during the 1960s. A study[18] showed that penetration of the salt had caused severe corrosion of the steel reinforcement in the concrete road bridge decks. The engineers' report concluded that:

> the cost of correcting faults which amount to little more than poor design
> and construction details has proved very high . . . Pelham bridge is by no
> means unique.

The experience at Pelham was corroborated by a report from the New York Department of Transport which stated[8] that in about 30 years 95 per cent of all New York's bridges would be deficient if maintenance remained at the same level as it was in 1981. The obvious deterioration had been seriously accelerated since the mid 1960s when the extensive use of road salt as a de-icer began.

Buildings, too, have had their problems. In Paris, the prestigious, award-winning French cultural showpiece, the Centre Pompidou, showed signs of severe corrosion in 1981[19] even though completed only in late 1976. Monitoring of the corrosion began in 1979 when the first effects were found on the 'inside-out' building of steel-framed construction with external steel Warren trusses supported between corner columns. Fire protection was achieved by mineral fibre blankets encased in stainless-steel sheeting on the trusses. The nodes were protected with cement, plastic coated to prevent water penetration. It appears that somehow water must have penetrated the node protection, for during the winter of 1978–79 freezing of the trapped water caused spalling. As a result of danger to the public from falling insulation the coatings were removed leaving the steelwork protected only by the 7.5 micron thick metal-sprayed zinc coating.

While bridges are by no means the only structures susceptible to corrosion, they serve to provide good example of the catastrophic effects of the most trivial corrosion problem. On 15 December 1967 the Point Pleasant bridge in Ohio collapsed killing 46 people.[20] It was found that the cause of the disaster was a stress corrosion crack 2.5 mm deep in the head end of an eyebar. Here, the metal had a low resistance to fracture once a notch had been initiated, failure resulted, and the bridge collapsed.

Sometimes, corrosion problems can be caused by the most unexpected agencies. In December 1979 the City of Westminster, London, reported[21] that it had a problem with falling lamp posts. It was suggested that the main culprits for the corrosion that had occurred at the base of the posts were dogs. The city's pets were daily depositing about 2000 litres of urine, mostly at the base of the city lamp posts and this caused a great increase in the rate of corrosion. This surprising cause was attributable to one of many problems experienced with the Charing Cross railway bridge in central London, for at a certain point the continuous visits from dogs had caused severe crevice corrosion in a part of the structure which was impossible to maintain.

A problem with major ramifications beyond those of corrosion alone has been the recently highlighted difficulties over acid rain. This serious air pollution can travel over national frontiers for many hundreds of miles and, under some conditions, can lead to rain as acid as vinegar. The resulting corrosion alone has been serious enough, but the ecological consequences have been nothing short of catastrophic, and it has been largely attributed to the massive emissions of sulphur and nitrous oxides from Europe's coal- and oil-burning power stations, though the automobile is by no means blameless. In a report by the Organisation for Economic Co-operation and Development[22] a 50 per cent reduction in sulphur emissions by 1985 would cost $4.6 billion, but the estimated savings on improved water quality, health, crops, damage to buildings and stonework, and corrosion would more than compensate for this expenditure.

It is rarely that corrosion could be considered to be actually beneficial. The expansion of rusting nuts and bolts as a result of the build-up of corrosion product in the threads prevents them from working loose, a property that is as inconvenient as it is useful. It has sometimes been held that a thin film of rust on the moving parts of an engine prevents them from pressure welding together and seizing prematurely, though the use of more modern materials makes this less common.

By far the greatest use for corrosion is in batteries. This is not entirely within the scope of this book and will only be touched upon in Chapter 4, but it should be remembered that many of the chemical reactions which occur in batteries are corrosion reactions. One recently developed battery which promises to have a profound impact, is simply an efficient form of a corrosion cell in which aluminium is dissolved by salt water.[23] A battery suitable for the motor car and which has the capacity for many kilometres of travel requires just the replacement of an aluminium plate at surprisingly long time intervals. The cost is quite low, given that aluminium is the third most common element on Earth.

Recently, a new method of cutting steel immersed in sea water has been developed.[24] Initially, it was proposed as a method for removing the tips of growing cracks by trepanning a hole with the crack tip as its centre. (Trepanning is a term which describes the use of a cylindrical saw to remove a circular coupon from a surface. The circular coupon of metal can be recovered as evidence that the crack tip has been removed.) The new method involves rotating a hollow metal electrode which has non-conducting abrasive particles embedded in its working end, while it is spring-loaded against the steel surface. The electrode is made cathodic with respect to the steel and the surrounding sea water acts as both coolant and electrolyte. The resulting hole, which can be 10 mm diameter or more, is smooth-walled and stress-free, effectively blunting the crack and retarding its continued growth.

As for the aesthetics of corrosion, it was mentioned briefly that the controlled corrosion of arms and armour could be used with startling effect. It has always been considered by many that the green patina on copper roofs and other ornamental objects is desirable, rather than unsightly. A well-known lover of rust, sculptor Anthony Caro, spent many years producing abstract steel sculptures. Initially they were painted brown, the colour of rust, but he later allowed them to rust naturally and then preserved it by varnishing. Few people, however, would subscribe to the view that, 'Rust is beautiful'. Corrosion of any kind is rarely useful, and its consequences are almost always expensive, but most importantly of all, it can often be avoided.

1.4 CONCLUSIONS

In spite of overwhelming evidence that corrosion is causing a vast amount of financial damage to economies, wasting large amounts of natural resources and causing a great deal of human suffering, society seems unable or unwilling to learn the lessons of the past. The answer does not lie entirely within the domain of engineering education, where many attempts have been made to increase levels of awareness, but also within the close confines of the accountants' ivory towers. Budgetary regulations which forbid expenditure on corrosion protection or the proper selection of material because it is of no immediate financial benefit to the project in question, are anti-social and extremely short-sighted. Too often the problem is passed on to others who have to sort out the long term problems caused by short term cost-cutting exercises. Until society as a whole assumes greater responsibility for corrosion prevention the situation will not greatly improve.

1.5 REFERENCES

1. Royal Naval Engineering College (RNEC) archive
2. Villiers A 1967 *Captain Cook*. Hodder and Stoughton, p 68
3. Pliny 1938 *Natural history of the world* (book 34). Heinemann
4. Fancourt L D 1970 *Llanthony priory – history and guide*. Published privately
5. Editorial 1980 *Corrosion Prevention and Control* 27(3): 1
6. Editorial 1985 *Corrosion Prevention and Control* 32(3): 41
7. Hoar T P (Chairman) 1971 *Report of the committee on corrosion and protection.* HMSO
8. Editorial 1982 *Corrosion Prevention and Control* 29(1): 1
9. Editorial 1981 *Corrosion Prevention and Control* 28(4): 1
10. Editorial 1982 *Corrosion Prevention and Control* 29(2): 1
11. Editorial 1981 *Corrosion Prevention and Control* 28(6): 1
12. Dabbs W F 1974 Choice of materials, in *Cars made to last*. Symposium of the Institute of Metal Finishing
13. McArthur H 1981 Motor vehicle corrosion – safe at any age? *Corrosion Prevention and Control* 28(3): 5–8
14. Chandler H E 1984 Update on corrosion control coatings for autos. *Metal Progress* Feb: 39–43

15. Editorial 1981 *Corrosion Prevention and Control* **28**(3): 1
16. Editorial 1980 *Corrosion Prevention and Control* **27**(4): 1
17. Editorial 1983 *Corrosion Prevention and Control* **30**(3): 1
18. Evans R A, Lee D M Corrosion fight on Pelham Bridge, Lincoln, *Highway Engineer* **28**(10): 16–21
19. Editorial 1981 *Corrosion Prevention and Control* **28**(6): 1
20. Highway accident report 1971 *Collapse of US35 highway bridge Point Pleasant, West Virginia, December 15, 1967.* NTSB-HAR-71-1
21. Editorial 1979 *Corrosion Prevention and Control* **26**(6): 1
22. The Organisation for Economic Co-operation and Development, 1981 *The cost and benefits of sulphur oxide control.* OECD, Paris
23. Fitzpatrick N, Scamans G 1986 *New Scientist* **111**(1517): 34–37
24. Editorial 1985 *Metals and Materials* **1**(12): 725

2 ENABLING THEORY FOR AQUEOUS CORROSION

Science arises from the discovery of identity amidst diversity.

(W J Jevons, 1874)

Corrosion is an interdisciplinary subject, in other words, it combines elements of physics, chemistry, metallurgy, electronics and engineering. Many of us who work in the field of corrosion often have a background in one or other of the mainstream subjects but not in all. Thus an electrochemist does not always appreciate the metallurgical or engineering aspects of corrosion, while metallurgists and mechanical or structural engineers have no reason to understand completely the electrical principles behind corrosion testing. Similarly, a student often finds it difficult to understand corrosion because he lacks the knowledge of one or more of these subjects. This chapter is designed to provide the necessary theory from all relevant subjects to enable you to understand corrosion better. Readers may find that some of the subject matter which follows is elementary and should study only those sections with which they are not familiar.

2.1 ENERGY: THE RULE OF LAW

It is hard to think of a natural process which is not governed by energy changes. Corrosion is a problem caused by nature: it affects material substance and is governed by energy changes. This chapter is therefore concerned with the fundamentals of energy and substance and we shall begin by examining the former.

The study of energy changes is called **thermodynamics**, a very precise subject with many definitions, variable quantities (called parameters) and equations. The discussion presented here is restricted to some basic concepts.

We shall often refer to **systems**, which can be defined as any particular mass of material in which we are interested. Around the system is an imaginary boundary wall separating it from the environment, which is known as the **surroundings**.

Let us consider two important statements:

Enabling theory for aqueous corrosion

The Law: *Energy can neither be created nor destroyed.*

The Rule: *All spontaneous changes occur with a release of free energy from the system to the surroundings at constant temperature and pressure.*

The first statement is the First Law of Thermodynamics and is extremely important when considering the changes which occur when metals corrode. The second statement is one form of the Second Law of Thermodynamics. When corrosion occurs in nature it is a spontaneous process and therefore occurs with a release of free energy. Left to its own devices nature is extremely concerned always to minimise energy. Water runs downhill, cups of coffee go cold and spinning tops slow down and stop: all these are consequences of the Second Law of Thermodynamics which is most simply stated in the form, *heat will not flow, of its own accord, from a cold place to a hot one.* Man can interfere to keep the water at a high level, to keep the coffee hot or the top spinning, but it is necessary for him to supply energy to the system in some form or another. We know that many metals will corrode given sufficient time, so it would appear that nature minimises the energy of metals through corrosion.

There are many forms of energy but the driving force for corrosion comes from **chemical energy**. This is partly derived from energy stored within the chemical bonds of substances which is called the **internal energy** of the system. Only a proportion of this internal energy is available as the useful energy which powers our engines or the destructive agent which causes a corrosion reaction. This available energy is called the **free energy,** of which we shall see more in Chapter 4.

An important concept which helps to explain the rates of corrosion reactions is the **Transition State Theory**. Consider the following equation:

$$A + B \rightarrow C + D \qquad\qquad [2.1]$$

The equation is a shorthand which can be taken to mean, 'Two species, A and B, known as the reactants, interact together in such a way as to form two new species, C and D, known as the products.' In order to produce the new species it is essential that A and B do not just come into contact with each other, but physically join together, forming an intermediate species, AB. In reality, this may happen for only the briefest instant, and then only when the reactants possess sufficient energy and the correct orientation for the joining to occur. AB is called the **transition state**, and it is the re-organisation of the transition state which leads directly to the products, C + D.

We can now use a diagram called an **energy profile** to describe the free energy changes which occur during the reaction (Fig. 2.1). The y-axis in the diagram is free energy, G; energy *changes* will appear as $\triangle G$, in which the Greek capital delta is used conventionally to mean 'a change in'. The x-axis is called **reaction co-ordinate** and can be thought of as the extent to which the process has progressed. It is not necessarily a time scale, but may be thought of as such in the first instance. Theory tells us that the transition state must be of higher free energy than the sum of the free energies of the separate species, A and B. Traditionally, this amount is given the symbol

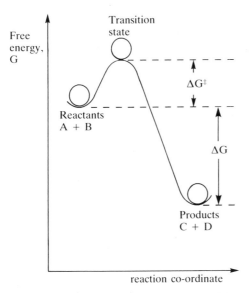

Fig. 2.1 Energy profile for the reaction which converts A+B into C+D via the transition state.

$\Delta G^{\ddagger}$. We are considering a spontaneous reaction, so the energies of the products, C and D, must be less than the energies of the reactants, A and B, by an amount ΔG. Once formed, the transition state can either revert to reactants or can progress to the products. By changing into C and D, the transition state is able to reach a lower free energy situation so this is the favoured of the two possibilities.

In simplest form, the rate of a corrosion reaction can be expressed thus:

$$\text{rate} = \text{rate constant} \times [\text{reactants}] \qquad [2.2]$$

The quantity in square brackets is a measure of the amount of substance. This will be explained in the next section. The rate constant can be shown to be related to the size of the free energy barrier:

$$\text{rate constant} = C \exp\left(-\Delta G^{\ddagger}/RT\right) \qquad [2.3]$$

where C and R are constants, and T is the absolute temperature. Inspection of the equation shows that as T increases, so also does the rate constant (and hence the rate) but when the size of the barrier ($\Delta G^{\ddagger}$) is increased, the rate constant decreases. Equation [2.3] is a modified form of an important equation called the **Arrhenius equation**, named after the scientist who first described this successful theory.

Now consider the reverse process. Overall, it cannot be spontaneous, because there is an increase in free energy upon conversion of C and D into A and B. Furthermore, there is a bigger free energy barrier for the new reactants, C and D, to cross. The Transition State Theory says that the reverse process *is* possible, but occurs at a much reduced rate, represented by an equation similar to equation [2.3] in which the activation free energy has been increased from $\Delta G^{\ddagger}$ to $(\Delta G^{\ddagger} + \Delta G)$, as in Fig. 2.1. The reverse process is possible only on the molecular scale where the energies of an individual C and D pair may be such as to allow the formation of the transition state. (Remember that our rule applies to an *overall* free energy change for a bulk system.) The reverse process occurs at a rate far less than the rate of the forward process, so the net reaction observed on the large scale will always appear to be a steady conversion of A + B into C + D. In order for the reverse process to occur in the bulk system, energy must be supplied to the system (e.g. in electrolysis).

The use of energy profiles is of considerable assistance in the understanding of corrosion processes and will be frequently used throughout our discussions of corrosion theory. We shall now progress to a description of substances.

2.2 MATTERS OF SUBSTANCE

All substances are ultimately composed of **atoms**, **molecules** or **ions**. Currently, 103 different kinds of atoms, or **elements**, have been discovered. Many of these are extremely rare or can be made only in highly specialised nuclear reactions. Several surprising statements can be made about the elements.

Fact: Just *five* of the 103 elements, oxygen, silicon, aluminium, iron and calcium, make up 91 per cent of the matter in the Earth's crust.

Fact: The amounts of oxygen and silicon together make 74.3 per cent of the mass.

Fact: Of the 103 elements, the great majority are metals.

Since corrosion is a problem which afflicts metals, the third fact adds new emphasis to the need for an understanding of corrosion.

Corrosion is about substance and energy. We began by discussing energy and shall continue with an examination of the properties of substances, starting with some definitions which allow us to state the amount of substance.

Most people have agreed to measure the amount of substance in kilograms, incorrectly called a weight, but uniquely described as mass. Today, the phrase, **amount of substance**, is used specifically to refer to the number of atoms, molecules or ions in a given quantity of material, **NOT** a mass measured in kg. Amount of substance is one of the seven fundamental units of the SI system of measurement which has now been accepted across the world and which is used in this book. The unit is called the **mole**, abbreviated mol, and is expressed thus:

> **One mole of a substance is the molecular weight of that substance expressed in grams.**

The reason for the need to distinguish a number of atoms from a mass is obvious to anyone who has been fooled by the child's conundrum, 'Which is heavier: a kilo of lead or a kilo of feathers?' Lead is intrinsically more dense than a feather but the term 'weight' has a different implication and the answer to the question, of course, is that they both weigh the same. At the root of the answer lies the fact that the basic particle of lead, the lead atom, has more mass than the basic particle of a feather, let's say a carbon atom, and for a given number of atoms the lead will weigh much more. Thus we see the need to define the number of atoms in a given amount of substance, as well as its mass. Unfortunately the rephrasing of the conundrum 'Which is heavier, a mole of lead or a mole of feathers?' makes more sense and less fun. The major significance of this unit is that **1 mol of material A and 1 mol of material B have equal numbers of molecules but different masses**.

It is unusual to find isolated atoms in nature. The oxygen we breathe is better described as molecular oxygen, for it consists of **molecules** in which two oxygen atoms have combined together. The water we drink and which aids many corrosion reactions is composed of molecules in which two atoms of hydrogen have combined with one atom of oxygen. Thus discrete atoms are often of less importance in corrosion than the molecules derived from them. By undergoing such combinations, the atoms are able to attain lower energy states.

Question: *How do I work out the number of moles of oxygen in a given mass of gas?*

First take the base unit of oxygen gas, the oxygen molecule which consists of two atoms of elemental oxygen. Next find the atomic weights of all the constituents of the base unit (in this case, oxygen) and sum them. A table of atomic weights can be found in many science texts. Rounding off the weights to the nearest whole number, oxygen has an atomic weight of 16 and a molecular weight of 32. This number, expressed in grams, gives the mass of 1 mol of oxygen, i.e. 32 grams. The same method is used for liquids as well as gases. A water molecule consists of $2H + O$, and so the molecular weight $= 2 + 16 = 18$. Hence a mole of water is 18 grams.

An atom can be thought of as a diffuse cloud of **electrons**, tiny particles with negligible mass and a single unit of negative charge, surrounding a positively charged nucleus. The latter contains virtually all the mass and is made up of **neutrons** and **protons**. It is the number of protons which determines the element and is called the **atomic number**. The number of electrons always balances the number of protons because atoms are always electrically neutral: despite being composed of charged particles, they have no net charge.

The electrons in atoms are arranged into different levels of energy which are called, **orbitals**. It is those electrons in the orbital of highest energy which determine the reactive properties of the material. When reactions occur between atoms or molecules it is changes in the distribution of elec-

Enabling theory for aqueous corrosion

trons between the atoms concerned which determine the net thermodynamic energy changes. There are three ways in which atoms combine:

(a) by **sharing** electrons to form **covalent** bonds;
(b) by **exchanging** electrons to form **ionic** bonds;
(c) by **sharing** electrons to form **metallic** bonds.

Although covalent bonds are extremely important in the study of materials, they are much less important in corrosion processes and need not be discussed here.

Ionic bonding occurs when electrons are exchanged. One atom may lose an electron to another, in which case, the first has become an **ion** with a net positive charge and the second has changed into an ion with a net negative charge. The former is called a **cation**, the latter, an **anion**. If the process occurs spontaneously, both ions are lower energy states than the atoms from which they were made. In the solid state the ions are held together by the electrostatic attraction of positive to negative charge. The highly ordered arrangements which are formed are known as **crystal lattices**.

We shall refer frequently to sodium chloride. When dissolved in water it produces an extremely aggressive corrosion medium similar to sea water. It is formed when metallic sodium reacts with toxic chlorine gas to yield a harmless, white, non-metallic solid better known as common salt. The sodium and chlorine atoms have become ions and the reaction illustrates an important fact about all ions: being electrically charged, they are quite dissimilar in properties to the atoms from which they are derived.

Using the chemical shorthand in which sodium is represented by Na, chlorine by Cl, and e^- is an electron, we write the reactions which describe the formation of salt as:

$$Na \rightarrow Na^+ + e^- \tag{2.4}$$
$$Cl + e^- \rightarrow Cl^- \tag{2.5}$$

Equation [2.4] means that a sodium atom gives up one electron to form a positively charged sodium ion, while eqn [2.5] means that a chlorine atom accepts an electron to form a negatively charged chloride ion. Reactions such as eqn [2.4] which generate electrons are known as **oxidation** reactions, while those such as eqn [2.5] which consume electrons are called **reduction** reactions. As we shall see below, all corrosion reactions are of the former kind.

For clarity, we have expressed the event as two separate equations, but it is common to write the equation which results from the addition of eqns [2.4] and [2.5], thus:

$$Na + Cl \rightarrow Na^+ + Cl^- \tag{2.6}$$

In this form there is no direct indication of the electron changes which have occurred, though they are easy to deduce. As in all equations, it is important that the quantities on both sides should balance. Chemical equations should always balance for mass and electrical charge. In a sense, it is rather meaningless to refer to sodium chloride as NaCl. The large, ordered array of ions in the crystal lattice structure requires a formula, $(NaCl)_n$, but in calculations it is helpful to use the formula as NaCl.

When ionic materials are dissolved in water the ions separate and are dispersed at random within the molecules of water. Whatever amount of salt we use to make the solution, the number of sodium ions is always equal to the number of chloride ions. This statement is more important than it might at first appear and is better called the **principle of electroneutrality**. Thus, whenever a positive ion is created, a negative ion must also be created.

Question: *How are the proportions of ions to water molecules calculated?*

This is a most important question for it introduces our method of measuring the amounts of substance in solution. Let's take sodium chloride again. To make a 3.5 per cent solution (in order to crudely mimic sea water) 35 grams of crystals are weighed out and dissolved in water, say, 250 ml. Further water is then added to make the total volume 1 litre. One way of expressing the solution's composition is to say that the concentration of sodium chloride is 35 grams per litre. However, as we saw earlier, this is not necessarily as useful as expressing the number of moles which have been dissolved in the water. For this, we need the molecular weight of sodium chloride: using the relevant atomic weights, Na = 23, Cl = 35.5, thus NaCl = 58.5. A mole of sodium chloride is 58.5 g, but the weight dissolved was 35 g = 35/58.5 moles = 0.6 mol. When 1 mol has been dissolved in a litre of water we say that the solution is molar, thus our 3.5 per cent sodium chloride solution is 0.6 molar, abbreviated 0.6 M.

We have still not answered the original question, however. To do this we need to calculate the number of moles in a litre of water. Since a litre of water weighs 1000 g and 1 mol is 18 g then there are 1000/18 = 55.55 mol of water in a litre. In the 3.5 per cent solution there is 0.6 mol of salt and we make the approximation that the salt will have taken the place of 0.6 mol of water molecules in order to maintain a constant volume of 1 litre. There is one final point to make before the answer is obtained. A solid assembly of ions was dissolved assuming a formula NaCl. It is now that the usefulness of the mole becomes even more apparent, for 1 mol of sodium chloride dissolves to give 1 mol of sodium ions and 1 mol of chloride ions. The ratio of sodium ions to chloride ions is exactly 1. Thus, the answer to the original question is that the ratio of either ion to water molecules is 0.6/(55.55–0.6) = 0.011.

More precise texts use a parameter called an **activity** rather than a concentration, and coupled with this is a term called **molality** – a slightly different way of measuring the concentration of solutions and not to be confused with molarity. It is considered that in an introductory text such as this, molalities and activities are an unnecessary complication and they will not be discussed further. The excellent work by Atkins[1] describes all the terms precisely.

The understanding of the bonding of metals is rather more difficult to achieve. In practice this is not necessary at our level of discussion and we shall think of metals as an assembly of atoms in an ordered crystal structure.

Question: *How can a metal be crystalline? It does not look at all like a diamond or a grain of sugar, neither does it have similar properties.*

The property which determines whether a material is crystalline is internal ordering of atoms, usually over a large number of atomic distances. The atoms within metals are ordered in very regular patterns and hence metals **ARE** crystalline in nature, despite their apparent dissimilarity to more obvious crystalline materials. It is the special type of bonding in metals which gives them their characteristic properties and distinguishes them from other crystalline, non-metallic substances.

When a metal atom undergoes a corrosion reaction it is converted into an ion by a reaction with a species present in the environment. Using the symbol M for a metal atom contained within its solid structure, we can represent corrosion by the following equation:

$$M \rightarrow M^{z+} + ze^- \tag{2.7}$$

The integer, z, usually has the values 1, 2 or 3. Higher values of z are possible, though rare. **Equation [2.7] is the most general form of a corrosion reaction** and states that metal atoms can lose more than one electron. Of the possible values of z, 2 is most common. The value of z is called the **valency** and it is not uncommon for metals to have more than one valency. For example, iron, the most important engineering material of all when used in steel, has 2 and 3 as common valencies, a factor which greatly complicates study of the rusting process.

Equation [2.7], though correct, does not represent a complete process, according to the principle of electroneutrality. A positive ion has been generated from a neutral species, but a negative ion has not. (An electron is not an ion and cannot move about in solution, i.e. it is insoluble.) The equation is called a **half-reaction** because of its incompleteness but is nevertheless very useful. A further consequence of the principle of electroneutrality is that we cannot create free electrons: they must be 'consumed' elsewhere. This, of course, is achieved by generating a negative ion to balance the positive ion. A corrosion process is thus more completely described by an equation in which negative ions as well as positive ions are formed. For example, eqns [2.4] and [2.5] are half-reactions; eqn [2.6] describes the complete process and free electrons are not generated in this equation. The formation of anions in corrosion processes need not involve the metal at all and may take place at a location quite remote from the corrosion of the metal. We shall discuss the nature of the second half-reaction of a corrosion process in Chapter 4.

A fundamental definition of corrosion will now be made:

CORROSION IS THE DEGRADATION OF A METAL BY AN ELECTROCHEMICAL REACTION WITH ITS ENVIRONMENT.

Several points concerning this definition are worthy of emphasis:

(a) Corrosion relates to metals. This means that only a half-reaction such as eqn [2.7] can be a true corrosion reaction. The second half-reaction, though it describes a process without which corrosion cannot occur, is not itself a corrosion reaction.

(b) By using the word *degradation* we assume that corrosion is an undesirable process. There are circumstances in which this is not true, in which case the process is not usually referred to as corrosion. Chapter 1 discussed this in more detail.

(c) The degradation of the metal involves not just a chemical but an electrochemical reaction, that is, electron transfer occurs between the participants. Since electrons are negatively charged species, the transport of which constitutes electrical current, such reactions are influenced by electrical potential (see section 2.7).

(d) The environment is a convenient name to describe all species adjacent to the corroding metal at the time of the reaction, eqn [2.7]. The topic of environment and its control is the subject of Chapter 13.

The most important material in aqueous corrosion processes, apart from metals themselves, is water. We shall devote the next section to a more detailed look at the properties of water and ions in solution.

2.3   PROCESSES IN SOLUTION

Water is a neutral molecule in which two atoms of hydrogen are combined with one of oxygen. It does, however, undergo limited **dissociation** into a **hydrogen ion** and a **hydroxyl ion**:

$$H_2O \rightleftharpoons H^+ + OH^- \qquad [2.8]$$

Equation [2.8] is an example of what is generally referred to as an **equilibrium**, the sign $\rightleftharpoons$ being used to denote this. At any given temperature and pressure, the degree to which this process occurs is constant. If it were possible to remove some of the ions from the right-hand side of the equation, more water molecules would dissociate to redress the imbalance in the equilibrium. Such an equilibrium process can be expressed mathematically using the **Law of Mass Action** which is stated as follows:

For the equilibrium process

$$A + B \rightleftharpoons C + D \qquad [2.9]$$

then

$$\frac{[C].[D]}{[A].[B]} = \text{constant} \qquad [2.10]$$

Square brackets express the concentration of the species they enclose. When equilibrium exists, the constant is represented by K, whilst the substitution of non-equilibrium values for [A], [B], [C] and [D] in eqn [2.10] requires the use of symbol J.

Applying this law to the equilibrium eqn [2.8] we have

$$\frac{[H^+].[OH^-]}{[H_2O]} = \text{constant} \qquad [2.11]$$

It can usually be assumed that the concentration of water in dilute solution is constant. In the last section, the effect of adding 0.6 mol of salt to 55.55 mol of water made only a small change. In more dilute solutions, the effect becomes even smaller. Thus, for dilute solutions, we write $[H_2O]$ = 55.55 and obtain a new constant in eqn [2.12]:

$$[H^+].[OH^-] = \text{constant} \qquad [2.12]$$

We define standard conditions of temperature and pressure as 25 °C and 1 atmosphere. The constant in eqn [2.12] has been measured experimentally as 10^{-14} under standard conditions. This value, and the relationship eqn [2.12] are extremely important, for they form the basis of a scale of **acidity**. Many metals are dissolved by acids and acidity is thus closely related to corrosion. The common factor in the properties of acids is the presence in aqueous solution of the hydrogen ion. The 'opposite' of acid is **alkali**, by which is meant that acids are neutralised by alkalis, and that alkalinity is associated with hydroxyl ions. As eqn [2.8] shows, water is a source of both acid and alkali in equal quantities and must therefore represent a neutral substance.

The accepted method of defining acidity is by means of a term called **pH**. It is measured on a scale from 0 to 14 and is defined as follows:

$$pH = - \lg [H^+] \qquad [2.13]$$

At first sight this may appear a little complicated but it soon becomes very easy to understand and use successfully. The term in square brackets is the concentration of hydrogen ions expressed as a molarity. Thus a concentration of 10^{-8} M when substituted into eqn [2.13] gives a pH of 8, while one of 10^{-2} M leads to a pH of 2. A concentration of 2×10^{-4} is slightly more difficult to convert:

$$\begin{aligned}
pH &= - \lg (2 \times 10^{-4}) \\
&= - (\lg (2) + \lg (10^{-4})) \\
&= - (0.3 - 4) \\
&= 3.7
\end{aligned}$$

Returning to eqn [2.8], it is apparent that every water molecule produces one hydrogen ion and one hydroxyl ion. Therefore, in pure water:

$$[H^+] = [OH^-] \qquad [2.14]$$

Substitution into eqn [2.12] yields

$$[H^+]^2 = 10^{-14} \qquad [2.15]$$

i.e. $\quad [H^+] = 10^{-7} \qquad\qquad\qquad\qquad\qquad\qquad$ [2.16]

Thus the pH of pure water is 7. On the scale of pH, if the concentration of hydrogen ions is greater than 10^{-7} M then the pH is in the region 0–7 which is acid, while a concentration of hydrogen ions of less than 10^{-7} gives a pH in the region 7–14 which is alkaline.

It is possible to define the concentration of alkali (hydroxyl ion) in the same way, i.e. $pOH = -\lg [OH^-]$. However, it is not often necessary to use this quantity because, by taking negative logarithms to the base 10 of eqn [2.12] we get:

$$pH + pOH = 14 \qquad [2.17]$$

and pOH is always readily calculated from pH. You should try the questions at the end of the chapter to gain experience in these quantities. Vogel[2] contains a full discussion of much of the material in sections 2.2 and 2.3. Figure 2.2 summarises the different pH and pOH values in a memorable way.

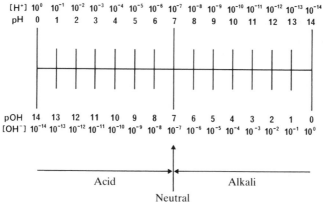

Fig. 2.2 The scales of pH and pOH (after Vogel[2]).

A point worth emphasising at this juncture is that pH 7 is neutral only under the standard conditions described. As the temperature goes up, so more dissociation occurs and the pH of neutral water falls. This can be extremely important for corrosion reactions in a chemical plant operating at high temperature and pressure.

Some other comments about the properties of solutions should be made, for it is important to appreciate that an ionic material such as sodium chloride in solution possesses very different properties from that of the solid. No longer are the ions bound together tightly by strong electrostatic forces. The dissolution process completely destroys the crystal lattice of the solid state. If the solution is dilute it may be that the sodium ions are separated from the chloride ions by relatively large distances. The ionic material is said to be **dissociated** when the ions have been separated in solution.

The addition of a second soluble ionic material to the solution merely adds new types of ions which become randomly dispersed among all the other species present. It is quite wrong to think of a solution merely in terms of the individual solids which were added to the water. Undissociated solid may be present in the solution, though the substance is not truly in solution because the distribution of ions in the solid and in the solution is not homogeneous: we say that the solid is a different **phase** to that of the solution. Although these points may seem obvious to some readers, it is a common error which students make to consider a solution as a mere mixture of one or more solids in water, rather than a homogeneous dispersion of a number of different ions. There is still a force of attraction between the positive and negative ions while in solution, but this force is greatly reduced by the presence of the water molecules.

2.4   METALS: MIGHT AND BLIGHT

Engineers are rightfully more than ready to exploit the might of metals but it is usually the task of the corrosion technologist to patch up the blight of corrosion. We have discovered so far that corrosion is the destruction of metals and that as many as 80 per cent of all known elements are metals. With so many metals to choose from, engineers use surprisingly few in significant amounts and of these it is alloys which are frequently preferred to pure elements.

The first elements to be used in a practical sense were gold and copper, being two metals which exist uncombined in nature. Gold was too soft to be of use for tools or weapons, but copper was hammered into crude tools about 8000 BC. By 4000 BC the Egyptians found that it could be obtained by heating a suitable ore in a kiln, while about 3000 BC, it was discovered that by alloying copper with tin much more durable weapons could be made than if either of the pure metals was used. Since then it has become common knowledge that alloying produces a wider range of useful properties than use of the pure metals themselves.

Metals are exceedingly useful for a number of reasons. Apart from being typically opaque and lustrous, they have such characteristic properties as ductility, high strength/weight ratio, and the ability to conduct heat and electricity. It is obvious that metals are extremely important in modern engineering, yet many can be badly affected by corrosion. What is it about them which makes them so useful yet so vulnerable? Unfortunately the answer to such a crucial question cannot be provided in a few paragraphs.

We have already noted that the bonding of metals in solid structures is different from that in non-metals: this is largely why metals have such characteristic properties. The most significant difference in the bonding mechanism is that, while the electrons in covalent and ionic substances are usually strongly localised around their parent atoms or ions, in metals a proportion of the electrons are free to move throughout the material. This gives metals the property of electrical conduction.

But what of the solid structure of metals? Many books have been written on this subject alone and it is impossible to give more than the briefest coverage here. However, some aspects are vital to the understanding of the corrosion mechanisms and their control. These will be discussed next.

2.5   METALS IN THE MELTING POT

The solid state is one of order and permanence. The paper you are reading does not disintegrate before your eyes because the atoms of which it is composed are held firmly within their molecules, and the molecules too are fixed in space by other bonding forces which maintain short, inter-atomic distances. The same is not true for liquids; shape is not retained because the energies of the atoms, molecules or ions are sufficient to overcome the forces which would have held them together in the solid state. Although they remain closely packed, there is little resistance to free movement within the bulk. All liquids can be made to form solid, ordered structures if their energies are reduced sufficiently by lowering the temperature. Substances melt when the species which make up the solid are given sufficient energy to overcome the bonding forces; conversely they freeze when that energy is taken away.

Metals are not different from other materials in this respect. Since the extraction, purification and fabrication of metals usually involves solidification from a **melt** – a quantity of molten metal – it is important to consider the events which take place during solidification. Imagine a pot (crucible) containing a sample of molten, pure metal. If the temperature of the sample is plotted as time proceeds, a graph rather like that shown in Fig. 2.3 is obtained. Initially, at A in the diagram, the crucible of molten metal contains the metal atoms jumbled together in a loosely packed assembly. At the temperature represented by the freezing point, B, the metal atoms begin to organise themselves into highly ordered arrays. At a given temperature for a given metal, the same array is always formed, although a variety of array patterns are found amongst the many different metals. The solidification process occurs with a release of energy, which is commonly called the **latent heat of fusion**. Under ideal conditions, the release of this energy maintains pure metals at a constant temperature for the duration of the solidification process (B to C in the figure), compensating for the natural tendency of the system to cool down to the temperature of its surroundings.

The number of crystals which nucleate is dependent upon the rate of cooling. Rapid cooling causes many nucleation sites in the liquid; conversely, slow cooling yields only a few crystals which continue to grow relatively slowly. Rate of cooling metals during the **casting** (solidification) process is very important in determining the mechanical properties of the

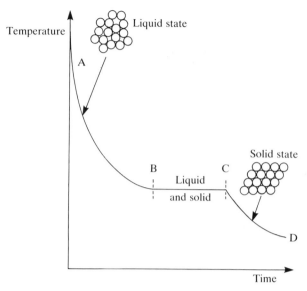

Fig. 2.3 A cooling curve for the solidification of a pure metal.

metal. As will later become apparent, it also significantly affects corrosion properties.

Figure 2.4 illustrates schematically the solidification of a pure metal. Once a crystal has been formed, Fig. 2.4(a), even though it may be only a tiny cluster of atoms, it is more natural for further solidification to occur on the established array rather than creating new crystals. Let us suppose that the crystal is a tiny cube. The rate of cooling is faster from the corners than from the edges. Likewise, cooling from an edge is faster than from a face. Thus the growth of the crystal is fastest at the corners which generate branches. These branches 'fill in' with solid material as the growth on the edges and faces tries to catch up with that at the corners. When a crystal has become branched (Fig. 2.4(b)) it is known as a **dendrite** and the process of solidification of metals is often referred to as **dendritic growth**. The orientation of each of the dendrites is different. Eventually the dendrites grow so large that their outermost atoms touch and movement becomes restricted, (Fig. 2.4(c)); the dendrites become fixed in their random orientations and the liquid which remains between the branches of the dendritic arms solidifies to give the final solid sample (Fig. 2.4(d)). At this point, represented by the end of the horizontal portion of the cooling curve, C in Fig. 2.3, cooling continues in the normal manner, in the solid state.

Once completely formed in the solid state, the individual crystals are known as **grains**. In the zone between any two grains, where the crystal pattern changes orientation, the atoms are mismatched to the lattices in the grains. These zones are called the **grain boundaries**. Crystal lattices are rarely perfect and the next section discusses some of the defects which are possible, together with their significance in corrosion.

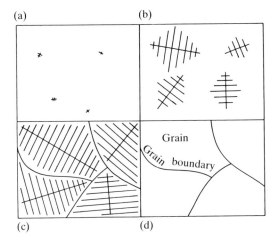

Fig. 2.4 Dendritic growth and solidification.
(a) Nucleation of crystals in the melt.
(b) Growth of crystals into dendrites.
(c) Complete solidification.
(d) Final grain structure.

2.6 DEFECTS IN METAL STRUCTURES

We have tended to assume that when metals solidify to form highly ordered crystal lattice structures there are no faults in the stacking arrangements. This is not the case. Metals always have imperfections, known as **defects**, in their lattice structures and these often have a considerable effect upon the corrosion properties of the metal.

In fact, we have already discussed one way in which the crystals are imperfect: grain boundaries are regions of mismatch between adjacent lattices, each with different orientations. Grain structure of metals arises from the solidification processes during the casting of the metal. It is also substantially affected by the mechanical treatment received during working and fabrication. The malleability of metals means that such processes can cause substantial deformation of the grains and fracture of the otherwise perfect parts of the lattice.

Question: *What types of defects occur within the grains themselves?*

Figure 2.5 shows three sorts of defects which can occur with individual atoms.

(a) **Vacancies**, where there is an atom missing from a lattice site.

(b) **Substitutional defects**, where a foreign atom occupies a lattice site which would have been occupied by a host atom.

31

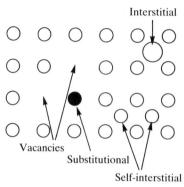

Fig. 2.5 Point defects in a crystal lattice.

(c) **Interstitial defects**, where an atom occupies a site which is not a normal lattice site and is squeezed in between atoms of the host lattice. The interstitial atom may be either a host atom or a foreign atom.

Single atom defects, or **point defects** as they are more correctly known, are very significant in the theory of alloying in which, though they are defects in a perfect lattice, they are used to great advantage in improving the mechanical properties of metals. They also play a role in some corrosion mechanisms, notably hydrogen embrittlement, selective attack, oxidation and hot corrosion, all of which rely upon the diffusion of species through the metal lattice.

A second type of defect occurs within the grain structure when planes of atoms, rather than individual atoms, are not perfectly fitted into the lattice. These are known as **line defects**. One example of a line defect is the **dislocation**. The mobility of both dislocations and point defects is intimately concerned with such properties as strength, hardness and toughness. Two important types of dislocation are:

(a) **Edge dislocations** where an 'unfinished' plane of atoms is present between two other planes, (Fig. 2.6(a)).

(b) **Screw dislocation** where a plane is skewed to give it a different alignment to its immediate neighbour, (Fig 2.6(b)).

Dislocations and their role in corrosion mechanisms are dealt with in Chapters 9 and 10.

A final class of defects which affect metals on a macroscopic scale is that of **volume defects** where relatively large parts of the volume of the metal do not conform to the overall structure of a perfectly formed material. This rightly implies that volume defects are a result of the processes occurring during the manufacture.

(a) **Voids** are holes within the material. They may be caused by a number of mechanisms, such as the entrapment of air, the release of gas during the process of pouring the metal into its mould or the presence of trace amounts

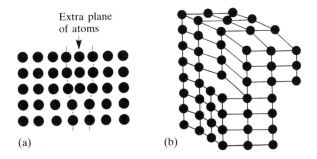

Fig. 2.6 Two types of dislocation in crystal lattices:
(a) Edge.
(b) Screw.

of moisture which turn to steam in contact with hot liquid metal. Voids can also be caused by inter-dendritic shrinkage during solidification.

(b) **Cracks.** These are much more severe than the mere mismatch of grain boundaries for they allow ingress of corrodants on a macroscopic scale. They are caused during casting, usually as a result of uneven rates of cooling and the setting up of stresses within the cast. Cracks can also be created during forging and are very common in and adjacent to weldments.

(c) **Inclusions** are particles of foreign matter embedded in the solid and therefore not part of the crystal lattice structure of the metal itself. The elements which make up the inclusions may solidify first and then become entrapped as individual particles within the dendrites of the metal as these in turn solidify. Alternatively, they may be the last to solidify after the host metal dendrites have been formed and the inclusions are again trapped in the grain boundaries.

In the ideal situation, the manufacturer's product should be free from all volume defects, but in practice this is very difficult to achieve in large-scale production. Obviously, if very small amounts are required then it is possible to obtain a material relatively free from such defects, but in the engineering world at large, it is common to find such defects in materials. Volume defects play an important role in corrosion mechanisms.

2.7 ELECTRICAL SCIENCE

Electricity is the passage of charged particles between two defined points. Normally this occurs through a metal wire connected to the two points in question. The charged particles are electrons, the flow of which is **current**. Electricity can also pass through aqueous solutions, but the electrical charge in this case is carried by positive and negative ions.

The amount of charge carried by an electron is known accurately and is measured in **coulombs**, for which the SI symbol is C. When an amount of charge is passed at a constant rate we talk about the flow of current which is given the symbol I and is measured in **amperes**, symbol A. Thus, when a charge of 1 C passes a given point in 1 s an electrical current of 1 A is flowing.

The driving force which causes the charged particles to move is analogous to that which causes water to flow downhill, except that we do not imagine the current flowing downhill in the physical sense of the word. We do, however, say that when there is a difference of **potential** between two points then the current will flow from the high potential to the low potential. We talk of a potential gradient and of the current flowing *down a potential gradient*.

Electrical potential is measured in **volts**, symbol V. Often, a driving force from a cell, or other source is called an **electromotive force**, abbreviated e.m.f.

The flow of electric charges is impeded by a quantity called **resistance**, and between any two points there is always some resistance to the passage of the current. In metal wires where the electron flow is extremely efficient the resistances involved are low, though not always negligible. In solution, the resistance is much bigger and may need to be taken into account. The passage of ionic current is a quite different process from the passage of electron current. Although the ions are charged to either one, two or sometimes three units of electron charge, the mass transported is greater by a factor of about 2000: the resistance to the motion of such massive particles is therefore much greater than for electrons. Electrical resistance is measured in **ohms**, symbol Ω.

We use a relationship known as Ohm's Law to equate potential, current and resistance. Hence:

$$\textbf{volts} = \textbf{amperes} \times \textbf{ohms} \qquad [2.18]$$

or, using the commonly accepted symbols:

$$V = I.R \qquad [2.19]$$

Charge is related to current by the expression:

$$\text{coulombs} = \text{amperes} \times \text{seconds} \qquad [2.20]$$

i.e. $\qquad C = A.t \qquad [2.21$

It should be obvious to the reader that the resistance to electron flow offered by a conductor will increase with the length, l, of the conductor, but decrease with the area of cross-section, A. The proportionality constant is called the **resistivity**, or **specific resistance** and written, ρ. Thus:

$$R = \rho.\frac{l}{A} \qquad [2.22]$$

By inspection, the units of resistivity are seen to be Ωm.

It is possible to define **conductance** as the inverse of resistance; logically, it is called the **mho** and has units, Ω^{-1}, though this unit has recently been called the **siemens**, symbol S. Likewise, **conductivity**, κ, is the inverse of resistivity, i.e. $\kappa = 1/\rho$, and the units of conductivity are Ω^{-1} m^{-1}, or S m^{-1}. Conductivities of aqueous solutions are an important consideration in the understanding of corrosion processes.

The ideas which we have put forward so far have implied that we can remove charge from one point and 'pile it up' at another. By using a circuit, or loop, around which charge is transported, charge is never accumulated at one point. In physics, the study of electrostatics often involves the storage of charge, and devices which operate by storing charge are very important in all areas of electronics. However, the passage of continuous **direct current** is only achievable when there is a circuit.

It is now relevant to make a historical point with reference to current. When a decision was made about the sign of the charge which constitutes electrical current in wires, it was decided that it should be positive. With the benefit of hindsight we know that the wrong choice was made: **electron current is negative**. Unfortunately, by the time this was realised, it had become conventional in electrical science to designate current flow as a movement of positive charges around a circuit. Many circuit diagrams in textbooks still indicate conventional current in this way.

In this text we shall refer almost exclusively to electron flow because most of the discussions will consider how electrons are produced in corrosion reactions. Discussions concerning ionic current will involve the passage of both positive and negative ions. When the term **conventional current** is used it will refer to the electrical engineering concept of current as a flow of positive charges around a circuit. Consistent with this is the concept of a positive current being driven from a high potential to a low potential, i.e. a positive potential difference. A consequence is, therefore, that **electrons flow from a higher negative potential to a lower negative potential**.

It is unfortunate that the legacy of that incorrect choice remains, for it is possible for students setting up experiments to confuse the role of the terminals of electrical instruments, or to find the application of positive and negative signs somewhat muddled. In some areas of corrosion it is vital to know the role of each terminal, and to distinguish positive from negative. One Royal Navy ship suffered severe corrosion damage to the hull because the positive and negative terminals of electrical equipment had been wired wrongly. Our discussions will try to clarify these aspects as they arise, but some comment is appropriate at this stage, particularly since there seems to be a discrepancy between an electrical engineer's terminology and that of a corrosion engineer.

Most laboratory corrosion measurements can be made using simple d.c. electrical circuits, of which Fig. 2.7(a) is an example. (Although recent specialist developments in corrosion science utilise alternating currents, all discussion in this book will be about direct current.) A source of e.m.f. is represented by the symbol (1) while symbol (2) is used for an ammeter. It is conventional to use the shorter line of the battery symbol for the negative terminal and the longer line for the positive. The negative pole, also known as the **cathode**, provides the electron current which flows around the circuit towards the attraction of the positive pole, or anode. In batteries, the

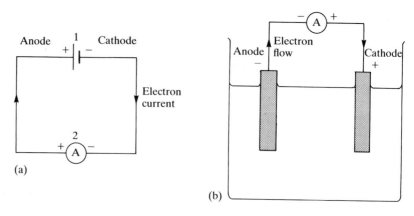

Fig. 2.7 Electrical science conventions:
(a) Anode, cathode, positive and negative definitions in a battery (1), and the direction of electron current flow which results in a positive current reading on an ammeter (2).
(b) Anode, cathode, positive and negative definitions in a corrosion cell and the direction of electron current flow in the circuit. With an ammeter connected as shown, a positive current reading will be indicated.

electrons are generated at this cathode by anodic electrode processes such as the oxidation reactions discussed in section 2.2. When the negative (low or black) terminal of the ammeter is connected to the negative (cathode) of the battery, as in Fig. 2.7(a), the ammeter will register a positive current because the **conventional current flows from the positive (high or red) terminal to the negative (low or black) terminal, while electrons flow in the reverse direction.**

In an aqueous corrosion cell, depicted in Fig. 2.7(b) (and explained in section 4.2), the electrode which produces the electrons for the external circuit is still the negative, but in this case it is referred to as the **anode** because convention leads us to think of the anode as supplying the positive current (positive ions) through the solution. It is this change of terminology which often causes confusion. *If you have not yet read Chapter 4 you will not understand why this change of terminology is necessary. You will find it useful to re-read this section after you have studied sections 4.2–4.5 at which time you will be familiar with the processes which occur at the anode and cathode in both corrosion cells and batteries.* Meanwhile, three rules which summarise the situation are:

> **Electrons flow from negative to positive through the metallic circuit.**
>
> **Conventional (positive) current flows from the anode of a d.c. power supply into the metallic circuit.**
>
> **Conventional (positive) current flows from the anode of an electrolytic or corrosion cell into the electrolyte.**

Laboratory corrosion studies, then, often involve measurements of current and potential. Current measurement is achieved with an ammeter which is inserted into the circuit, in series, so that the entire current flow in the circuit passes through the meter. Remember that ammeters are constructed to give positive deflections (or digital readings) when electrons pass through from the negative (low or black) terminal to the positive (high or red) terminal.

The use of ammeters in corrosion current measurements should always be treated with caution because their finite resistance adds to the resistance within the corrosion cell and can affect the reactions which occur. Instruments known as **zero resistance ammeters** (ZRAs) are used for accurate measurements. The experiments described in this book will be adequately carried out using a modern multimeter, used in the current mode.

The measurement of potential difference between two points is performed with a **voltmeter** which is connected across the points of interest, **in parallel**. The aim of such a measurement is to allow as little current as possible to be diverted from the circuit. This is achieved by having an extremely high resistance within the voltmeter. All traditional voltmeters draw some current from the circuit and will inevitably lead to a small error in the measurement, but the greater the internal resistance, the smaller the error. Many modern electronic instruments have resistances of the order of giga-ohms and are quite suitable for most work. Older instruments should be treated with more caution. Errors in the measurement of potential differences have been overcome by the use of **potentiometers** which are able to measure potential yet draw no current. These are not as simple to use as voltmeters, but should be used for accurate work. However, the modern electronic multimeter used in the volts mode is adequate for use within the experiments of this text. Remember that positive potential differences are indicated from the high (positive or red) terminal to the low (negative or black) terminal.

The interested reader will find ample further discussion of this subject in the bibliography, but all readers should now be equipped to progress to a study of the theory of corrosion.

2.8 GLOSSARY

acid	a solution in which the concentration of hydrogen ions exceeds the concentration of hydroxyl ions
alkali	a solution in which the concentration of hydroxyl ions exceeds the concentration of hydrogen ions
anion	ion with negative charge

Enabling theory for aqueous corrosion

cation	ion with positive charge
corrosion	the degradation of metals by an electrochemical reaction with the environment
defect (in metal)	any arrangement of the atoms in a solid which does not match the ideal lattice structure
dendrite	a characteristically ordered crystal grown from a melt
dissociation	break-up of ionic solid into discrete ions
electromotive force	electrical potential: the source of electricity
environment (in corrosion)	the chemical species which is immediately adjacent to a metal surface and which at a given temperature and pressure influences corrosion reactions
free energy	the portion of a substance's internal energy which is available to perform work
ion	an atom with electrical charge
mole	SI unit of substance: molecular weight in grams
molecular weight	sum of weights of atoms which comprise the molecule
neutral (solution)	when the concentrations of hydrogen and hydroxyl ions are equal
oxidation	a reaction which generates electrons
reduction	a reaction which consumes electrons
surroundings	all things distinct from the system
system	any particular amount of substance of interest
transition state	the arrangement of atoms intermediate between reactants and products
valency	the characteristic number of electrons lost when a metal atom becomes an ion

2.9 WORKED EXAMPLES

E2.1 What is the molecular weight of hydrated copper(II) sulphate, $CuSO_4.5H_2O$?

Ans: The atomic weights are:
$Cu = 63.5$, $S = 32$, $O = 16$, $H = 1$,
therefore the molecular weight is
$63.5 + 32 + (4 \times 16) + (5 \times 18) = 249.5$

E2.2 How do you make a solution of copper (II) sulphate which is 0.4 M?

Ans: A solution which is 1 M contains 249.5 g per litre of solution.
A solution which is 0.4 M contains $0.4 \times 249.5 = 99.8$ g per litre.

E2.3 The elements nickel and oxygen both have a valency of two. When nickel is heated in air it reacts with oxygen to form the ionic material, nickel oxide. Write equations to describe the processes involved.

Ans: The nickel undergoes an oxidation reaction:

$$Ni \rightarrow Ni^{2+} + 2e^-$$

The oxygen undergoes a reduction reaction:

$$O + 2e^- \rightarrow O^{2-}$$

The two processes together lead to:

$$Ni + O \rightarrow Ni^{2+} + O^{2-}$$

Note: According to our definition, the oxidation of nickel above is a corrosion reaction.

E2.4 Write a corrosion reaction for iron.

Ans: Iron has two valencies: 2 or 3
Thus there are two possible reactions:

$$1. \quad Fe \rightarrow Fe^{2+} + 2e^-$$

$$2. \quad Fe \rightarrow Fe^{3+} + 3e^-$$

E2.5 Under standard conditions, what is the concentration of (a) hydrogen ions (b) hydroxyl ions in a solution of pH 8.4?

Ans: By definition, $\quad 8.4 = -lg\,[H^+]$
i.e. $\qquad\qquad 0.6 - 9 = + lg\,[H^+]$
Thus $\qquad\quad lg[H^+] = lg\,(10^{0.6} \times 10^{-9})$
And $\qquad\qquad [H+] = 4 \times 10^{-9}$ M

$$\text{Since} \quad [H^+].[OH^-] = 10^{-14}$$
$$[OH^-] = \frac{10^{-14}}{4 \times 10^{-9}}$$
$$= 2.5 \times 10^{-6} \text{ M}$$

E2.6 A source of e.m.f. of 0.2 volts drives an electrical current of 2 mA through a solution of sodium chloride for 1 hour.
(a) How much charge has been passed?
(b) Assuming there is no resistance elsewhere in the circuit, what is the resistance represented by the solution?
(c) By what means does the charge pass through the solution?

Ans: (a) The charge passed is the current (A) × time (s)
$$= 0.002 \times 60 \times 60$$
$$= 7.2 \text{ C}$$
(b) From Ohm's Law, $\quad V = I.R$
so the resistance is voltage/current
$$R = 0.2/0.002$$
$$= 100 \ \Omega$$
(c) The charge is carried by the ions in the solution, i.e. Na^+, Cl^-, H^+, OH^-.

2.10 PROBLEMS

P2.1 Potassium chloride is an ionic substance, the aqueous solution of which is commonly used in corrosion experiments.
(a) What is its molecular weight?
(b) How do you make a solution which is 0.1 M?
(c) What is meant by a 'saturated' solution?

P2.2 What is the pH of a solution containing 10^{-3} M hydrogen ions?

P2.3 Nitric acid is a liquid with the formula HNO_3. When added to water it dissociates into a hydrogen ion and a nitrate ion, NO_3^-.
(a) What is the molecular weight of nitric acid?
(b) If 6.3 g is contained in a litre of water, what will be the concentration of hydrogen ions?
(c) What will be the pH of this solution?

P2.4 When iron rusts it forms a chemical compound with atoms of oxygen. If iron can have a valency of either two or three, while oxygen only has a valency of two,
(a) Give two possible ratios of iron/oxygen atoms in rust.
(b) Suggest two possible formulae for rust.

P2.5   You are given three liquids:
1. A solution containing 0.1 M sodium chloride
2. A solution containing 0.5 M hydrochloric acid
3. Tap water

Which solution will conduct electricity best?

2.11 REFERENCES

1. Atkins P W 1978 *Physical chemistry*. Oxford University Press, Ch. 8
2. Vogel A I 1981 *Textbook of quantitative inorganic analysis* (4th edn). Longman, p 36

2.12 BIBLIOGRAPHY

Anderson J C, Leaver K D, Alexander J M, Rawlings R D 1982 Materials science (2nd edn) Nelson.

Atkins P W 1978 *Physical chemistry*. Oxford University Press

Carroll A D, Duggan J E, Etchells R 1978 *Physical science, first level* (1st edn) Hutchinson TECtexts

Higgins R A 1981 *Materials for the engineering technician.* Hodder and Stoughton

Morley A, Hughes E 1970 *Principles of electricity in SI units* (3rd edn) Longman

Rosenberg J L 1980 *Theory and problems of college chemistry* (6th edn). Schaum's outline series; New York: McGraw-Hill

Vogel A I 1981 *Textbook of quantitative inorganic analysis* (4th edn). Longman, Ch. 2

3 PRACTICAL AQUEOUS CORROSION

It is a capital mistake to theorize before one has data.

(Sir Arthur Conan Doyle: *The Memoirs of Sherlock Holmes*)

Most of the experiments described in this chapter are suitable for use in college or school science laboratories. They are in their simplest form, and when students or teachers are familiar with them, may be refined using dedicated apparatus. In some cases, basic metallurgical equipment is advantageous, for example, grinding and polishing facilities. The experiments can be performed either by students or as demonstrations for classes. In the latter case, they form excellent visual aids which can be used in parallel with lesson material. The authors have used the experiments described for a number of years and are fully convinced of the value of even the simplest experiment in understanding corrosion. They are described early in the book on the basis that 'doing is learning' and at this stage it is not necessary to understand the theory behind the practice. Comments have been added where necessary, and a discussion occurs in section 3.3 at the end of the chapter.

Corrosion was defined in section 2.2 as the degradation of a metal by an electrochemical reaction with the environment. From a practical point of view there are three keywords in the definition: metal, electrochemical and environment. Each of the experiments illustrates the importance of one or more of these keywords and an understanding of all three is necessary for an appreciation of the mechanisms of corrosion and their control.

3.1 COMMON APPARATUS

Some items of common apparatus will be discussed here to avoid duplication later.

Power supply
A source of direct current is required. Commercial devices are available which are both potential and current controllable. Otherwise, an ordinary 12 V car battery charger is quite suitable. Dry cell batteries may also be

used if they can be mounted in a holder which enables easy connections to be made.

Inert electrodes

Electrodes are conducting media which provide an interface between the electron current in wired electrical circuits and the ionic current of aqueous solutions. Inert electrodes are best made of platinum foil or gauze, approx 100 mm^2, which should have platinum wires attached, leading away from the solution. Electrical connections can be made to these wires. (The connections should not be allowed to get wet.) A cheaper but satisfactory alternative is to use carbon rods to which suitable connections can be made, either by crocodile clips or a more permanent method. Some carbon rods disintegrate slowly and contaminate the solutions, but for quick, simple experiments, they are quite adequate.

Working electrodes

Copper plate is preferred in a number of experiments which are designed to explore comparatively small differences in the system. The electrodes should be as similar as possible or false observations may be made. It is desirable to prepare a number of copper plate electrodes, approx $150 \times 50 \times 3$ mm, which should be kept scrupulously clean.

Multimeters

Many experiments require measurements of current and potential. Modern electronic multimeters are versatile and easy to use. They can be used in current, voltage and resistance modes. More traditional ammeters are also suitable. However, errors in potential readings may arise when voltmeters are used in parallel. This is less likely with modern voltmeters which have much greater internal resistance.

Use of solutions

A common solution used in corrosion experiments is 3.5 per cent sodium chloride. This solution is best made up in bulk and is obtained by dissolving 35 g of laboratory grade reagent in a litre of water. When preparing solutions, use de-ionised or distilled water whenever possible. (These forms of water will be denoted by the term 'pure' in the descriptions to distinguish them from tap water.) Add the appropriate chemical and stir with a glass rod until dissolved.

Use of acids

When reference is made to acids, for example, nitric acid, it is the commonly supplied concentrated reagent which should be used. If dilution is necessary, details will be given. Dilution of acids should always be performed by addition of acid to water, **NOT** water to acid. In the case of sulphuric acid, this rule is essential; however, in expt 3.12 water is added to nitric acid in contravention of this general rule, but if done carefully, is quite safe. *Safety glasses should always be worn when handling acids.*

Laboratory glassware

Common laboratory glassware such as beakers, test tubes, Petri dishes, glass rods and tongs should be available. A large quantity of sodium bicarbonate should be on hand in case of acid spillages. Paper towels and rags are useful for spillages of aqueous solutions and for drying specimens.

3.2   THE EXPERIMENTS

Experiment 3.1

Aim: To demonstrate the different conductivities of liquids.

Apparatus
A. Power supply.
B. Three 250 ml beakers.
C. Two inert electrodes.
D. Ammeter, switch and connections.
E. Liquids: pure water, white spirit, and stock aqueous sodium chloride.

The apparatus is connected as shown in Fig 3.1. The power supply should be capable of providing a voltage of 2 to 5 V.

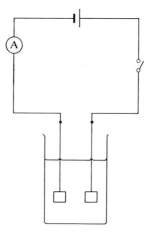

Fig. 3.1   Apparatus for expt 3.1

Procedure
1. Pour 200 ml white spirit into a beaker.
2. Immerse the electrodes.
3. Close the switch.
4. Record the current.
5. Open the switch.

6. Remove the beaker and test liquid. Clean and dry the electrodes.
7. Repeat steps 1–5 using pure water.
8. Repeat steps 1–5 using stock aqueous sodium chloride.

Note: You should have measured zero current using white spirit or pure water.

Question
Why does the aqueous sodium chloride conduct electricity while the other liquids do not?

Experiment 3.2

Aim: To demonstrate the importance of the environment in the process of rusting.

Apparatus
A. 6 test tubes and a stand.
B. 6 uncoated steel nails, 50 mm long.
C. Sealant.
D. 1 cork or bung.
E. Chemicals: sodium nitrite, sodium chromate.

Procedure
1. Place each tube in the stand and add a nail to each one.
2. Leave the nail in tube no. 1 exposed to the air.
3. Half cover the nail in tube no. 2 with tap water.
4. Completely cover the nail in tube no. 3 with tap water.
5. Fill tube no. 4 with boiled tap water. Add bung and seal so that no air can get into the tube.
6. Prepare aqueous solutions of sodium nitrite and sodium chromate, each 10 g in 100 ml tap water.
7. Completely cover the nail in tube no. 5 with aqueous sodium nitrite.
8. Completely cover the nail in tube no. 6 with aqueous sodium chromate.
9. Leave at least overnight and preferably for a week. Observe the effects of the different environments upon the nails.

Note: Sodium nitrite and sodium chromate in solution are called inhibitors. These will be discussed in Chapter 13.

Exercise
Correlate any corrosion you have observed with the different environments which you have created.

Experiment 3.3

Aim: To illustrate relative energy levels of different metals.

Apparatus
A. Voltmeter.
B. 1 250 ml beaker.
C. Selection of various electrodes (any convenient geometry or size) such as copper, iron, zinc, tin, lead, magnesium, silver, platinum, graphite rod.
D. Electrical connections: crocodile clips for quick and easy attachment to metals.
E. Stock aqueous sodium chloride.
The arrangement of the apparatus is shown in Fig. 3.2.

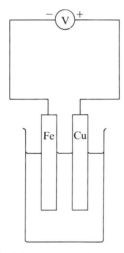

Fig. 3.2   Apparatus for expt 3.3

Procedure
1. Pour 200 ml of stock sodium chloride solution into the beaker.
2. Use copper and iron as the two electrodes. Connect iron to the black (negative) terminal of the voltmeter, and immerse in the beaker. Take care that the electrodes do not touch, and that the electrical connections do not make contact with the solution.
3. Note the magnitude and sign of the potential when it has become reasonably steady.
4. Keep the copper in position, but exchange the iron for each of the other available metals in turn.
5. Compile a table of potentials placing them in numerical order from the most positive to the most negative.

Note: You have shown that different materials have different energy levels. These levels are measured as a series of potentials which develop spontaneously when dissimilar metals are in electrical contact in an aqueous environment. The potentials are called galvanic potentials and the series is the galvanic series.

Questions
(a) What is the significance of the potentials you have found?
(b) What does the change of sign tell you?
(c) Draw a circuit diagram and examine the flow of electrons in the various cases. What is happening at the metal surface in solution?

Experiment 3.4

Aim: To illustrate the difference in energy levels between a metal and its oxide.

Apparatus
A. 1 250 ml beaker.
B. 2 copper electrodes.
C. Bunsen burner and tongs.
D. Stock solution of aqueous sodium chloride.

Procedure
1. Using the tongs, hold one copper electrode in the Bunsen flame until heavily oxidised (blackened).
2. Cool the copper and use it as one of the electrodes in the apparatus of Fig 3.2.
3. Measure the magnitude and sign of the potential between the electrodes.

Note: The potential you have measured is a difference in energy levels between the copper and its oxide.

Questions
(a) Draw a circuit and indicate the direction of the current flow. Which of the two electrodes supplies electrons into the wires of the circuit?
(b) Which of the two electrodes is of higher energy?

Experiment 3.5

Aim: To illustrate the effect of temperature upon energy levels.

Apparatus
A. 2 250 ml beakers.
B. 2 copper electrodes.
C. Large filter paper.
D. Hotplate.
E. Stock solution of aqueous sodium chloride.
The apparatus in Fig. 3.3 is for use in expts 3.5–3.7.

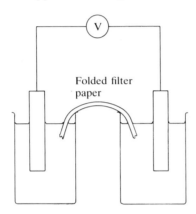

Folded filter
paper

Fig. 3.3   Apparatus for expts 3.5–3.7

Procedure
1. Set up the apparatus of Fig 3.3.
2. Immerse the copper electrodes in the solutions.
3. Heat the solution in one beaker using the hotplate.
4. When the solution is almost boiling, switch off the heat.
5. Connect the voltmeter and measure the magnitude and sign of any potential difference between the two electrodes.

Notes: (i) The filter paper is folded into a strip and bridges the two beakers. It must be saturated with the solution from the beakers before any attempt is made to measure potential.
(ii) A bunsen burner may be used to heat the solution but care should be taken not to burn the filter paper. It may be necessary to raise the non-heated solution on a small stand.

Question
Which of the two electrodes is in a higher energy state:
the hot or the cold copper?

Experiment 3.6

Aim: To illustrate the generation of energy difference because of changes in the concentration of oxygen in the environment.

Apparatus
A. 2 250 ml beakers.
B. 2 copper electrodes.
C. Large filter paper.
D. Supply of nitrogen or argon.
E. Stock solution of aqueous sodium chloride.

Procedure
1. Set up the apparatus as in Fig. 3.3 using copper electrodes.
2. Bubble nitrogen or argon through one of the solutions for 10–15 minutes.
3. Measure the magnitude and sign of any potential difference.

Note: The aim of bubbling gas through the solution is to reduce greatly the level of dissolved oxygen in that solution.

Question
Which electrode is at a higher energy: the one with more oxygen in the solution, or the one with less oxygen in solution?

Experiment 3.7

Aim: To show how a change in concentration of the ions in solution affects the energy levels of metals.

Apparatus
A 3 250 ml beakers.
B 2 copper electrodes.
C. Large filter paper.
D. Chemicals: copper(II) sulphate.

Procedure
1. Prepare an aqueous solution of copper(II) sulphate, 20 g in 200 ml.
2. Add 100 ml of solution to each beaker.
3. Set up the apparatus as in Fig. 3.3, using the copper sulphate solution instead of sodium chloride.
4. Immerse the copper electrodes in the solutions.

5. Connect the voltmeter and monitor any potential differences between the two electrodes.
6. Fill the third beaker with tap water and pour into one of the solutions of copper sulphate.
7. Observe any changes of potential.

Question
Which electrode is at a higher energy: the one in the more dilute or less dilute solution?

Experiment 3.8

Aim: To illustrate the effect of mechanical work on the energy of a metal.

Apparatus
A. 1 piece of mild steel 30 × 30 × 10 mm.
B. Metal punch.
C. Chemicals: copper(II) ammonium chloride, hydrochloric acid.

Procedure
1. Grind the steel specimen to remove coarse scratches.
2. Using the punch, mark the specimen so that an easily identifiable symbol is stamped into the surface of the metal to a depth of about 1 mm.
3. Grind the specimen until the identification mark is invisible. Polish to achieve a fair mirror finish.
4. Prepare a solution of copper(II) ammonium chloride, 12 g in 100 ml.
5. Prepare an acidic solution of copper(II) ammonium chloride, 12 g in 100 ml, to which is added 5 ml hydrochloric acid.
6. Slowly pour the non-acidified solution over the surface of the specimen, allowing the solution to run to waste.
7. Immerse the specimen in the acidified solution and leave for 1 hour.
8. Remove the specimen. The copper coating should be easily rubbed off under a running tap. The specimen should show the original identification mark.

Notes: (i) The treatment in step 6 must precede that with the acidic solution because it provides an underlying loosely adherent layer and facilitates easy removal of the copper coating in step 8.
(ii) This experiment is the basis of a forensic examination technique which assists in the identification of stolen property.

Experiment 3.9

Aim: To illustrate regions of different energy within metals.

Apparatus
A. 1 convenient cast brass object.
B. Commercial brass polishing fluid and cloth.
C. Cotton wool.
D. Chemicals: iron(III) chloride, hydrochloric acid.

Procedure
1. Polish the specimen until a mirror finish is obtained.
2. Prepare an acidified aqueous solution of iron(III) chloride, 5 g in 100 ml, to which is added 2 ml hydrochloric acid.
3. Dip a cotton wool ball into the solution and rub gently over the polished surface for about 15 seconds.
   CAUTION: *Avoid skin contact with the etchant. Use rubber gloves.*
4. Wash under the tap and dry.

Notes: (i) The procedure is known as etching and is commonly used in metallurgical laboratories for examination of the grain structure of metals.
(ii) Sand cast brass objects work well in this experiment because they usually have a grain size large enough to be seen with the eye: many grain structures are visible only under a microscope.

Question
Which part of the metal is of higher energy: the area within a grain or the grain boundary?

Experiment 3.10

Aim: To use chemical colour changes as an indicator of energy levels.

Apparatus
A. 2 Petri dishes.
B. 2 uncoated steel nails, approx 50 mm long.
C. Chemicals: agar-agar, sodium chloride, phenolphthalein, potassium ferricyanide, copper(II) ammonium chloride, hydrochloric acid, ethanol or clean methylated spirits.

Procedure

1. Prepare an aqueous solution of potassium ferricyanide, 1 g in 100 ml.
2. Prepare a solution of phenolphthalein, 1 g in 100 ml ethanol (or clean methylated spirits), diluted with 100 ml water.
3. Boil 2.5 g agar-agar and 7.5 g sodium chloride in 250 ml de-ionised water until a uniform suspension is obtained. Add 2 ml of the solution of potassium ferricyanide and 2 ml of the phenolphthalein solution.
4. Pour some of the liquid agar-agar mixture into a Petri dish and allow to cool. Before the gel sets, carefully immerse a nail.
5. Prepare a solution of aqueous copper(II) ammonium chloride, 12 g in 100 ml, to which is added 5 ml hydrochloric acid.
6. Immerse a nail in the acidified solution for about 5 minutes.
7. Remove the nail and carefully dry.
8. Abrade one half of the nail to remove the copper coat.
9. Repeat step 4 using the partially plated nail.
10. Leave for 12 hours.

Notes: (i) Phenolphthalein is a solution which turns pink in the presence of hydroxyl ions.
(ii) The potassium ferricyanide solution turns blue in the presence of iron(II) ions, Fe^{2+}.

Exercise
Describe the colour differences in both cases and account for the changes which occur in the gel.

Experiment 3.11

Aim: To use chemical colour changes as an indicator of energy levels.

Apparatus
A. 1 piece of mild steel plate, any convenient size or geometry.
B. Dropping pipette.
C. Agar-agar indicator solution from expt 3.10.

Procedure
1. Pipette about 1–2 ml of the agar-agar indicator solution used in step 4 of expt 3.10, on to the cleaned surface of the steel plate so that a well-rounded droplet is obtained.
2. Allow to stand for an hour and observe the colour changes which occur inside the droplet.

Note: If the stock solution of agar-agar has gelled, simply reheat and stir to obtain a uniform suspension.

Exercise
Correlate the colour changes which have occurred with possible differences in the oxygen levels within the solution.

Experiment 3.12

Aim: To illustrate the formation of surface films which protect metals against corrosion.

Apparatus
A. 2 250 ml beakers.
B. 1 piece of clean mild steel sheet, $100 \times 50 \times 2$ mm approx.
C. Screwdriver, tongs.
D. Chemicals: nitric acid.

Procedure
1. Carefully pour 100 ml nitric acid into one beaker and fill the second with tap water.
2. Gently stand the steel plate in the acid.
3. Allow the specimen to stand undisturbed for a minute.
4. **VERY** carefully add 100 ml of water to the beaker, pouring down the side of the beaker so as not to disturb the acid.
5. Using the screwdriver, tap or, if necessary, scratch the surface of the steel as it stands in the solution. As soon as the reaction starts, use the tongs to remove the specimen and place it in the beaker of water.

Notes: (i) *Take care in case of spillage when the reaction starts. Sodium bicarbonate should be on hand to spread on any acid spillage. Safety glasses should be worn. The experiment should be carried out in a fume cupboard. Carefully dispose of solutions with copious quantities of tap water.*
(ii) Nitric acid is an oxidising agent, which means that it is capable of transforming metals into oxides.

Questions
(a) Why does the metal not corrode very vigorously in the concentrated acid?
(b) Why does it corrode when it is knocked in the dilute acid?

Experiment 3.13

Aim: To illustrate the formation of surface films which protect metals against corrosion.

Apparatus
A. 1 piece of clean, polished mild steel, 100 × 50 × 2 mm approx.
B. 2 × 250 ml beakers.
C. Glass rod.
D. Chemicals: nitric acid, copper(II) sulphate.

Procedure
1. In one beaker dissolve 10 g copper(II) sulphate in 100 ml water.
2. Pour 100 ml nitric acid into the second beaker.
3. Ensure that the steel is scrupulously clean and polished. Dip it vertically in the beaker of copper sulphate so that the solution half covers it. Leave for 1 minute.
4. Remove the steel slowly from the solution and allow excess solution to drain off.
5. Dip the steel vertically into the nitric acid so that the acid level is just above the level reached by the copper sulphate solution. Leave for 5 seconds.
6. Carefully remove from the acid solution, allow excess acid to drain off, and immediately dip the steel into the copper sulphate.
7. Remove the steel at once from the copper sulphate solution and immediately strike the surface with the glass rod.

Notes: (i) The specimen quality is very important in this experiment. The first immersion causes the steel to turn pink. The surface should be uniformly coloured: if it is not then the specimen is of doubtful quality and the success of the experiment is unlikely.
(ii) The immersion in the nitric acid causes the colour to disappear.
(iii) The second immersion in the copper sulphate is crucial. It must be done carefully and quickly, and neither the steel nor the solution must be agitated in any way. Contact of the specimen with the beaker must be avoided. The copper sulphate must **NOT** reach a higher level than did the nitric acid. If the steel turns pink or brown before it has been struck then the experiment has failed. Try again with a new specimen.

Questions:
(a) What is the pink coloration?
(b) Why does it disappear in nitric acid?
(c) Why does it not reappear during the second dip into the copper sulphate?
(d) Why did the colour reappear only after the touch of the glass rod?

Experiment 3.14

Aim: To illustrate the formation of surface films which protect metals against corrosion.

Apparatus
A. 1 piece of aluminium sheet, 100 × 50 × 2 mm approx.
B. Screwdriver.
C. Dropping pipette.
D. Chemicals: mercury(I) nitrate, nitric acid.

Procedure
1. Prepare an aqueous solution of mercury(I) nitrate with 1 g in 100 ml. **CAUTION:** *Toxic solution. Use rubber gloves.*
2. Pipette a drop of solution on to the clean surface of the aluminium to give a drop of about 20 mm diameter.
3. Using the pointed corner of the screwdriver, score the sheet, commencing on one side of the drop and passing through it to the other side.
4. Dry the metal with an absorbent paper towel and observe the metal for approximately 10 minutes.

Experiment 3.15

Aim: To illustrate the crevice corrosion of stainless steels.

Apparatus
A. 1 250 ml beaker.
B. 1 piece stainless steel, 100 × 50 × 2 mm approx.
C. 1 rubber band.
D. Chemicals: iron (III) chloride, hydrochloric acid.
E. Stock sodium chloride solution.

Procedure
1. Wrap the rubber band lengthwise around the specimen so that it is taut.
2. Prepare an acidified solution of iron(III) chloride, 5 g in 100 ml to which 2 ml of hydrochloric acid is added.
3. Pour 100 ml of stock sodium chloride solution into the beaker and add 5 ml of the acidified iron(III) chloride solution.
4. Stand the specimen in the beaker and leave for 24 hours.
5. Remove the specimen, wash under tap and dry.
6. Remove the rubber band.

Note: The best material is a 430 type stainless steel, or other member of the 400 series, but a sample of type 304 will probably give the desired result. A simple test which may help in identification is to measure the magnetic property of the stainless steel: alloys in the 400 series are magnetic, but those in the 300 series are not.

Exercise
(a) Comment upon the performance of stainless steel in the given electrolyte.
(b) Give a reason for the behaviour you have observed.

Experiment 3.16

Aim: To illustrate the effect of the combination of stress and a specific environment upon 70/30 brass.

Apparatus
A. 1 250 ml beaker, tongs.
B. 2 identical pieces of hard drawn 70/30 brass tube, 20 mm long, 20 mm diameter.
C. Chemicals: mercury(I) nitrate, nitric acid.

Procedure
1. Heat treat one piece of tube for 1 hour at 500 °C.
2. Prepare an acidified aqueous solution of mercury(I) nitrate, 1 g in 100 ml, to which 1 ml nitric acid is added.
3. Make sure that the two specimens are distinguishable.
4. Immerse both specimens in the solution and leave for 10–15 minutes.
5. Using tongs, remove the specimens and inspect for cracks.

Notes: (i) **CAUTION:** *the solution is very toxic and should be handled carefully. Use rubber gloves throughout.*
(ii) The heat treatment is to relieve stresses which are residual from the manufacturing process.
(iii) If the residual stress is sufficiently high the tube may disintegrate. Gentle pressure in the tongs may be enough to achieve this.

Question
What is the silvery coating on the tubes after removal from the solution?

3.3 DISCUSSION

Corrosion is a chemical reaction of a metal with its environment as a result of which electric current flows. All the environments which have been considered in the experiments have been aqueous, but this does not mean that corrosion is absent if there is no water present. There are many corrosion reactions in what can be considered as dry environments. Furthermore, remember that corrosion can occur in the atmosphere where the presence of water vapour, together with an ionic material from any available source, is enough to cause corrosion for the same reasons as if the metal were immersed. The presence of both water and an ionic material is complementary: current can only be carried through water by free ions, while the water causes dissociation of the ionic solid to produce the necessary free ions. **Expt 3.1** was designed to show that electric current flows in solutions only if they possess ions, aqueous sodium chloride for example. If there are no ions, as in white spirit, or the very low quantities of ions in pure water, there is no current flow and no electrical circuit.

Expt 3.2 is an introduction to corrosion reactions, designed to show how variations in the environment cause very different degrees of corrosion. The requirement for ions, oxygen and moisture to be present is clearly demonstrated. An interesting lack of corrosion is obtained in the presence of inhibitors, the action of which is discussed in Chapter 13.

In the first experiment a current was driven around the circuit using a source of potential, but in the remaining experiments this was absent – or was it? Of course, it was not absent, merely disguised. In **expt 3.3** a source of potential was created by connecting two different metals together. Remember in Chapter 2 it was shown that electrical potential can be considered to be the same as energy, creating a 'force' which drives electric current from a high energy state to a low one. If it occurs spontaneously, this is in keeping with the rule that nature is striving to achieve low energy states. Corrosion is a spontaneous process and we stimulated it by connecting dissimilar metals in expt 3.3. The amount of energy released was different for each combination, and by examining the direction in which the current flowed in the circuit you should have been able to tell which electrode was supplying electrons into the external circuit. Remember the corrosion reaction illustrated by eqn [2.7] which shows how the metal generates electrons for the external circuit. In any of the experiments, the metal which does so must be corroding.

Some of the most common corrosion mistakes made by engineers relate to galvanic effects and dissimilar metal couples. This is the subject of Chapter 5 and will be discussed again in the theory of Chapter 4.

It should be clear that it is not necessary to have two different metals connected together in order to obtain corrosion. The object of the sequence of expts 3.4 to 3.9 is to show how corrosion may occur on the surface of a single piece of metal.

Question: *Then why do the experiments use two electrodes?*

It is necessary to use two electrodes in order to measure the potentials more easily. The two identical electrodes are linked, as one piece of metal, via the external circuit. By using identical electrodes we can explore the effects produced by small changes to the environment or surface condition of the electrode. Once this is achieved, the energy differences lead to potential differences, which in turn cause corrosion.

Expt 3.4 showed the difference of energy between a metal and one of its chemical compounds, in this case the oxide. Remember, a metal behaves like an assembly of atoms, but in its oxide it exists as ions. You should have discovered that the metal is in a higher energy state than its oxide and provides the electrons for the circuit, i.e. it corrodes. Thus, in any piece of metal which is partially coated with oxide, the metal will form a local corrosion cell with the oxide and corrode. If there are no pores in the film and the oxide layer completely covers the metal, this cannot occur.

An obvious way of giving a material energy is to heat it. In **expt 3.5** a metal was heated and compared with another piece of the same metal which was cool. You should have measured a potential, the sign of which told you that the hot electrode had more energy than the cold one. The implication of this is that in a metal structure where there are temperature variations, local corrosion cells can be established in which the hot areas begin to corrode compared to the cold ones.

In **expt 3.6** a similar effect was created by the expulsion of oxygen from one of the solutions. The result obtained should have been that the electrode in the solution with less oxygen was supplying the electrons and therefore corroding. Again, we can say that any system where there is a variation in the concentration of oxygen is a possible corrosion cell. This is a vital conclusion with respect to what happens inside crevices or in other areas which have restricted access to air. Thus we see again that not only does water play an important role in corrosion, but oxygen does too. Crevice corrosion is discussed in Chapter 7.

Expt 3.7 showed how changes in the concentrations of ions in the environment also generate potentials. A change in the concentration of liquid passing down a metal pipe, for example, could produce a corrosion problem for a process engineer.

Expt 3.8 is interesting as an act of 'magic'. An identification mark, apparently removed, has been made visible again. The reason why this can be done is that when the mark is punched into the metal, the material is mechanically worked and some of the energy is transmitted to the areas immediately below the mark. Even though the mark is removed the energy remains in the material below the mark. An etching process causes the areas with more energy to corrode preferentially and the mark reappears.

Expt 3.9 is not an obvious corrosion experiment but the method is fundamental in metallographic analysis and works because grain boundary areas are of higher energy than areas within grains. Atoms at grain boundaries are mismatched to the regular crystal lattice structure (nature would prefer them to be neatly located in ideal sites) and such defects have higher energy. The contact with a corrosive solution, the etchant, causes the grain boundary areas to corrode in preference to the areas within the grains and the result is that a mirror smooth surface is transformed into one in which the topography is visible. The experiment should have shown the large grain structure common in cast brasses. Grain boundary corrosion is discussed again in section 6.1.

The use of indicators in **expt 3.10** to indicate corrosion reactions is a colourful experiment. The blue colour which develops shows that iron is corroding and the regions which turn pink correspond to the lower energy areas. With a nail, you should observe that the head and the tip are most likely to corrode because they have been subjected to more mechanical work during the manufacturing process. When half the nail has been coated with copper we have a small galvanic cell and the steel end corrodes turning blue, while the coppered half goes pink. An interesting variation is described in **expt 3.11** in which the colour changes beneath the droplet show that the area at the centre of the drop is the zone of corrosion. This is a result of the geometry of the droplet: the diffusion path of oxygen through the environment is longer at the centre of the drop than at its edge. (A detailed discussion occurs in section 7.2.) The colour changes are therefore indicative of accessibility by oxygen and reinforce the conclusions of expt 3.6 which were that areas with less oxygen are likely to corrode.

In expt 3.12 the lack of corrosion of steel in nitric acid is examined. Because nitric acid is an oxidising agent, it causes a very thin layer of non-reactive oxide to form on the surface. This effectively excludes the environment and there is no appreciable corrosion, even when the acid is diluted. (The film cannot be generated in dilute acid alone.) However, the film is very fragile and a small tap is usually enough to cause rapid corrosion to occur.

The most visually dramatic of all the experiments and one which is worth practising is **expt 3.13**, an extension of the previous one. Immersion of the steel in the copper sulphate solution produces a thin plating of pink copper on the surface. A detailed discussion of what is happening in this reaction occurs in section 4.1. When the plate is then immersed in nitric acid, the copper dissolves and the steel is given a very thin and fragile protective oxide film. The second immersion in copper sulphate does not plate copper because the protective coating acts as a barrier. When the steel is removed from the solution mechanical damage to the inert film allows copper ions access to the steel beneath and the pink copper plates out immediately.

A similar effect is illustrated by **expt 3.14** in which the surface oxide film on aluminium is damaged while underneath a corrosive medium. In air, aluminium rapidly oxidises to repair any damage to the adherent, pore-free corrosion-resistant layer. If the film is disrupted while the metal is beneath the surface of a corrosive solution, then the resulting reaction is rapid, and in this case spectacular.

The type of corrosion known as crevice corrosion has already been mentioned. **Expt 3.15** shows clearly the fallibility of believing that stainless steel is the answer to all corrosion problems. Stainless steel is particularly susceptible to corrosion in areas where there is restricted access by oxygen. The mechanisms are discussed in section 7.1.

Another form of corrosion which is the subject of Chapter 10 is illustrated in **expt 3.16**. Here, the combination of a specific environment and a specific material in which there is residual stress can lead to catastrophic failure. The example chosen is one of the most rapid for ease of demonstration, but the problem is widespread throughout the engineering world. The stresses which remain in a tube after fabrication are sufficient to cause the rapid failure in the experiment. Elimination of the stress by the heat treatment renders the material much less susceptible.

Now that the scope of corrosion has been examined in the laboratory and you have begun to see some of the reasons why metals corrode, we shall look in detail at the theory of aqueous corrosion.

4 THE THEORY OF AQUEOUS CORROSION

> The whole of science is nothing more than a refinement of everyday thinking.
>
> (Albert Einstein: *Out of my later years*)

In nature, most metals are found in a chemically combined state known as an **ore**. These may be oxides, sulphides, carbonates or other more complex compounds, and because many have been present in the Earth's crust since it was formed, we may presume that their chemical condition is somehow 'preferred' by nature. This preference is a result of fundamental laws which make up the study of thermodynamics, and using its terminology we say that ores and other such compounds are in low energy states. In order to separate a metal such as iron from one of its ores, iron oxide, say, it is necessary to supply a large amount of energy. This is usually done by heating (with a reducing agent) in a blast furnace to about 1600 °C. Metals in their uncombined condition are therefore usually high energy states. This is illustrated in Fig. 4.1 where an energy profile has been used to chart the thermodynamic progress of a typical metal atom, from being combined in an ore, separated as a metal atom, and recombined as a corrosion product. In the figure our rather vague term, energy, has been more specifically called the free energy, in continuance of our discussion in section 2.1.

Thermodynamic laws tell us that there is a strong tendency for high energy states to transform into low energy states. It is this tendency of metals to recombine with elements present in the environment that leads to the phenomenon known as corrosion. A good definition of corrosion is therefore:

CORROSION IS THE DEGRADATION OF A METAL BY AN ELECTROCHEMICAL REACTION WITH ITS ENVIRONMENT

The low energy corrosion product is obviously not the same as an ore, although it may be similar, as in the case of iron oxide and rust, and the energies of the ore and the corrosion product may well be comparable.

Question: *Why, then, do metals exist at all in the uncombined state?*

The free energy difference between a metal and its corrosion product, ΔG in Fig. 4.1, represents only the *tendency* of the metal to corrode: it tells us

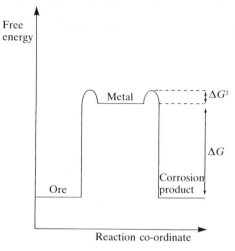

Fig. 4.1 A thermodynamic energy profile for metals and their compounds.

nothing about the *rate* at which the metal corrodes. This is because of the presence of an energy barrier between the metal and the corrosion product. Metal atoms must surmount this barrier before they may corrode and more energy must be supplied for this to occur. In our example, the energy barrier is called the free energy of activation, represented by the symbol $\Delta G^{\ddagger}$. It is the size of the free energy of activation which determines the rate of a corrosion reaction, the rate constant of which we shall denote by k_{corr}. The rate of a corrosion reaction, v, can be expressed as

$$v = k_{corr}. \text{[reactants]} \qquad [4.1]$$

where

$$k_{corr} = A \exp(-\Delta G^{\ddagger}/RT) \qquad [4.2]$$

A is an undefined constant, R is the universal gas constant and T the absolute temperature.

It is well known that hydrogen and oxygen have a strong tendency to form water. The relatively high energy states of the gases can be converted into the low energy state of water with an explosive release of energy, but no measurable reaction occurs in a mixture of the gases until a spark is introduced. A source of energy has then been found to drive the molecules over the activation energy barrier, and without it the rate of reaction is extremely slow even though its thermodynamic tendency is large. The next section examines in more detail the effect of thermodynamics on corrosion and the benefits obtained from its study.

4.1 THERMODYNAMIC ASPECTS OF CORROSION REACTIONS

We have seen that free energy is the single factor which determines whether or not corrosion will take place spontaneously. This is not surprising:

All interactions between elements and compounds are governed by the free energy changes available to them.

In section 2.1 it was stated that, for a reaction to occur spontaneously, there must be a net release of free energy. In this treatment, individual free energies of species are expressed as G and the net change of energy in a reaction is given by ΔG. In line with our notion that natural changes involve a transition from a high to a low free energy state, it is conventional to denote energy *given out* by a *negative sign*, while a net *absorption of energy* by the system is given a *positive sign*. Thus,

For a spontaneous reaction to occur, ΔG must be negative.

In Fig. 4.1 you can see that for the metal to corrode spontaneously, the change in G must be negative. At room temperature most chemical compounds of metals have lower (more negative) values of G than the uncombined metals, hence our original premise that:

Most metals have an inherent tendency to corrode.

Question: *Why do gold, platinum and other precious metals not corrode?*

The theory we have developed so far gives two explanations. First, the energetics may not be favourable. Consider the following reactions and the respective free energy change per mole:

$$Mg + H_2O + \tfrac{1}{2}O_2 \rightarrow Mg(OH)_2 \quad \Delta G^\circ = -596 \text{ kJ/mol} \qquad [4.3]$$

$$Cu + H_2O + \tfrac{1}{2}O_2 \rightarrow Cu(OH)_2 \quad \Delta G^\circ = -119 \text{ kJ/mol} \qquad [4.4]$$

$$Au + \tfrac{3}{2}H_2O + \tfrac{3}{4}O_2 \rightarrow Au(OH)_3 \quad \Delta G^\circ = +66 \text{ kJ/mol} \qquad [4.5]$$

We shall use the symbol ($^\circ$) to represent parameters at 298 K and 1 atmosphere pressure, i.e. standard state parameters.

The free energy data show clearly that copper and magnesium have negative values of ΔG°, while that for gold is positive. Thus, thermodynamics alone tells us that copper and magnesium are expected to corrode naturally in wet, aerated atmospheres, while gold is not.

A second answer to the question is that the free energy of activation may be too great, thus making the rate very slow. A detailed discussion of corrosion rates will be reserved for later sections of this chapter.

It is not just the precious metals which can remain uncorroded in nature. Iron objects have been found remarkably preserved after centuries of immersion at the bottom of peat bogs. This is almost certainly because of the total exclusion of oxygen which, as we found in expt 3.2, is often necessary for corrosion to occur. The example is a good one because it reminds us that the environment is just as important a consideration as the metal.

There may always be special circumstances why corrosion does not occur when expected, but the general principles outlined above form a useful guide.

Experiment 4.1

Place a piece of iron in a beaker containing a solution of 10 per cent copper sulphate. After about 10 minutes you will see that the iron becomes coated with copper, an effect we shall call replating. If the solution is analysed it will be found to contain iron ions.

This example of the corrosion of a piece of metal immersed in an electrolyte has been carefully chosen because it illustrates a number of points and aids our development of a theoretical approach. Of course, there are countless combinations of metal and environment in which corrosion may be observed, yet it is possible to reduce most of these to a set of common features. In section 2.2, a corrosion reaction was defined by the general equation, [2.7], but it was pointed out that this was only one half of an overall process in which the electrons produced in eqn [2.7] are consumed by another reaction. In aqueous corrosion, these two reactions occur at the solid/liquid interface; the electron producing reaction is an anodic reaction, also called an oxidation process. (Note that the term used in this sense is not concerned with a reaction involving oxygen.) The electron consuming reaction is one of reduction and is termed, cathodic. It is important to realise that both anodic and cathodic reactions may occur on the same metal surface.

Question: *What is happening in the beaker in experiment 4.1?*

We can observe that the iron is corroding; in other words, the metal is being degraded into ions according to the simplified equation,

$$Fe \rightarrow Fe^{2+} + 2e^- \qquad [4.6]$$

In contrast, copper ions are replated from the solution thus:

$$Cu^{2+} + 2e^- \rightarrow Cu \qquad [4.7]$$

This cathodic reaction is the reverse of a corrosion reaction, though not all cathodic processes are. Other cathodic reactions will be discussed below. The complete effect is represented by the addition of the two reactions:

$$Fe + Cu^{2+} \rightarrow Fe^{2+} + Cu \qquad [4.8]$$

Notice that although eqn [4.7] would be expected to have a positive value of ΔG, it is a reduction in free energy overall (i.e. for eqn [4.8]) which drives the reaction. Since we know that this reaction is occurring spontaneously then $\Delta G < 0$.

All corrosion is dependent upon temperature. This is because the free energy states of the species involved are dependent upon temperature. Thus it is important to be able to calculate the value of ΔG at any given temperature. This can be done by the important thermodynamic equation:

$$\Delta G = \Delta G^{\circ} + RT \ln J \qquad [4.9]$$

J is defined in this way:

For a reaction: $A + B \rightarrow C + D$

$$J = \frac{[C].[D]}{[A].[B]} \qquad [4.10]$$

The expression for J contains any arbitrary (non-equilibrium) values corresponding to the non-equilibrium free energy change, ΔG, in eqn [4.9]. If the system reaches a point where there is no net change of free energy, we say the system is in equilibrium and $\Delta G = 0$. Then, $J = K$, where K is the equilibrium constant. This leads to:

$$\Delta G^{\circ} = -RT \ln K \qquad [4.11]$$

As you have read in section 2.3, the terms in the square brackets in eqn [4.10] represent the concentrations of the species in the solution. (More precise work requires the use of a quantity called an activity but this will not be used here.) Concentrations are expressed in molarities (moles per litre).

In the case of the copper/iron system above we can use eqn [4.8] to substitute for J in eqn [4.9], thus:

$$\Delta G = \Delta G^{\circ} + RT \ln \frac{[Fe^{2+}].[Cu]}{[Cu^{2+}].[Fe]} \qquad [4.12]$$

It can be shown that the terms involving the concentration of a solid substance can be replaced by unity.

Remember that we are considering the corrosion of iron in a solution of copper sulphate, and it is the free energy change which is driving the reaction.

Question: *How can I measure these energy changes and hence study the corrosion in more detail?*

Experiments 3.3 to 3.7 showed that these free energy differences are measurable as electrical potentials and flow of current. Thus electrical measurements are one way of studying corrosion in more detail. The exact methods involved will be discussed later, but at this stage we can be content to define the expression which relates electrical potential with free energy. This equation is due to the great scientist, Michael Faraday, who expressed the work done (the free energy change of the corrosion process) in terms of the potential difference and the charge transported:

$$\Delta G = E \cdot -zF \tag{4.13}$$

Equation [4.13] is known as **Faraday's Law**. The symbol F represents the charge transported by one mole of electrons and has the value of 96,494 coulombs per mole. E is the measured potential (volts) and z is the number of electrons transferred in the corrosion reaction, two in the case of the iron reaction above. A negative sign is necessary to indicate the conventional assignment of negative charge to electrons.

Again, using the superscript (°) to represent standard conditions, we can rewrite eqn [4.13] as

$$\Delta G^{\circ} = -zFE^{\circ} \tag{4.14}$$

Treating the concentration of solid substances as unity, we can substitute into eqn [4.12] to obtain

$$-zFE = -zFE^{\circ} + RT \ln \frac{[Fe^{2+}]}{[Cu^{2+}]} \tag{4.15}$$

Rewriting eqn [4.15]

$$E = E^{\circ} - \frac{RT}{zF} \ln \frac{[Fe^{2+}]}{[Cu^{2+}]} \tag{4.16}$$

This equation can be written more generally as

$$E = E^{\circ} - \frac{RT}{zF} \ln \frac{[\text{products}]}{[\text{reactants}]} \tag{4.17}$$

Equation [4.17] is known as the **Nernst Equation** and is an extremely important equation which has great theoretical and practical significance. As a consequence, it is common to introduce the numerical values into the equation. Using the standard temperature, 298 K, and $R = 8.3143$ J mol^{-1} K^{-1}, together with conversion into logarithms to the base of 10, eqn [4.17] becomes

$$E = E^{\circ} - \frac{0.059}{z} \lg \frac{[\text{products}]}{[\text{reactants}]} \tag{4.18}$$

E is the non-equilibrum potential generated by performing the reaction, with [reactants] being the reactant concentration and [products] being the product concentration. If these concentrations were chosen to be the equilibrium values, there would be no driving force ($\Delta G = 0$) and E would be zero.

Surprisingly, it is easier to understand the events which occur when a metal corrodes in an electrolyte by considering the anode and cathode reactions as occurring on separate metal surfaces. This will be discussed next.

4.2 THE BASIC WET CORROSION CELL

It is usually possible to identify different regions of a corroding metal surface at which the anodic and cathodic reactions occur, and these are logically termed anodes and cathodes. There are many reasons for anodes and cathodes to be present on a metal surface. Some of these were investigated in the experiments of Chapter 3, and much of this book deals with the matter in depth. For the present, we shall consider a system in which a single anode and cathode are considered to be present in a system known as the basic wet corrosion cell, Fig. 4.2. Four essential components can be identified:

(a) **The anode.** The anode usually corrodes by loss of electrons from electrically neutral metal atoms to form discrete ions. These ions may remain in solution or react to form insoluble corrosion products. (The latter is a common anode reaction. If it occurs it may block further metal dissolution and the corrosion is retarded. In such a case, the surface is said to be passivated – a subject about which much will be said later.) As we have seen, the corrosion reaction of a metal M is usually expressed by the simplified equation:

$$M \rightarrow M^{z+} + ze^- \tag{2.7}$$

in which the number of electrons taken from each atom is governed by the valency of the metal. Commonly, $z = 1, 2$ or 3.

(b) **The cathode.** The cathode does not normally corrode, although it may suffer damage under certain conditions which will be discussed later. Two important and common reactions which may occur at the cathode, depending upon the pH of the solution are:

(1) pH < 7: $$H^+ + e^- \rightarrow H \text{ (atom)} \tag{4.19}$$

$$2H \rightarrow H_2 \text{ (gas)} \tag{4.20}$$

(2) pH $\geqslant$ 7: $$2H_2O + O_2 + 4e^- \rightarrow 4OH^- \tag{4.21}$$

(For an explanation of pH, read section 2.3.) Other cathode reactions are possible. Remember that the only criteria are that the reaction must consume the electrons produced by the anode process and that the energy change must be favourable. In expt 4.1 we saw that the cathode process was one of metal replating. Other cathodic reactions will be described as it becomes necessary.

(c) **An electrolyte.** This is the name given to the solution, which must, of necessity, conduct electricity. Very pure water is not normally considered to be an electrolyte: the conductivity of typical commercial de-ionised water is about $1-10$ mS m^{-1}. Under most practical conditions, however, an aqueous environment will have a sufficient conductivity to act as an electrolyte. 'Soft' tap water has conductivity typically in the region of

10–20 mS m^{-1}, compared with a value for 3.5 per cent sodium chloride solution of 5.3 S m^{-1}. (Note that it is not intended to present an exact definition of an electrolyte in this paragraph, although the one given will serve us well at this stage. In hot corrosion, for example, there need be no water present, for molten salts act as electrolytes.)

(d) **Electrical connection**. The anode and cathode must be in electrical contact for a current to flow in the corrosion cell. (Obviously, a physical connection is not necessary when the anode and cathode are part of the same metal.)

Given the system described in Fig. 4.2, corrosion will be possible on the anode if there is a difference in free energies between the anode and cathode. As was expressed in eqn [4.13], this energy difference is manifest as an electrical potential, which can be measured by the inclusion of a voltmeter in the electrical circuit of Fig. 4.2. This potential is precisely what it says – a tendency for corrosion. When a circuit is completed between the electrodes, the potential drives a current composed of the electrons produced by the reaction, eqn [2.7]. Thus the corrosion can best be monitored by using a galvanometer to measure the flow of current in the wet corrosion cell.

All aqueous corrosion reactions can be thought of in terms of the example of the simple wet corrosion cell we have just described. Even when part of the same metal surface, anodes and cathodes are usually identifiable. (Several of the experiments in Chapter 3 illustrate how this is possible.) If we concede that, for corrosion to occur, the four components must be present, then we can state that:

> **The removal of any one of the four components of the simple wet corrosion cell will stop the corrosion reaction.**

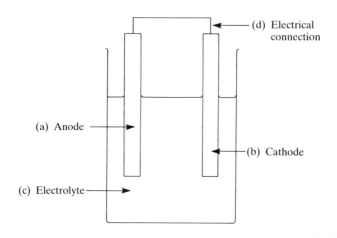

Fig. 4.2 The basic wet corrosion cell.

This is our own First Law of Corrosion Control. The control of corrosion is obviously a very important subject, to which much space is devoted in this book. Many of the measures adopted in the later chapters are just ways of applying this law.

4.3 STANDARD ELECTRODE POTENTIALS

In the basic wet corrosion cell, discussed above and illustrated in Fig. 4.2, we saw that potential differences between the anode and the cathode could be measured by simply inserting a voltmeter into the circuit. While this is quite acceptable for the most basic of laboratory measurements, more accurate work requires the addition of a third electrode to the cell. This is necessary because of the need to define an absolute value of the potential of an electrode. The problem in the measurement of potentials is that only a potential *difference* can be measured. Thus, for example, if in Fig. 4.2 the anode were iron and the cathode copper, the potential difference between the iron and the copper might be measured as a constant value, although the individual potentials of the iron and the copper might be varying (equally).

It would be very desirable to be able to predict the potential of a single metal electrode in an electrolyte, but it is, of course, impossible to measure potentials absolutely, since all measurements are a comparison of one potential with another. This problem is overcome in traditional scientific fashion by defining a standard electrode against which all other measurements can be made. Then, by defining its potential as zero volts, the measurement of an 'absolute' potential of a test electrode can be made.

In expt. 3.3 a simple galvanic series was built up by making a series of measurements using copper as a reference electrode, but copper was an arbitrary selection: in fact, hydrogen is chosen as a standard of reference for rigorous scientific work. By definition, it is given an electrode potential of exactly zero volts. The electrode potential of another element is then compared with that for hydrogen and is called the standard electrode potential for that element. A table similar to the one obtained in expt. 3.3 is then compiled for different metals.

Question: *How can a gaseous element be used for a standard electrode?*

A cell is constructed according to the illustration of Fig. 4.3. An inert metal electrode of platinum is used, over which is passed a steady stream of hydrogen gas at exactly 1 atmosphere pressure and 298 K. The electrode is immersed in a solution in which the concentration of hydrogen ions is exactly 1 M. This half of the cell is separated from the half cell to be tested by a porous membrane designed to minimise the cross-contamination of the electrolytes which occurs as a result of the flow of ions. The test electrode is a combination of a metal and a solution of its ions, also of concentration

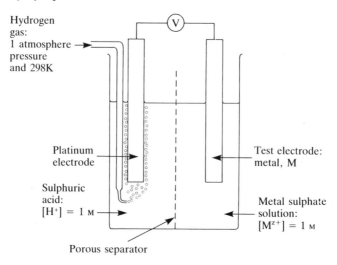

Fig. 4.3 The determination of standard electrode potentials.

1 M. The whole cell must be at the standard temperature of 298 K.

The measurement of standard electrode potential can now be made. If, for example, we wish to measure the standard electrode potential of iron then the measured potential difference between the hydrogen electrode and the iron, under the exact conditions defined above, will be the required value. The half-cell in which the hydrogen reaction takes place is called the **Standard Hydrogen Electrode**, often abbreviated SHE.

Question: *How does the Nernst equation relate to this?*

We can guess that the iron will be an anode compared with a solution of hydrogen ions (iron dissolves in acid). The equations for the reactions are:

(a) iron dissolves:

$$Fe \rightarrow Fe^{2+} + 2e^-$$ [4.6]

(b) hydrogen gas formed:

$$2H^+ + 2e^- \rightarrow H_2 \text{ (gas)}$$ [4.19/20]

(c) overall reaction:

$$Fe + 2H^+ \rightarrow Fe^{2+} + H_2 \text{ (gas)}$$ [4.22]

Substituting into the Nernst equation, eqn[4.17], we have:

$$E = E^o - \frac{RT}{zF} \ln \frac{[Fe^{2+}] [H_2]}{[Fe] [H^+]^2}$$ [4.23]

The terms [H$^+$] and [H$_2$] have been made equal to 1, while [Fe] has been approximated as unity, so eqn [4.23] simplifies to:

$$E = E^\circ - \frac{0.059}{2} \lg [Fe^{2+}] \hspace{3cm} [4.24]$$

Furthermore, in the correct setting up of the cell, the concentration of iron ions is also made equal to 1 M. This reduces the lg term in eqn [4.24] to zero and leaves E = E°. In other words, the measured potential difference *is* the electrode potential of the iron under standard conditions, and we say that for the reaction, eqn [4.6], the standard electrode potential is given by E°. Notice that eqn [4.6] is an oxidation reaction. Note also that, because of Faraday's Law, eqn [4.13], if the measured value of potential is positive then $\Delta G < 0$, indicating a spontaneous reaction. This is, indeed, the case: E° is found to be +0.44 V, supporting the well-known property of iron that it dissolves spontaneously in acid. Notice, also, that the values substituted into eqn [4.17] need not be equilibrium values: the concentrations may take any values (here, usefully made unity), to which the potential will correspond.

In the same way as we have described the measurement of E° for iron, the standard electrode potential of other metals can be measured and a table of values compiled. In so doing, it is conventional to list the entries as *reduction* reactions, and the table produced is of **standard reduction potentials**. This means that the value obtained for the oxidation of iron, +0.44 V, is listed as the reduction potential, −0.44 V. A representative selection of standard reduction potentials is given in Table 4.1. Note that oxidation is the exact reverse of reduction. Sometimes, standard oxidation potentials are tabulated; these values are the *same in magnitude* as those in Table 4.1, but of *opposite sign*. For this reason, the values are often referred to as **standard**

Table 4.1 Standard reduction potentials[1]

Electrode reaction	E° (*volt*)
$Au^+ + e^- = Au$	+1.68
$Pt^{2+} + 2e^- = Pt$	+1.20
$Hg^{2+} + 2e^- = Hg$	+0.85
$Ag^+ + e^- = Ag$	+0.80
$Cu^{2+} + 2e^- = Cu$	+0.34
$2H^+ + 2e^- = H_2$	0.00
$Pb^{2+} + 2e^- = Pb$	−0.13
$Sn^{2+} + 2e^- = Sn$	−0.14
$Ni^{2+} + 2e^- = Ni$	−0.25
$Cd^{2+} + 2e^- = Cd$	−0.40
$Fe^{2+} + 2e^- = Fe$	−0.44
$Cr^{3+} + 3e^- = Cr$	−0.71
$Zn^{2+} + 2e^- = Zn$	−0.76
$Al^{3+} + 3e^- = Al$	−1.67
$Mg^{2+} + 2e^- = Mg$	−2.34
$Na^+ + e^- = Na$	−2.71
$Ca^{2+} + 2e^- = Ca$	−2.87
$K^+ + e^- = K$	−2.92

redox potentials. *Some texts show opposite signs for the values listed in Table 4.1. This text uses the IUPAC convention which ascribes the signs stated.* (To avoid unnecessary confusion, the novice student is urged to avoid any texts using the opposite convention. The treatment we have developed has no need even to consider the possibility of using alternative signs.)

Question: *How is the electrochemical series different from the galvanic series?*

The electrochemical series is determined under standard conditions. The values are absolute for each element and are independent of the electrolyte used. They may be substituted into the Nernst equation to obtain a prediction of the potential of a corrosion reaction under non-standard conditions.

The galvanic series, on the other hand, is true only for specified conditions of electrolyte, pressure or temperature. Unlike the electrochemical series, it can include the relative performance of alloys, a great advantage to the designer. It is therefore of considerable practical significance, whereas the electrochemical series is more useful in theoretical contexts.

Question: *The standard hydrogen electrode is a rather complicated arrangement to use in the laboratory. Isn't there a simpler way of measuring potentials?*

Yes. Scientists have measured the reduction potentials so accurately that there is now no practical need to use the hydrogen electrode. Other robust and very stable standard electrodes have been devised which can be used conveniently in laboratory measurements. We shall now look at these, and ways of carrying out the measurements.

4.4 REFERENCE ELECTRODES

The most common reference electrode in the laboratory is the **Standard Calomel Electrode**. The construction is designed to give a constant and well-defined potential against which other potential measurements can be made.

Calomel is an old name for mercury(I) chloride, Hg_2Cl_2, and it is the combination of this compound, mercury and a solution of chloride ions which provides the very stable and reproducible potential. For convenience of use in laboratories the chloride solution used is a saturated solution of potassium chloride. In this case the standard calomel electrode is called the **Saturated Calomel Electrode**, an example of which is shown in Fig. 4.4(a). Care should be taken not to confuse standard with saturated: the abbreviation, SCE, is often used to indicate the latter, though it could be taken to mean either. Since there are other standard electrodes which are not saturated and which have different potentials (see below) it is safest to be quite clear which electrode is being used.

The saturated calomel electrode consists of a platinum wire in contact

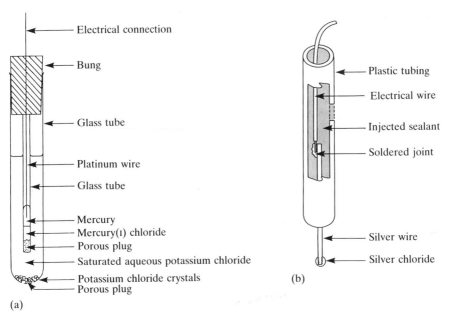

Fig. 4.4 Two designs of reference electrode:
(a) A saturated calomel reference electrode.
(b) A silver/silver chloride reference electrode.

with a small sample of mercury and mercury(I) chloride. This is held within a thin glass tube by a porous plug, and the glass tube is contained inside a larger glass tube, again with a porous plug at the end. The larger tube contains the saturated potassium chloride solution. The porous plugs allow the passage of ions (and hence current) without causing significant cross-contamination of the potassium chloride and the electrolyte of the test cell. If this were to happen then the potential of the calomel cell would not be constant, and the potential of the test electrode might also change. A simple electrical connection to the platinum wire allows the electrode to be introduced into a corrosion cell such as will be considered in section 4.8.

Variations of the calomel electrode exist in which different concentrations of potassium chloride are used. This is because the solubility of the salt is not constant over a range of temperature, and a solution which is saturated at one temperature is not saturated at a higher one. However, since most experimental work is performed at 298 K, there is little error involved with the use of the saturated solution and the electrode described here is probably the most widely used. It has the advantage that in use it is a simple matter to inspect the electrolyte visually: if crystals of potassium chloride are present in the tube then the electrode is probably functioning correctly. If crystals are absent then the electrolyte is possibly diluted and the electrode may not be providing the required reference potential. The porous plugs are not infallible and regular checks should be made to ensure that they are still allowing free passage of ions.

A second type of reference electrode exists which is gaining favour for laboratory usage because it can be made quite small – more so than is convenient for the calomel electrode. It is made with a silver wire, the end of which carries a coating of silver chloride. This robust coating can be easily applied by dipping the wire into a melt of the silver salt. The wire is then enclosed in a suitable glass tube or capillary, and an electrical connection made.

Silver/silver chloride electrodes are particularly suitable for use in sea-water environments, and are often used as reference electrodes in cathodic protection systems (see Ch. 16.) A more refined electrode suitable for laboratory use is depicted in Fig. 4.4(b). Two or three dips of the silver wire in molten silver chloride are sufficient to give a satisfactory coating, and by threading the end of the wire the tendency for the coating to fall off is reduced. Before it can be used, the electrode must be activated by immersion in electrolyte while connected to a piece of metal which is anodic to the electrode, say, aluminium. The immersion need only be for about 15 minutes which is a sufficient time to produce an extremely thin electrically conducting layer of silver.

The reference potentials of some common standard electrodes are given in Table 4.2.

Table 4.2 Standard reference electrode potentials[2]

Electrode	Electrolyte	Potential (V)
Calomel (SCE)	Saturated KCl	+0.2420
Calomel (NCE)	1.0 M KCl	+0.2810
Calomel	0.1 M KCl	+0.3335
Silver/silver chloride (SSC)	1.0 M KCl	+0.2224
SSC	Sea water	+0.25 approx
Copper/copper sulphate (CSE)	Sea water	+0.30 approx
Zinc	Sea water	−0.79 approx

Question: *How do I use reference electrodes in corrosion measurements?*

You need to set up a three-electrode cell. This will be described in section 4.8.

4.5 CELL POTENTIAL

Modern batteries are merely refined corrosion cells in which the electrical current produced by the corrosion reaction is used to drive an external device. The **Daniell cell** was a very early form of battery which consisted of copper and zinc immersed in solutions of their salts, Fig. 4.5. The external circuit shown in the figure is not part of the cell, but will be used in an experiment below. The use of the Daniell cell as an example to aid devel-

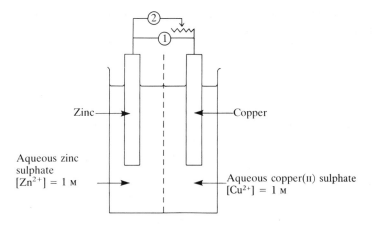

Fig. 4.5 The Daniell cell and associated apparatus for expt 4.2: (1) and (2) represent digital multimeters.

opment of our theory is convenient because it is simple in construction and common in teaching laboratories. Although its practical uses as a battery are now few because of the availability of more modern designs, it can be thought of as a wet corrosion cell, as previously described in Fig. 4.2.

Let us begin by adopting a standard form of symbols by which the cell can be represented, rather than in a diagram. This is done as follows:

$$Zn \mid Zn^{2+} \parallel Cu^{2+} \mid Cu$$

This symbolism shows the two electrodes at the extreme left and right, with the ions, in which they are immersed, between them. The double line represents some form of separator for the two different ionic species. Notice that we have written it in the same sense as the figure, i.e. with the zinc on the left.

Let us suppose that the zinc is the anode, and that the two electrode reactions are:

$$Zn \rightarrow Zn^{2+} + 2e^- \qquad [4.25]$$
$$Cu^{2+} + 2e^- \rightarrow Cu \qquad [4.7]$$

making the overall reaction

$$Zn + Cu^{2+} \rightarrow Zn^{2+} + Cu \qquad [4.26]$$

We can use these reactions to find the theoretical potential which can be obtained from the cell.

Using the Nernst equation, eqn [4.18], for each of the cell half-reactions we can write:

$$E_{(Zn/Zn^{2+})} = E^o_{(Zn/Zn^{2+})} - \frac{0.059}{2} \lg [Zn^{2+}] \qquad [4.27]$$

and
$$E_{(Cu^{2+}/Cu)} = E^o_{(Cu^{2+}/Cu)} - \frac{0.059}{2} \lg \frac{1}{[Cu^{2+}]} \qquad [4.28]$$

If we let $E_{(cell)} = E_{(Zn/Zn^{2+})} + E_{(Cu^{2+}/Cu)}$, and $E^o_{(cell)} = E^o_{(Zn/Zn^{2+})} + E^o_{(Cu^{2+}/Cu)}$, then by adding eqns [4.27] and [4.28] we get:

$$E_{(cell)} = E^o_{(cell)} - \frac{0.059}{2} \lg \frac{[Zn^{2+}]}{[Cu^{2+}]} \qquad [4.29]$$

Calculated in this way, the cell potential is the sum of the oxidation potential of the anode and the reduction potential of the cathode. Let us simplify matters by using ion concentrations of 1 M. The lg terms vanish, leaving $E_{(cell)} = E^o_{(cell)}$. Remember, by using this form of the Nernst equation, [4.18], we have decided to consider only standard conditions. Thus, for 1 M electrolyte,

$$
\begin{aligned}
E^o_{(cell)} &= E^o_{(Zn\ oxidation)} + E^o_{(Cu\ reduction)} \qquad [4.30]\\
&= (+0.76\ V) + (+0.34\ V)\\
&= +1.10\ V
\end{aligned}
$$

In many texts, a convention has been adopted which must be explained. Whenever we write down the symbols for a cell, or sketch one in a diagram, there is always one electrode on the left and one on the right. It has been agreed to subtract the potential of the electrode on the left from that of the one on the right. In Fig. 4.5 we drew the zinc on the left and the copper on the right, so we can write:

$$E^o_{(cell)} = E^o_{(copper)} - E^o_{(zinc)} \qquad [4.31]$$

This convention requires that we substitute **reduction** potentials into eqn [4.31]. Writing the two half-reactions as reduction reactions:

$$Zn^{2+} + 2e^- \rightarrow Zn \qquad E^o = -0.76\ V \qquad [4.32]$$

$$Cu^{2+} + 2e^- \rightarrow Cu \qquad E^o = +0.34\ V \qquad [4.7]$$

whereupon, according to the *convention*,

$$
\begin{aligned}
E^o_{(cell)} &= (+0.34\ V) - (-0.76\ V)\\
&= +1.10\ V
\end{aligned}
$$

The equivalence of the two approaches has been demonstrated. The cell reaction is found by eqn [4.7] − eqn [4.32] This gives:

$$Cu^{2+} - Zn^{2+} \rightarrow Cu - Zn \qquad [4.33]$$

which, on rearrangement, gives eqn [4.26].

An important check is now made to see if the cell reaction is possible in the direction in which we have written it. By simply substituting the value +1.10 V into Faraday's eqn, [4.13], we see that a negative value of ΔG^o is obtained, and the way that we have described the cell leads to the prediction of a spontaneous reaction. If we had written the cell with the copper on the left, we would have obtained a negative cell potential, whichever method was chosen, and we would have arrived at the conclusion that the cell reaction was not spontaneous. Thus we always know by inspection which is the anode and which the cathode.

Question: *What if I did not use ion concentrations which were exactly 1 M?*

Then the potential of the cell will be modified by the Nernst equations, [4.27] to [4.29]; the lg terms will no longer be zero. Note that the Nernst equation, as we have used it, becomes inaccurate at high concentrations of electrolyte. This is because of our use of concentrations rather than activities. A further small problem is the presence of a so-called junction potential which is a potential difference across the interface of the two different electrolytes. At this stage it is sufficient to be aware of it: we shall not discuss junction potentials in this chapter since they do not interfere with the experiment that follows.

Experiment 4.2

Set up a Daniell cell and circuit as shown in Fig. 4.5. It is useful to use two modern digital meters (labelled 1 and 2 in Fig. 4.5) because they can be used to measure both potential and current at the flick of a switch. They also have an extremely high impedance when used in the volts mode, which greatly reduces errors in the measurements of potential and eliminates the need to use the more cumbersome potentiometer.

First switch meter 1 to the volts mode with meter 2 switched off. No current will flow through meter 2 so meter 1 will measure a potential quite close to $+1.1$ V, confirming our calculations. Now switch meter 2 to the current mode. Instantly you will see that the potential measured by meter 1 has diminished as current flows in the external circuit. You can use the rheostat to alter the resistance of the circuit and show that the more current that flows, the more the potential diminishes.

Question: *Why does current flow reduce the potential?*

The potentials used in eqn [4.31] are for equilibrium (no current) conditions and are calculated using the Nernst equation. When a current is drawn the potential of the anode rises, that is, it becomes less negative. Similarly, the potential of the cathode falls as it becomes less positive. The cell potential is thus reduced. This is another consequence of thermodynamics which in chemistry is known as Le Chatelier's Principle:

A system will always react to oppose a change imposed upon it.

The cell potential can be thought of as an ability to supply current. As soon as current is drawn, the ability is reduced. If this were not so then we could go on drawing as much current as we liked from a battery without penalty – effectively creating a limitless supply of free energy.

The answer to the question is more fully explained in later sections, and in particular, section 4.10.

We have, until now, been considering the thermodynamic implications of corrosion reactions, but we have discovered that when a current flows the electrode potentials change. This brings us back to the point that thermodynamics tells us only about the *tendency* of a system to corrode; an equilibrium situation is required and this necessitates no net flow of current. Corrosion reactions, however, cause current to flow so in order to investigate them fully, we must also consider the kinetics of corrosion reactions.

4.6 THE KINETICS OF CORROSION REACTIONS

Consider two pieces of metal, one of 10 square millimetres and the other of 1 square millimetre, such that they both corrode in separate cells and each produces a current of 10 electrons per second. It is easy to see that the smaller piece will corrode 10 times faster than the larger piece. This illustrates the necessity for using a parameter called **current density** in which the *area* of the corroding metal is taken into account.

Throughout the discussions which follow we shall use I to represent an absolute current (amps) and i to represent current density (amps/square metre). We shall always imply a flow of electrons as being the current: conventional currents will not be used.

Now consider a hypothetical situation in which we place a piece of copper, say, of unit area, in a beaker of pure water. Immediately a situation parallel to the energy profile of Fig. 4.1 and redrawn in Fig. 4.6(a), will apply. There is sufficient available energy in the environment for a steady flow of copper atoms to 'pass over the energy barrier' and proceed to the ionic form. The copper begins to dissolve (corrode) and the concentration of copper ions in the water, which was initially zero, will slowly increase:

$$Cu \rightarrow Cu^{2+} + 2e^- \qquad [4.34]$$

Question: *How can this be a corrosion cell? It isn't a bit like the arrangement in Fig. 4.2.*

A single piece of metal placed in an electrolyte can act as its own anode, cathode and electrical connection. Individual areas of the metal can be anodic to others because of variations in the solid structure of the metal, or environmental differences over the surface as a whole (see Ch. 6).

As we have seen, the tendency of the copper to corrode decreases as the current increases from zero, and the value of ΔG diminishes, together with the potential, in accordance with Faraday's Law. The thermodynamic energies of metal atom and ion tend to approach each other.

We have already discussed in the early sections of this chapter how the rate of reaction diminishes as the activation free energy barrier increases. As soon as copper ions are present in solution, there is a possibility for them to 'pass back over the energy barrier' and plate back out on to the metal.

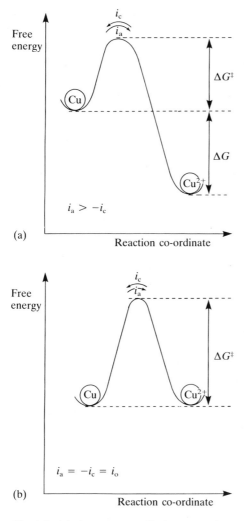

Fig. 4.6 (a) An energy profile for copper in pure water: $i_a > -i_c$.
(b) An energy profile for copper in equilibrium with a solution of its divalent
ions: $i_a = -i_c = i_0$.

The rate of this is governed by the activation free energy in the reverse
direction, a quantity initially greater than that for the forward reaction.
However, this free energy barrier is reduced in size as the energies of the
two species approach each other and the backward reaction of copper ions
plating out increases. The rate of the forward reaction, on the other hand,
decreases because its activation free energy increases. The situation is thus
obtained that the rate of the decreasing forward reaction becomes equal to
the rate of the increasing backward reaction and equilibrium is established,
Fig. 4.6(b).

79

Experiment 4.3

Place a clean piece of copper, together with a saturated calomel electrode, into a solution of about 1–2 litres of 3.5 per cent sodium chloride and connect them to the terminals of a digital meter, copper to the 'high' terminal. Switch to measurement of volts (or millivolts). Record the values of potential on an x/t chart recorder (or manually if none available). Continue readings for about 1–2 hours until the potential has settled to a fairly constant value.

This potential is called the **free corrosion potential** and is given the symbol E_{corr}. You should obtain a value for copper of about −0.240 V. It is close to being an equilibrium situation, although there is in fact a very slow corrosion of the copper. The full significance of free corrosion potentials will be dealt with in section 4.8.

Using i_a and i_c to denote the current densities of the respective forward (anodic) and backward (cathodic) reactions, we can rewrite eqn [4.34] as:

$$\overset{i_a}{\underset{i_c}{\rightleftharpoons}}$$

$$Cu \rightleftharpoons Cu^{2+} + 2e^- \qquad\qquad [4.34a]$$

When the state of equilibrium is reached, $i_a = -i_c$, and no *net* current flows. There *is* a current flowing, but it is *equal and opposite in both directions and cannot be measured*. It is called the **exchange current** and is denoted by I_o, or i_o if, as in this case, the area is unity. Remember that as soon as corrosion starts, the equilibrium situation no longer exists and $i_a > -i_c$.

When the copper is immersed in the water the first event is the passage of metal ions to the electrolyte side of the metal–electrolyte interface. The metal ions (cations) initially remain associated with the metal solid and the electrons they have left behind. Thus we can imagine the metal as a negatively charged solid, surrounded by a coating of cations. (It is often confusing to students that the anode is the negatively charged electrode. This pictorial mnemonic is very useful.) This non-homogeneous distribution of ions which has resulted from the immersion of a metal in an aqueous electrolyte is commonly referred to as the double layer.[3] It is illustrated schematically in Fig. 4.7.

The double layer consists of two parts:

(a) A compact layer, the Helmholtz layer, closest to the surface in which the distribution of charge, and hence potential, changes linearly with the distance from the electrode surface.

(b) A more diffuse outer layer, the Gouy–Chapman layer, in which the potential changes exponentially.

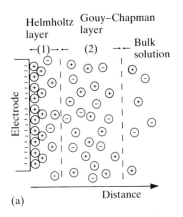

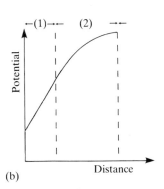

Fig. 4.7 The double layer:
(a) Distribution of ions as a function of distance from an electrode behaving as an anode.
(b) Variation of potential with distance for the model in (a).

The constitution of the double layer will parallel the changes of potential which occur on the electrode and will also reach an equilibrium condition corresponding to the energy profile of Fig. 4.6(b). When the equilibrium is destroyed by an increase in either the forward or backward reactions, a new dynamic equilibrium is established in which a continuous flow of anions and cations in the electrolyte adjacent to the electrodes performs the current-carrying requirement of the electrode reactions.

We now have an alternative way of visualising the effect described above, in which the corroding electrode loses its ability to corrode, as well as a good example of the literal use of the term potential. The corrosion potential of a metal is determined by the amount of negative charge developed on it when placed in an electrolyte. When corrosion occurs, the electrons produced by the anodic metal are conducted away to the (remote) cathode. The negative charge on the metal is reduced and the corrosion potential is diminished. Thus, for example, we can imagine that zinc acquires more negative charge than iron when immersed in an aqueous medium. This correlates with the electrode potentials of -0.76 V and -0.44 V for zinc and iron respectively (no current condition). If either were to corrode independently, negative charge would be conducted away and each would become less negative. If they are connected together, electrons are passed from the zinc (which becomes less negative) to the iron (which becomes more negative). For further discussion of this most important concept, see section 4.10.

Faraday's Law of Electrolysis states that:

$$Q = zF.M \qquad [4.35]$$

where Q is the charge created by the ionisation of M mol of material. Differentiating with respect to time we get:

$$dQ/dt = zF .dM/dt \qquad [4.36]$$

Now the rate of flow of charge is current, I, and if we consider the passage of charge across unit area of cross-section then we can use current density, i. Then, dM/dt becomes J, the flux of substance, and eqn [4.36] becomes:

$$i = zFJ \qquad [4.37]$$

The flux of substance is another name for corrosion rate per unit area. *Hence we confirm the important concept that current density and corrosion rate can be equated.* The ability to determine a corrosion rate by measurement of current density is a most important finding. However, in practical terms, to say that a metal is corroding at a rate of 0.003 A m^{-2} is rather meaningless. Engineers prefer to consider an average rate of deterioration per unit area expressed as the average depth of corrosion over a given area in a given time. For example, a corrosion rate of 2.5 mm per year means that in one year, the metal will have corroded, on average across the whole of its exposed area, to a depth of 2.5 mm. This unit of measurement is often abbreviated as mmpy. In the USA, a corrosion rate expressed in mpy means milli-inches per year. E18.4 in section 18.1 is a worked example to show how the conversion from current density to mmpy is carried out. In certain forms of corrosion, such as crevice or pitting corrosion, this method of considering corrosion rate is dangerous because an average corrosion rate is meaningless; corrosion can be very rapid and penetrating over very small areas of a large exposed surface.

4.7 POLARISATION

When a metal is not in equilibrium with a solution of its ions, the electrode potential differs from the free corrosion potential by an amount known as the **polarisation**. Other terms having equivalent meaning are **overvoltage** and **overpotential**. The symbol used for polarisation is η. Polarisation is an extremely important corrosion parameter which enables useful statements to be made about the *rates* of corrosion processes.

Consider the process described in eqn [4.34a]. In the last section we saw that corrosion rate and current density are directly related. At the beginning of this chapter we said that corrosion rate, v, could be expressed:

$$v = k_{corr}.[reactants] \qquad [4.1]$$

where

$$k_{corr} = A \exp (-\Delta G^{\ddagger} /RT) \qquad [4.2]$$

(*A* is constant). From these two equations we see that

$$v = A \exp \left(\frac{-\Delta G^{\ddagger}}{RT} \right) . [\text{reactants}] \qquad [4.38]$$

At equilibrium, the rate of the forward (anodic) reaction is i_a, and equals the rate of the reverse (cathodic) reaction, i_c. (Remember that $i_o = i_a = -i_c$.) It is usually possible to treat the concentration of reactants (e.g. the solid metal, for the forward reaction) as constant, and we shall incorporate the term into a new constant, A_o. Thus, if we consider the rate of the forward reaction for which the activation free energy is $\Delta G^{\ddagger}$, we can write eqn [4.38] as:

$$i_a \text{ (at equilibrium)} = i_o = A_o \exp \left(\frac{-\Delta G^{\ddagger}}{RT} \right) \qquad [4.39]$$

When the forward reaction is faster than the reverse reaction ($i_a > -i_c$) and an overall corrosion process occurs, equilibrium is destroyed and the free energies of the metal and its ions are at different levels, as in Fig. 4.6(a).

The deviation from the equilibrium potential, the polarisation, is the combination of an anodic polarisation on the metal and a cathodic polarisation of the environment. (Compare Fig. 4.6(a) and (b): the energy of the metal has increased and that of the environment has decreased.) These potential deviations away from the equilibrium value may or may not be equal, and we shall assume for the moment that they are not. The changes are redrawn in Fig. 4.8.

If the total polarisation is η then we can define the anodic polarisation as $\alpha\eta$ and the cathodic polarisation as $(1-\alpha)\eta$. Note that in Fig. 4.8, the

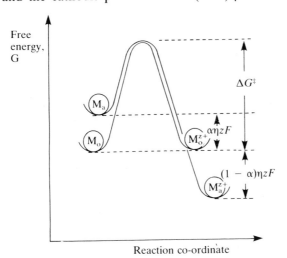

Fig. 4.8 An energy profile for an anode at equilibrium, represented by the curve M_o/M_o^{z+} and a similar profile for an anodic activation polarisation of η, represented by the curve M_a/M_a^{z+}.

polarisations have been converted into energies by multiplication by the factor, zF, as in the Faraday equation, [4.13]. This enables us to determine the new activation energy for the anodic reaction which can be seen to be $(\Delta G^{\ddagger} - \alpha\eta \, zF)$, because the energy state of the metal has increased and the activation energy reduced. Thus we can write:

$$i_a = A_o \exp \left(\frac{- \Delta G^{\ddagger} + \alpha\eta \, zF}{RT} \right) \qquad [4.40]$$

$$= A_o \exp \left(\frac{-\Delta G^{\ddagger}}{RT} \right) . \exp \left(\frac{\alpha\eta \, zF}{RT} \right) \qquad [4.41]$$

(Note that, similarly, the activation energy of cations, M^{z+} being converted into M, i.e. the cathodic reaction, has increased by an amount $(1-\alpha) \, \eta zF$. We shall continue to consider only the anodic reaction for the moment.)

Substituting eqn [4.39] into eqn [4.41]

$$i_a = i_o \exp \left(\frac{\alpha\eta \, zF}{RT} \right) \qquad [4.42]$$

Now let $A' = \alpha zF/RT$ then

$$i_a = i_o \exp (A' \, \eta) \qquad [4.43]$$

Taking logarithms we get

$$\ln i_a = \ln i_o + A'\eta \qquad [4.44]$$

Rearranging eqn [4.44]

$$\ln (i_a/i_o) = A'\eta \qquad [4.45]$$

converting to base 10 logarithms and rearranging eqn [4.45]

$$\eta = \frac{2.303}{A'} \lg (i_a/i_o) \qquad [4.46]$$

Letting $\beta_a = 2.303/A'$ we now have the important result that

$$\eta_a = \beta_a \lg (i_a/i_o) \qquad [4.47]$$

where $\beta_a = 2.303RT/\alpha zF$ $\qquad [4.48]$

Written in the form:

$$\eta = C \lg i + D \qquad [4.49]$$

the equation is known as the Tafel equation. The derivation above was for the anodic polarisation. Continuing with our use of the subscript a to represent anodic parameters, we obtain the specific equation: for the anodic reaction:

$$\eta_a = \beta_a \lg i_a - \beta_a \lg i_o \qquad [4.50]$$

An identical approach can be applied to the cathode reaction and leads to a similar equation in which i_c is substituted for i_a and a subscript c is used with the β value to denote the cathode reaction. Thus:

$$\beta_c = \frac{2.303RT}{(1-\alpha)zF} \qquad [4.51]$$

The constants β_a and β_c are called the anodic and cathodic Tafel constants.

Examination of the Tafel equation in its form of eqn [4.49] tells us immediately that a graph of η against $\lg i$ for either of the two processes will give a straight line with a slope equal to the respective β constant. The intercept, D, is given by $- \beta\lg i_o$ with the exchange current being a constant for a given material in a given electrolyte. In practice, it is found that β values usually fall within the range 0.03–0.3 volts per decade of current density.

As an example of the Tafel equation expressed in graphical form, let us use reasoning first put forward by Stern and Geary.[4] In order to plot the variation of polarisation with $\lg i$ for both anodic and cathodic reactions, we need to decide upon representative values for the Tafel constants and the exchange current density. If we choose, as representative values, $\beta_a = +100$ mV per decade, $\beta_c = -100$ mV per decade, and $i_o = 0.01$ A m^{-2}, substitution of these values into eqn [4.50] leads to data which, when plotted, take the form of Fig. 4.9(a). Thus, for example, when $i_a = 0.01$, $\eta = 0$. This is true also for $i_c = 0.01$. Remember that we are plotting η versus $\lg i$, not $\lg i/i_o$. The anodic polarisation varies as line (i_a) and the cathodic polarisation as line (i_c).

Examination of Fig. 4.9(a) shows that when the electrode is anodically polarised to $+100$ mV, the anodic current density is 0.1 A m^{-2} while the cathodic current has fallen to 0.001 A m^{-2}. Since we can only measure the difference between the anodic and cathodic currents, then the measured current,

$$\begin{aligned} i_{meas} &= i_a - i_c \qquad [4.52] \\ &= 0.1 - 0.001 \\ &= 0.099 \text{ A m}^{-2} \end{aligned}$$

As the polarisation is increased, so $i_{meas} \rightarrow i_a$. Using eqn [4.52] and substituting for i_a in eqn [4.47], we obtain:

$$\eta_a = \beta_a \lg \left(\frac{i_{meas} + i_c}{i_o} \right) \qquad [4.53]$$

Obviously when $i_{meas} \gg i_c$ then linear Tafel behaviour will be observed experimentally. However, at polarisations close to E_{corr} where i_a is comparable with i_o the measured value of the current density will be far removed from the true value of i_a and substantial deviations from linearity will be obtained. The same arguments apply whether anodic or cathodic polarisations are being used. Thus if we try to obtain the data experimentally, the graph of Fig. 4.9(a) will become Fig. 4.9(b). Extrapolation of the linear portions of the polarisation plots allows a determination of i_o.

Question: *But how do I carry out a laboratory experiment to quantify a corrosion reaction?*

Read the next section.

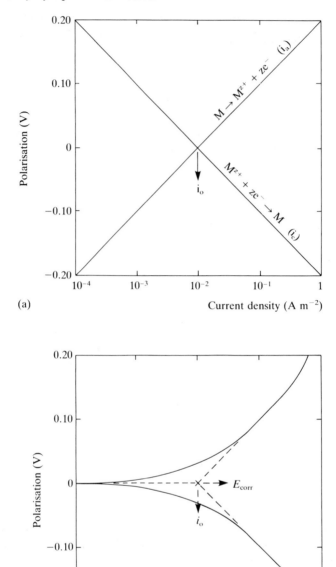

Fig. 4.9 Two Tafel plots:
(a) Theoretical Tafel plot.
(b) An idealised practical Tafel plot.

4.8   THE THREE-ELECTRODE CELL AND THE *E*/lg *i* PLOT

The three-electrode cell is the standard laboratory apparatus for the quantitative investigation of the corrosion properties of materials. It is a refined version of the basic wet corrosion cell and a typical example is illustrated in Fig. 4.10. It can be used in many different types of corrosion experiments. First we shall examine the components in more detail.

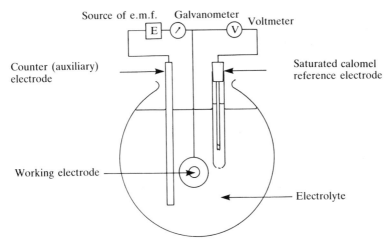

Fig. 4.10   The three electrode cell.

The **working electrode** is the name given to the electrode being investigated. It is useful, though not essential, if the electrode is designed to have a surface area of 100 square millimetres. Current measurements can then be more readily converted into current densities, which should be used in calculations. We use the term working electrode rather than anode because we are not limited to investigations of anodic behaviour only. As we shall see in a moment, cathodic behaviour can also be examined.

Practical working electrodes can be constructed in a variety of ways. One method is simply to mount a small specimen in cold-setting resin. Electrical connection must be made to the specimen, and this can be done before mounting. After mounting, the surface should be ground and polished, as for metallographic examination.

Although these specimens are adequate for most purposes, accurate work may require a more carefully designed method of mounting. Two designs which are commercially available are illustrated in Fig. 4.11(a) and (b). Specimens in the form of thin discs of approximately 15 mm diameter, or cylinders of about 10 mm diameter × 10 mm length, can be quickly assembled into working electrodes with well-defined characteristics.

The **counter (auxiliary) electrode** is the name given to the second elec-

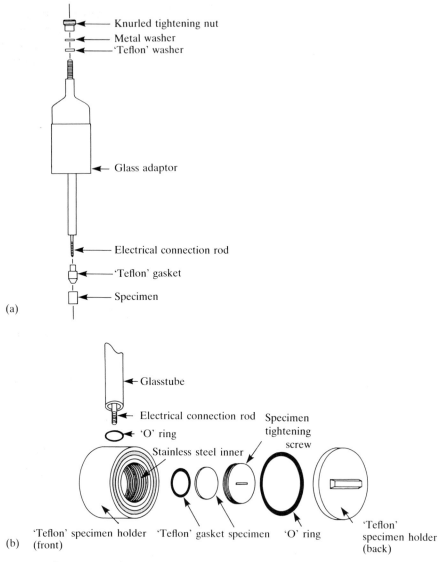

Fig. 4.11 Designs for working electrodes:
(a) Cylindrical specimens
(b) Disc specimens
(Courtesy EG & G, Princeton Applied Research Companv. Copyright © 1984 EG & G Princeton Applied Research).

trode which is present specifically to carry the current in the circuit created by the investigation. It is not required for measurements of potential. Usually, a carbon rod is used, but it can be any material which will not introduce contaminating ions into the electrolyte. Platinum or gold can also be used with success, especially if space is at a premium, when smaller electrodes can be used.

The **reference electrode** is present to provide a very stable datum point against which the potential of the working electrode can be measured. It cannot itself carry any more than the most negligible current. If it did it would participate in the cell reactions and its potential would no longer be constant, hence the requirement for the counter electrode. By far the most convenient reference electrode to use in such an experiment is a saturated calomel electrode.

The external circuit can be varied considerably. The essential components are the following:

(a) A current measuring device capable of reading milliamps or, preferably, microamps.

(b) A potential measuring device. It is important that no current is drawn during the act of measurement and traditionally potentiometers have been used for this purpose. The modern digital meter, however, can have an impedance of the order of giga-ohms, and may be used with as good an accuracy as a potentiometer.

(c) A source of potential that will serve to 'drive' the working electrode such that the desired cell reactions occur. A typical instrument is a **potentiostat** which is readily available commercially and has been used extensively by corrosion scientists. Potentiostats apply pre-determined potentials to the working electrode so that measurement of the cell current can be made. This is done by altering the current at the counter electrode to whatever value is necessary to maintain the set value of working-to-reference potential. A simple constant voltage source is not suitable.

Many instruments have built-in stepping facilities which allow the application of a range of potentials over a given period at a constant rate. Other instruments are available which combine all three of the above functions and perform all measurements by either automatic or manual control. However, a very simple and inexpensive rig is quite capable of accurate measurements.

The three electrodes are placed in a suitable glass vessel, of about 1 to 2 litres capacity containing the chosen electrolyte.

Question: *Does the electrolyte I choose affect any of the measurements?*

Yes, considerably. It is extremely important to consider the conductivity of the electrolyte, since, by carrying the ionic current, it plays such an important role in corrosion reactions. The use of a reference electrode is to enable the potential of a working electrode to be measured and it should be placed as close to the electrode surface as possible. This is because the potential which is measured will always include the potential difference across the electrolyte which occupies the space between the working electrode surface and the reference electrode. Because most corrosion measurements involve the use of direct currents, Ohm's Law applies, and this potential difference across the electrolyte can be estimated by means of eqn [2.19], i.e. $V = IR$. Not surprisingly, this potential is often referred to as

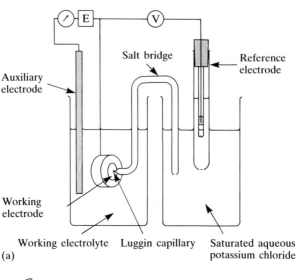

Auxiliary electrode

Salt bridge

Reference electrode

Working electrode

Working electrolyte   Luggin capillary   Saturated aqueous potassium chloride

(a)

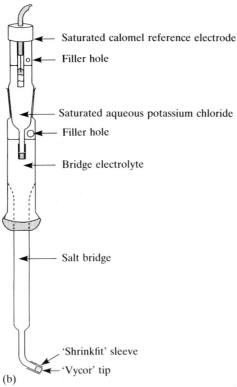

Saturated calomel reference electrode

Filler hole

Saturated aqueous potassium chloride

Filler hole

Bridge electrolyte

Salt bridge

'Shrinkfit' sleeve

'Vycor' tip

(b)

Fig. 4.12 Designs of Luggin capillary for minimising the *IR* drop in a corrosion cell:
(a) Simple capillary made from glass tube salt bridge
(b) Specialist design incorporating ball and socket ground glass joint
(Courtesy EG & G Princeton Applied Research Company.)

the **Ohmic** or **IR drop**, and may be large if either the current or the resistance of the electrolyte is large. It is usually preferable to make the *IR* drop as small as possible, otherwise its contribution to the overall cell potential may be difficult to quantify. If a high conductivity electrolyte, such as 3.5 per cent sodium chloride solution or sea water, is used the effect will be small and the experimental apparatus described in Fig. 4.10 is adequate for most investigations. If more dilute solutions are necessary for the experiments then it is essential to use a more sophisticated design of reference electrode measurement. This is achieved by means of a device called a **Luggin capillary** and Fig. 4.12(a) illustrates such an arrangement.

The aim is to provide a high-conductivity path from the working electrode surface to the reference electrode. The traditional Luggin design utilises a glass capillary with a very fine tip which must be placed at the metal surface. The capillary leads away from the reaction vessel to a separate small receptacle in which a SCE is placed and the whole is filled with saturated potassium chloride solution. The tube which carries the electrolyte between the reference and test electrodes is known as the salt bridge.

The biggest problem with such an arrangement is from contamination of the low conductivity electrolyte by diffusion of the saturated potassium chloride. A better method is to use the equipment shown in Fig. 4.12(b) in which use is made of a special 'Vycor' ® tip which has extremely low leakage rates together with a low *IR* drop through the tip.

Experiment 4.4

Set up the equipment described in Fig. 4.10 using pure copper of measured, exposed surface area as a working electrode, and a solution of 3.5 per cent sodium chloride as an electrolyte. Adjust the potentiostat to read -400 mV. (It is common in the laboratory to refer to potentials in millivolts, rather than -0.400 V, although the latter figure is preferable in calculations and some graphical displays.) Scan through the potentials to $+400$ mV and take measurements of current at 10 mV intervals. Allow the cell to settle at each value of potential for one minute before moving on to the next. Observe the surface of the specimen and note any changes that occur.

Question: *What shall I do with the data I get from this experiment and what does it mean?*

Compile a table of applied potential (volts) and current density (amps per square metre), then plot a graph of *E* as ordinate and lg *i* as abscissa. You will have obtained both positive and negative current values. On your

---

® Vycor is a registered trade mark of the Corning Glass Corporation.

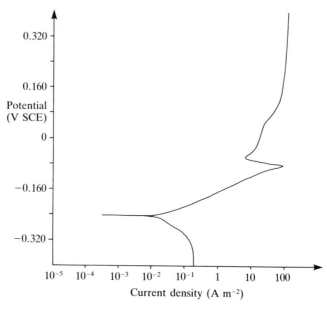

Fig. 4.13 Potentiodynamic scan for copper in 3.5 per cent sodium chloride solution.

graph, plot all the current density values as positive. You should obtain a graph similar to that illustrated in Fig. 4.13.

The $E/\lg i$ plot is one of the most common methods of examining the corrosion behaviour of materials. It has become common practice in $E/\lg i$ plots for all current densities to be treated as positive. This is really just a convenience for it reduces the size of the graphs and gives a much clearer indication of the value of potential when the current density changes from negative to positive. The portion of the graph for which you had measured negative currents, that is from -400 mV to about -240 mV, represents the copper behaving as a cathode. From -240 mV to more positive values of potential, the copper is behaving as an anode and it is during this part of the scan that you will have observed visible changes to the specimen because of the numerous reactions which occurred as the copper corroded.

Question: *But so far you have described only one copper corrosion reaction, eqn [4.34]. Are there more?*

Yes. The reaction which we have been considering, in which copper corrodes to its divalent ion, Cu^{2+}, has been much simplified. The corrosion reactions of copper, especially in a solution containing chloride ions, are quite complicated. For a complete description see reference 6, but the reactions may briefly be described as the formation of copper(I) (cuprous) and copper(II) (cupric) oxides, together with insoluble hydrated chlorides.

You should have observed the formation of surface corrosion products during the time that the specimen was in the anodic regime, i.e. -240 mV and upwards.

At potentials in the region of -0.05 to -0.08 V, a marked reduction in the corrosion current is observed. This represents partial protection of the metal because of the presence of the corrosion products. This effect is known as **passivation**. In this case, the passivation is quite small, as evidenced by the relatively small reduction in current. Much more will be said about passivation in sections 4.11, 7.3 and 16.6 where examples of strong passivation will be given.

Question: *What is the significance of the changeover from negative to positive currents?*

Considerable. In practical terms it represents *the state that the metal assumes under freely corroding conditions.* The value of potential is commonly called the **free corrosion potential** and the symbol used is E_{corr}. In your experiment you should have measured E_{corr} as about -0.240 mV. At this potential the specimen could be described as being in a pseudo-equilibrium condition, equivalent to that which it would have achieved with the potentiostat switched off. (Remember that the purpose of the instrument is to drive the potential of the specimen away from its rest potential.) In practice, equilibrium conditions are impossible to achieve; the metal surface acts as an assembly of many tiny anodes and cathodes, and corrosion occurs at a rate given by the theoretical anode current density, i_a ($=$ the exchange current density, i_o). For this reason, i_o is often replaced by the symbol, **i_{corr}, the corrosion current density**. This is the simplest way of quantifying the actual corrosion rate under freely corroding conditions, thus emphasising the importance of carrying out polarisation scans and Tafel-type plots of the data obtained.

Question: *If I carry out the scan in expt 4.4 in the reverse direction, that is, from $+400$ mV to -400 mV, will I achieve the same result?*

No, not in this case, because the formation of surface corrosion products affects the currents observed. You should not expect to observe the same behaviour by starting with a surface which is already heavily corroded (as it is at $+400$ mV) and scanning to more negative potentials. It is frequently best to begin all such scans at E_{corr} (when the surface will be as close as possible to naturally occurring conditions) and then to consider deviations from this value, whether these polarisations are anodic (more positive) or cathodic (more negative). If the experiment were performed with iron in dilute sulphuric acid (using an appropriate range of potential) the reverse scan would be the same as the forward scan. More will be said about reversing the direction of scan in section 7.3.

Question: *When the copper is in the region of potential from −240 to −400 mV, it is acting as a cathode in the aqueous corrosion cell. What is the anode and what corrodes in this cell at these potentials?*

The auxiliary electrode, in this case the carbon rod, becomes the anode. There is no corrosion in the manner of eqn [2.7] because the auxiliary electrode is chosen so that it will not contaminate the solution with ions which are not part of the investigation. Other reactions must take place at the electrode surface if the electrons required by the copper cathode are to be produced. A common reaction in the electrolyte is the generation of oxygen gas by the oxidation of water:

$$2H_2O \rightarrow O_2 + 4H^+ + 4e^-$$
[4.54]

Another possibility, again in the chloride containing electrolyte we have used, is the generation of chlorine gas:

$$2Cl^- \rightarrow Cl_2 + 2e^-$$
[4.55]

Question: *Is there a simpler way of measuring free corrosion potentials?*

Yes. When you measured a simple galvanic series in expt 3.3 you were actually comparing the free corrosion potentials of the specimens. You used a copper electrode as a reference electrode and assumed that its potential would not change significantly. A better idea would be to carry out the same experiment using a SCE instead. Specimens should be given ample time to reach the pseudo-equilibrium (usually about an hour in a 1–2 litre electrolyte solution).

Question: *You explained in detail that the Tafel equation predicts a linear relationship between E and lg i. My E/lg i plot isn't at all linear.*

There are several reasons why an $E/\lg i$ plot may be non-linear. We have already discussed the deviations from linearity which occur in the vicinity of E_{corr} when i_{meas} is significantly different from i_a or i_c. Three other reasons can be offered:

(a) If there is more than one electrode reaction. The corrosion products which formed on the surface of the specimen during your experiment serve to reduce the conductivity of the anode. This results in lower current densities than would be predicted by the Tafel equation and a deviation upwards on the $E/\lg i$ plot. This is a good reason why the determination of cathodic polarisation data is made by scanning either from the free corrosion potential to more negative potentials, or from the cathodic to the anodic regime. There is little point in obtaining cathodic data on a specimen which has already been badly corroded, unless there is a special reason for doing so.

(b) If the scan rate is too fast. If the specimen is not allowed time to react to each potential step then linearity cannot be expected.

(c) Because of diffusion polarisation. The theory which we have examined so far has considered the kinetics of the electrode processes only. Another major factor in the kinetics of corrosion is the transport of charge (diffusion) through the electrolyte and this will be discussed next.

4.9   DIFFUSION PROCESSES AND THE DOUBLE LAYER

When the rate of a corrosion process is examined, several different stages must be analysed. So far we have discussed the kinetics of the reactions occurring on the electrode surfaces. The transport of the charge through the solution is also important because, unlike the passage of electrons as current, which is deemed to occur in a negligible time compared with other processes, the current in the solution is carried by the ions themselves. These have atomic masses and therefore measurable mobilities.

In the analysis of rates of reaction it is an important principle that:

The rate of a reaction is determined by the slowest step.

A simple analogy is to consider what happens to the traffic on a motorway at the scene of roadworks. When the traffic is light, the road-works pose no obstruction and the rate of passage of vehicles along the road is the same as the rate at which they join it. However, when the traffic is heavy, the bottleneck caused by the restriction creates a long tailback, while the traffic which has passed by flows freely. The rate of flow of cars along the motorway as a whole is then the rate at which they pass the obstruction and it is this rate which is called the **rate determining step**.

In a corrosion cell, the step which is the slowest can vary at any given time, just as on our motorway. When small currents are involved the transport of ions through the solution is relatively easy and the activation process is the rate determining step. However, when large currents flow the cell is demanding a greater passage of charge than can be accommodated by the electrolyte. The speed of the ions becomes the slowest step and is thus rate determining. Under these conditions we refer to the process being diffusion controlled.

Figure 4.14 represents the variation of anion concentration with distance from an electrode which is to be the cathode of a corrosion cell. Under zero current conditions, labelled (1) in the figure, the concentration of ions, c_o, will be uniform throughout the electrolyte. When the cell is connected and current flows, (2) in the figure, the anions are repelled by the cathode, a drop in concentration results, and the current flows through the electrolyte as the (net) transport of charge from the cathode to the anode. A concentration gradient is established such that for concentrations, c, and distances, x, the gradient is dc/dx. Fick's First Law of Diffusion states that:

$$J = -D \ . \ dc/dx \qquad [4.56]$$

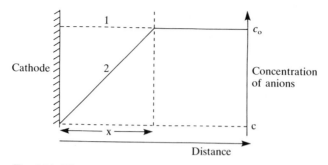

Fig. 4.14 The variation of anion concentration with distance from a cathode.

where J is the flux of substance and D is a diffusion coefficient. We have already seen, from Faraday's Law of Electrolysis that:

$$i = zFJ \tag{4.37}$$

thus, substituting eqn [4.56] into eqn [4.37]:

$$i = -zFD.dc/dx \tag{4.57}$$

To simplify the situation, assume that the concentration gradient is linear and can be replaced by $(c_o - c)/x$. Equation [4.57] becomes:

$$i = -zFD. (c_o - c)/x \tag{4.58}$$

The negative sign is conventional: it tells us that the current is being carried away from the cathode. We are interested only in the magnitude of the current. Thus, since c can never be negative the magnitude of the current is greatest when $c = 0$ and a maximum current, i_{max} is obtained:

$$i_{max} = -zFD. c_o/x \tag{4.59}$$

(Notice that when $c > c_o$ the implication is that the concentration of anions in the region of the electrode is increasing. The resulting sign change of i from negative to positive tells us that the current is reversed.)

Using the Nernst equation for condition 1 (no current):

$$E_1 = E^o + \frac{0.059}{z} \lg c_o \tag{4.60}$$

while for condition 2

$$E_2 = E^o + \frac{0.059}{z} \lg c \tag{4.61}$$

We have previously defined the polarisation as the change of potential away from the equilibrium (no current) condition. Thus $\eta = E_2 - E_1$ and subtracting eqn [4.60] from eqn [4.61] we get:

$$\eta = \frac{0.059}{z} \lg \left(\frac{c}{c_o}\right) \tag{4.62}$$

Using eqn [4.58] and eqn [4.59] it can be shown that

$$\frac{c}{c_o} = \left(1 - \frac{i}{i_{max}}\right)$$ [4.63]

Substituting eqn [4.63] into eqn [4.62] we obtain

$$\eta = \frac{0.059}{z} \lg\left(1 - \frac{i}{i_{max}}\right)$$ [4.64]

From this equation it can be seen that as $i \to i_{max}$ then $\eta \to$ −infinity. The overall effect of diffusion polarisation on the cathodic part of the $E/\lg i$ plot is easy to define and should be illustrated in the results you obtained in the experiment. For small currents, $i_{meas} \to 0$ because $i_c \to i_o$: non-linearity of the $E/\lg i$ plot is obtained. For intermediate currents, $i_{meas} \to i_c$, the Tafel equation holds and linearity is observed. As the current increases towards the limiting value the plot begins once more to deviate from linearity towards more negative values. In theory it should approach the limiting current density asymptotically. This is clearly shown in the potentiodynamic polarisation scan of Fig. 4.13. Both the anodic and the cathodic portions of the curve show limiting current density values which result from diffusion control at the respective potential ranges.

4.10 MIXED POTENTIAL THEORY

In section 4.5 the Daniell cell was considered as a corroding system in which, as current is drawn, the cell potential falls from its maximum theoretical value. We can use a graph of potential versus current density to illustrate in diagrammatic format the typical polarisation of both of the electrodes of a corrosion cell. By plotting current density on a logarithmic scale, the polarisation lines will be linear, in accordance with the Tafel equation. The diagrams which are drawn in the way which will be described below are commonly called **Evans diagrams**, after one of the founders of corrosion science, Ulick Evans.

Figure 4.15(a) shows such a diagram using a zinc anode (A) and a copper cathode (C), as we were using in the Daniell cell. First the free corrosion potentials of the individual metals in their solutions, $E_{corr(A)}$ and $E_{corr(C)}$ are recorded at their respective exchange current densities, the cathode having the more positive value of E_{corr}. As the zinc corrodes, according to the equation:

$$Zn \to Zn^{2+} + 2e^-$$ [4.25]

it is polarised upwards to more positive values by an amount η_a, the anodic polarisation, while the copper reaction

$$Cu^{2+} + 2e^- \to Cu$$ [4.7]

causes a cathodic polarisation, η_c, downwards to more negative values. For completeness it may be desirable to draw both the anodic and cathodic polarisations for each electrode. The slopes of these lines, the anodic and cathodic beta constants, are not necessarily the same, or even similar. The limiting cell current, i_L, will be obtained at the intersection of the anodic polarisation line of the anode and the cathodic polarisation line of the cathode. At this point, the cell potential is theoretically zero and the electrodes short-circuited.

Question: *How can this be so? You are saying that as the cell potential decreases, the current increases until, when the potential is zero the current is maximised. You can't have a current without a driving force.*

Correct. The flaw in the argument is that the potential does not actually reach zero and the resistance also falls rapidly to a minimum, but not to zero. Thus, at the point at which the electrodes seem to be short-circuited, the resistance is small and the cell potential is also small, though the current is significant (Ohm's Law, eqn [2.19]). This minimum cell resistance, often called the **internal resistance** in working cells, prevents the limiting current ever being achieved in practice. Instead, the maximum cell current is i_{cell}, see Fig. 4.15(a), where $i_{cell} < i_L$.

Question: *If the potential of the cathode is more positive than for the anode, why do the anode reactions have positive slopes in the diagram? Surely, as the anode corrodes it will become more anodic and the line will have a negative slope.*

Remember the simple explanation given towards the end of section 4.6. When a piece of metal is placed in an electrolyte, metal ions enter the solution and the metal is left with an excess of negative electrons. This is the source of the free corrosion potential. If the metal is now connected to another metal which is cathodic to it, the electrons will flow away from the anode, towards the cathode, and the anode will be less negative than it was before. Thus the potential will become more positive. Le Chatelier's Principle says that the system always reacts to oppose the change we try to impose upon it. When a metal corrodes it loses some of its thermodynamic desire for corrosion.

Question: *How do I calculate values for i_o?*

There is no theoretical method for doing this – only a practical one. You have to perform $E/\lg i$ measurements very carefully for each metal, both anodically and cathodically polarised. If good linearity of the resulting Tafel plot is obtained, then the extrapolated linear portions of the anodic and cathodic polarisations should intersect at the value corresponding to $\lg i_o$ and E_{corr}. (Refer to Fig. 4.9(b).)

The use of Evans diagrams such as that shown in Fig. 4.15(a) is not

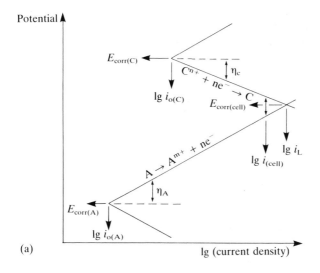

(a)

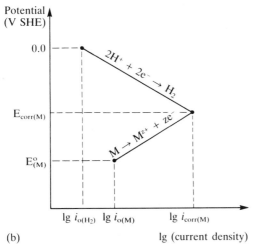

(b)

Fig. 4.15 Two mixed potential plots (Evans diagrams):
(a) The Daniell cell. The individual free corrosion potentials of the copper (C) and zinc (A) are changed to a cell potential, $E_{corr(cell)}$, when the metals are short-circuited. The limiting current density, i_L, is never achieved because of the finite internal resistance of the cell.
(b) A metal, M, corroding in an acid solution with the evolution of hydrogen. The microstructure of the metal causes it to act as its own anode and cathode; the cathode reaction is the reduction of hydrogen ions.

restricted to simple cases in which the cathode reaction is one of replating metal ions. Section 4.2 described two common cathode reactions which occur when a metal corrodes in aqueous solution: hydrogen evolution when the electrolyte pH is less than 7, and reduction of dissolved oxygen for pH ≥ 7. The well-known chemical property whereby metals dissolve in acids with the evolution of hydrogen is a corrosion reaction which can be described by an Evans diagram such as that shown in Fig. 4.15(b). The cathode reaction is now represented by eqns [4.19/20] and at the point where the polarisation line intersects that of the metal dissolution reaction, the corrosion rate is $i_{corr(M)}$. The value of $i_{o(M)}$ occurs at 0.0 V SHE, but at different current densities according to the metal. Obviously, the rate of corrosion of a metal in acid is governed by such factors as:

(a) the anodic polarisation line of the metal;
(b) the exchange current density of hydrogen evolution on the metal.

Thus, if $i_{o(M)}$ is moved to higher current densities for the same anodic polarisation line, the metal will corrode much faster. A good example of this is that of iron and zinc. Under identical electrolyte conditions, iron corrodes much faster in hydrochloric acid than does zinc, despite zinc's much more active position in the galvanic series (section 5.1). This is because $i_{o(H_2)}$ is much greater on iron than on zinc. The explanation is shown in Fig. 4.16. (Noble metals such as platinum and palladium have very high values of $i_{o(H_2)}$ but their greatly positive electrode reduction potentials explain their lack of corrosion.)

The hydrogen evolution line for zinc is shown as line a, while the anodic dissolution of zinc is line b. The free corrosion potential and corrosion

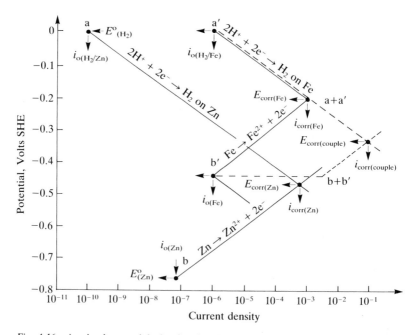

Fig. 4.16 A mixed potential plot for the bimetallic couple of iron and zinc. The diagram also explains the higher corrosion rate of iron than zinc in hydrochloric acid solution. Despite its more positive reduction potential, iron has a high exchange current density for the evolution of hydrogen.

100

current density of zinc are shown at the intersection of these two lines. The equivalent lines for iron (corroding independently of the zinc) are shown as a′ and b′ respectively. The corrosion potential and corrosion current density for the zinc/iron couple occur at the intersection of the cathodic and anodic lines for the couple. In order to obtain these, it is necessary to sum the two cathodic processes (a + a′) and the two anodic processes (b + b′). These are shown as the dashed diagonal lines in Fig. 4.16.

This type of diagram is extremely useful in both explaining and predicting corrosion rates in different environments. This example (in acid solution) was chosen so that the cathodic lines were representative of hydrogen evolution. In neutral or alkaline solutions, the cathode reaction is represented by eqn [4.21], but is strongly affected by the rate of oxygen diffusion to the metal surface and by stirring. As we shall see in Chapter 5, a different environment can alter considerably the plot obtained. In particular, experimental effects can sometimes cause complications (section 5.2). However, this should not detract from the usefulness of Evans diagrams, especially in situations where two or more metals may be coupled together. *The success of Evans diagrams is that they combine thermodynamics and kinetics to form a whole picture: the potential axis is the thermodynamic factor and the current axis is the kinetic factor.*

4.11 E/pH (POURBAIX) DIAGRAMS

In section 4.1 the Nernst equation was examined in detail and we saw how the potential of an electrode deviated from the standard electrode potential by an amount which depended upon the concentration of ions. We have also seen that E_{corr} can be used to represent the pseudo-equilibrium condition which exists when an electrode is corroding freely with no externally applied potential.

E_{corr} is essentially a parameter which is determined in the laboratory, but theory can also be used to predict a potential at which a similar condition exists. This concept was first used by Marcel Pourbaix who distinguished a corroding condition from a non-corroding condition as follows:

A metal is deemed to be in a corroding condition when the concentration of its ions in solution $\geq 10^{-6}$ M.

If the concentration of ions does not exceed this value then the metal is deemed to be in the immunity condition.

Until now we have considered all corrosion reactions to be represented by the equation

$$M \rightarrow M^{z+} + ze^- \qquad [2.7]$$

where the ionic species has been assumed to enter the solution as a soluble species. This may not be the case: many corrosion products are insoluble in the electrolyte – oxides, for example. Often, the insoluble product forms a film on the corroding surface which effectively prevents the electrolyte from coming into contact with the metal and greatly reduces the corrosion rate. This is the third electrode condition and is called **passivity**. Experiments 3.12, 3.13 and 3.14 illustrate this property, and the effect of film formation on current density was described in the discussion on expt 4.4. There it was noted that the corrosion of copper in chloride solutions is quite complex. Indeed, the corrosion of many metals, even in simple electrolytes, consists of more than one reaction. These reactions are often dependent on the pH of the electrolyte. (For an explanation of the term, pH, see section 2.3).

Pourbaix succeeded in correlating the dependence of pH and the potential of the electrode upon the condition of the electrode. The result of his work is a chart for each metal which shows the conditions under which a metal will be corroding, not corroding, or passivated in aqueous solution. The charts are called *E/pH* or **Pourbaix diagrams** and a comprehensive selection may be found in Pourbaix's *Atlas of Electrochemical Equilibria in Aqueous Solutions*.[5]

The construction of an E/pH diagram, rather like a table of logarithms, is based upon quite simple principles, but requires many calculations. It has the compensation that the task need only be performed once. As an example we shall consider the construction of the E/pH diagram for zinc in water. The description has been much simplified to assist the explanation, but the principles involved in the construction of the diagram are the same for all metals. For the complete description of the zinc/water system, you are referred to the work of Pourbaix.

When zinc corrodes in pure water, up to four species can be present over the complete range of potential and pH. Therefore five reactions must be written to describe the possible reaction processes for the interconversion of each of the species:

(a) The usual anode reaction

$$Zn = Zn^{2+} + 2e^- \tag{4.25}$$

(b) The formation of insoluble zinc hydroxide

$$Zn + 2H_2O = Zn(OH)_2 + 2H^+ + 2e^- \tag{4.65}$$

(c) The formation of a soluble zincate ion

$$Zn + 2H_2O = ZnO_2^{2-} + 4H^+ + 2e^- \tag{4.66}$$

(d) The dissolution of zinc hydroxide by acid

$$Zn(OH)_2 + 2H^+ = Zn^{2+} + 2H_2O \tag{4.67}$$

(e) The formation of zincate from zinc hydroxide

$$Zn(OH)_2 = ZnO_2^{2-} + 2H^+ \tag{4.68}$$

Those reactions which involve the generation of electrons (a), (b) and (c) must be influenced by variations of electrode potential, while those in which hydrogen ions are formed (b), (c), (d) and (e) will be controlled by

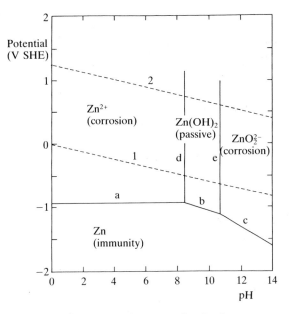

Fig. 4.17 E/pH (Pourbaix) diagram for zinc in water.

pH. Reactions (b) and (c) are controlled by both potential and pH.

Reaction (a) has already been studied in detail. It is independent of pH variation and influenced only by variation of E in accordance with the Nernst equation. Substituting the value of $[M^{z+}] = 10^{-6}$ M we can calculate the potential at which a metal begins to corrode.

$$E = -0.76 + \frac{0.059}{2} \lg (10^{-6})$$

i.e. $E = -0.76 - 0.177$

$$E = -0.937 \text{ V}$$

Remember that all potentials calculated from the Nernst equation are relative to the standard hydrogen electrode and are more accurately quoted as -0.937 V SHE.

We have thus established that a pseudo-equilibrium occurs at -0.937 V when the concentration of $Zn^{2+} = 10^{-6}$ M. On a graph of E as ordinate and pH as abscissa we can draw a horizontal line at -0.937 V.

Figure 4.17 shows an E/pH diagram for the system, and the line at -0.937 V is labelled line a.

Question: *How are the other lines obtained?*

For small values of pH reaction (a) is the only significant one, but as the electrolyte becomes more alkaline, the concentration of hydroxide becomes large enough to cause reaction (b) to predominate. If we take the equilibrium condition and substitute into the Nernst equation we can write:

$$E = E^o + 0.0295 \lg \frac{[Zn(OH)_2]. [H^+]^2}{[H_2O]^2 . [Zn]} \qquad [4.69]$$

As has been explained earlier, our simplified treatment using concentrations rather than activities requires that we substitute unity for terms other than ions in solution. This simplifies eqn [4.69] to:

$$E = E^o + 0.0295 \lg [H^+]^2 \qquad [4.70]$$

E^o is the standard electrode potential for the $Zn/Zn(OH)_2$ combination. This value has been measured as -0.439 V, and as we saw in section 2.3, $-\lg [H^+] = pH$. Equation [4.70] therefore becomes:

$$E = -0.439 - 2.(0.0295).pH \qquad [4.71]$$

Equation [4.71] is the equation of the line representing reaction (b) and labelled as such in Fig. 4.17. We can confirm its point of intersection with line a by substituting the value of -0.937 V for E in eqn [4.71]. Thus:

$$pH = \frac{-0.937 + 0.439}{-0.059} = 8.44$$

This tells us the equation of the line d, for at this value of pH, $Zn(OH)_2$ becomes the most stable species instead of Zn^{2+}. A vertical line is thus drawn at pH 8.44 (shown as line d in Fig. 4.17) to represent reaction (d) which is independent of potential.

Line d can be found by another method, for if the equilibrium constant for reaction (d) is known, we can write:

$$K = \frac{[Zn^{2+}]. [H_2O]^2}{[Zn(OH)_2]. [H^+]^2} \qquad [4.72]$$

i.e. $\lg K = \lg [Zn^{2+}] + 2pH \qquad [4.73]$

K has been measured by chemical means and found to have the value 7.58×10^{10} and we have defined $[Zn^{2+}] = 10^{-6}$, therefore substituting into eqn [4.73]:

$$10.88 = -6 + 2pH$$
$$16.88 = 2pH$$
$$8.44 = pH$$

As pH is increased still further it is found that the insoluble $Zn(OH)_2$ dissolves to form the so-called zincate ion. This occurs at pH $= 10.68$ and is vertical line e in Fig. 4.17. Line c can be determined in a manner identical to that described for line b.

The four regions of the diagram can be considered as being domains of immunity, corrosion or passivation. Each domain indicates a region in which one species is the most thermodynamically stable. If it is the metal which is the most stable species then it is considered to be immune to corrosion, but if a soluble ion is most stable then the metal should corrode. A region in which an insoluble corrosion product is the most stable species is considered to be passive. Here it is not so easy to predict whether the metal will corrode or not. There are many factors which determine whether a solid corrosion product will form a sufficiently protective film on the surface of the metal to prevent corrosion from taking place. These factors will become more apparent as the corrosion properties of different materials are discussed in other chapters.

Question: *What do the dashed lines in Fig. 4.17 mean?*

They represent two other reactions which are possible in aqueous solutions:

1. The reduction of hydrogen ions to liberate hydrogen gas:

$$2H^+ + 2e^- \rightarrow H_2 \qquad\qquad [4.19/20]$$

2. The oxidation of water to liberate oxygen gas:

$$2H_2O \rightarrow O_2 + 4H^+ + 4e^- \qquad\qquad [4.54]$$

Reaction 1 gives a Nernst equation:

$$E = E^o - 0.059pH \qquad\qquad [4.74]$$

assuming that the pressure of hydrogen is 1 atmosphere. E^o for hydrogen is, of course, 0.00 V, so the line intersects the y-axis at $E = 0$ for pH $= 0$ and at pH $= 10$ has a value of $E = -0.59$ V. Below line 1 hydrogen gas is the most stable species, while above it, the hydrogen ion is stable. Hydrogen gas is always liberated at a cathode and this will occur when the cathode is in the domain of potential and pH below line 1.

Reaction 2 gives an equivalent Nernst equation:

$$E = E^o - 0.059pH \qquad\qquad [4.74]$$

where E^o has been measured as +1.228 V. The line 2 therefore intersects the potential axis at this value and at pH $= 10$ has the value $1.228 - 0.59 = 0.638$ V. Above this line oxygen gas is the most stable species and will be liberated on an anode which lies in this region of potential and pH.

Question: *How can I make water either strongly acid or strongly alkaline without it effectively becoming another electrolyte?*

The question has highlighted one weakness of this theoretical approach to corrosion. In short, in order to cause an imbalance in either the hydrogen or hydroxide ions, it is necessary for another counter-ion to be present and the properties of water may be modified as a result. This may invalidate some of the calculations and their predictions. Often, it is possible to have a counter-ion present which does not significantly interfere, but this is not always so. In the case of copper and water, for example, the presence of chloride ions in the electrolyte *does* have a significant bearing on the electrode processes. Thus if we had attempted to confirm the measurements we made in expt 4.4 by reference to an E/pH diagram for copper and water, we would have found very poor correlation, because of the reactions which had not been taken into consideration. In this particular case it has been possible to construct a practical E/pH diagram.[6] Such diagrams are a step nearer the goal of total predictability.

A further problem is that the domains have been calculated using thermodynamic data and, as we have seen, the actual reactions are kinetically as well as thermodynamically controlled. Thus, unlike Evans diagrams, E/pH diagrams give no information about corrosion rates. There are many environmental factors, too, which cannot be encompassed by the E/pH diagram. Changes of flow rates, oxygen concentration, temperature and pressure are just some of the ways in which the environment continually influences the course of corrosion reactions, and what may be a simple system on paper, or even in the laboratory, becomes highly unpredictable in nature. Although the E/pH diagram would seem at first sight to be exceedingly useful to predict the course of corrosion reactions, there are limitations which must be borne in mind before it can be used to effect.

4.12 REFERENCES

1. Uhlig H H 1953 *Corrosion handbook*. John Wiley, p 1134
2. Evans U R 1981 *An introduction to metallic corrosion*. Edward Arnold, p 285
3. Skoog D A, West D M 1971 *Principles of instrumental analysis*. Holt, Rinehart and Winston, p 502
4. Stern M, Geary A L 1957 Electrochemical polarisation, *Journal of the Electrochemical Society* **104**(1): 56
5. Pourbaix M 1966 *Atlas of electrochemical equilibria in aqueous solutions*. Pergamon Press
6. Bianchi G, Longhi P 1973 Copper in sea water: potential pH diagrams, *Corrosion Science* **13**: 853

4.13 BIBLIOGRAPHY

Evans U R 1960 *The corrosion and oxidation of metals*. Edward Arnold
Fontana M G, Greene N D 1978 *Corrosion engineering*. New York: McGraw-Hill, Chs 2, 9 and 10
Fried I 1973 *The chemistry of electrode processes*. Academic Press
Scully J C 1966 *Fundamentals of corrosion*. Pergamon Press
Selley N J 1977 *Experimental approach to electrochemistry*. Edward Arnold
Shrier L L (ed.) 1979 *Corrosion* (Vol. 1) Newnes-Butterworths, Ch. 1
West J M 1980 *Basic corrosion and oxidation*. Ellis Horwood

5   DISSIMILAR METAL CORROSION

A good scientific theory should be explicable to a barmaid.

(Ernest Rutherford. Quoted by R W Clark in *The Greatest Power on Earth*)

Dissimilar metal corrosion is the name given to the corrosion which results when two different metals are coupled together to form a basic wet corrosion cell, as defined in section 4.2. It is also frequently called **bimetallic corrosion**, or **galvanic corrosion**, because the corrosion is primarily galvanic in nature. However, the last name is something of a misnomer because all aqueous corrosion is caused by the galvanic effect.

The corrosion problems associated with the coupling of dissimilar metals have been appreciated for over two hundred years (see Ch. 1) yet dissimilar metal corrosion is still rife throughout the engineering world. In the report to the Admiralty in 1763 it was suggested that when it was necessary to couple dissimilar metals they should be insulated from each other, but even when this sound advice is carried out, problems still arise.

Case 5.1:   Major restoration work has only recently been completed on the Statue of Liberty which was built in 1886 but which by 1980 had suffered serious weakening of its structure, partly through dissimilar metal corrosion. In its construction, the copper skin of the statue was supported by a network of iron ribs. At intervals, the skin was riveted to a copper band called a saddle which, in turn, was supported by a rib (see Fig. 5.1). The saddles were free to slip on the ribs to allow for contraction and expansion. In addition, the copper saddles were isolated from the ribs by a layer of tar, though the ribs and the skins were in contact. The tar deteriorated during the statue's hundred year lifetime and, with the inevitable ingress of water, galvanic corrosion ensued. The build-up of corrosion product caused saddle rivets which held the copper to the supports to pull out and resulted in serious deformation of the green patina-covered skin.[1]

Although the original design of the Statue of Liberty had made some provision for the possibility of dissimilar metal corrosion by insulating with tar, the case history is typical of what has probably been the most common

Dissimilar metal corrosion

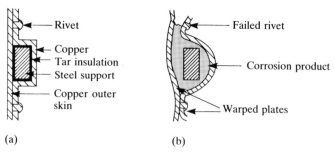

Fig. 5.1 The Statue of Liberty which has recently been given a major structural overhaul. Weakening of the iron support members had occurred because of bimetallic corrosion with the copper skin, despite the original use of insulated joints.

problem of corrosion between dissimilar metals: the coupling of copper and copper-based alloys with iron and steel. The cheapness and availability of these engineering materials has inevitably resulted in their use in many situations in which for corrosion resistance they are quite incompatible.

Case 5.2: The owner of a 'classic' sports car spent many hours restoring his prize possession to showroom condition, but when renewing the plastic fasteners for his soft top, decided to fix them to the steel body with small brass screws because they 'would not go rusty'. Within weeks, rings of rust had broken through the paint coat immediately surrounding the fasteners.

Such a story is typical of the basic errors made, even with the best of intentions, through ignorance. On many occasions, the maintenance engineer is fully aware of the galvanic effect, but, faced with the requirement for a rapid repair and the non-availability of compatible materials, uses the wrong combination and remarks, with a shrug of his shoulders, 'Oh, that'll be all right'. Sometimes, dissimilar metal corrosion results from a quick, thoughtless action.

Case 5.3: An aircraft handler experienced difficulty in operating a magnesium alloy fuel pipe coupling (Fig. 5.2) when refuelling his aircraft. The application of graphite grease proved to be a very short-term remedy, for the galvanic action between the graphite and the alloy caused seizure of the coupling when it next rained.

The failure to appreciate the problems posed by the very noble and electrically conducting material, graphite, has been the cause of many cases of galvanic corrosion. Even in its amorphous form, carbon is a significant electrical conductor which is almost as likely to cause galvanic corrosion cells as graphite. Deposition on to metal surfaces of soot from smoke stacks is bad enough, but is often exacerbated because sulphur oxide emissions which usually accompany the soot cause 'acid rain'. The galvanic action which results in such cases is rapid and severe.

Fig. 5.2 A magnesium alloy aircraft fuel pipe coupling which seized because of galvanic action after it had been lubricated with graphite grease.

Numerous other case histories involving bimetallic corrosion will be found throughout this book, but, with care, it is possible to use dissimilar metal combinations with relatively little adverse effect. Metals which are widely separated in the galvanic series (discussed in the next section and illustrated in Fig. 5.5) should not be coupled – combinations with copper and copper alloys are always risky. However, the alliance of aluminium and steel can sometimes be made.

Figure 5.3(a) shows how insulating plastic sleeves and washers enable the safe use of aluminium nuts and bolts to fasten steel plates to aluminium, but insulation is not always necessary. In the marine industry, the stability of ships has been improved by reducing top weight with aluminium alloy superstructures instead of steel. Normal welding techniques cannot be used to join aluminium alloy to steel so riveted joints were originally used, and corrosion protection was provided by suitable paint systems. This method has not proved entirely satisfactory and corrosion in the joint region has been observed. A more recent development has been to use an explosively bonded transition joint. Three sheets of metal, one of aluminium alloy (usually a 5000 series, aluminium/magnesium alloy), one of commercial purity aluminium and the third of steel, are joined by an explosion which creates millions of atmospheres' pressure between the layers. The bonded sheets are then cut into bars of suitable dimensions which are used in the transition between aluminium and steel structures by making two fillet welds: one of the steel in the joint to the steel of the ship's deck, and the other of the alloy in the joint to the aluminium alloy superstructure. Figure 5.3(b) illustrates the jointing method. Notice that the transition joint material must be wider than the sheets being joined to ensure maximum

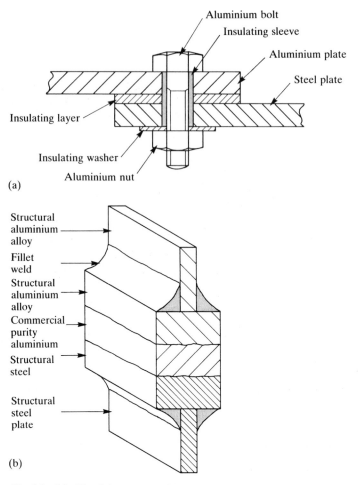

Aluminium bolt

Insulating sleeve

Aluminium plate

Steel plate

Insulating layer

Insulating washer

Aluminium nut

(a)

Structural aluminium alloy

Fillet weld

Structural aluminium alloy

Commercial purity aluminium

Structural steel

Structural steel plate

(b)

Fig. 5.3 (a) Aluminium nuts and bolts can be used to secure steel plates to aluminium if plastic insulating sleeves and washers are used.
(b) The use of an explosively bonded transition joint for the fastening, by welding, of aluminium alloy to steel.

strength. Figure 5.4 shows the Royal Navy patrol vessel, HMS *Peacock*, just one of many modern ships which have been constructed using transition joints.

Case 5.4: The use of explosively bonded transition joints requires careful fabrication techniques to avoid adverse effects in the aluminium/steel interface, but with properly applied protective paint the joints are reported to perform well and to show little signs of corrosion.[2,3] Unfortunately, it has been reported[4] that in some shipyards, the manufacturers' recommendations for welding and fabrication techniques are not always adhered to, especially in complicated geometrical arrangements. This, combined with the deficiencies which exist in all

Fig. 5.4 A patrol vessel of the Royal Navy, HMS *Peacock*, which has an aluminium alloy superstructure welded to a steel hull by means of an explosively bonded transition joint.

but the best paint coatings, seems to indicate that corrosion problems will occur.

5.1   THE GALVANIC SERIES

The principles of dissimilar metal corrosion and the galvanic effect have already been discussed in some detail in Chapter 4 because they are fundamental to the understanding of all forms of corrosion. Experiment 3.3 described the establishment of an order of corrosion tendency, better known as a *galvanic* series. Such a series is of great practical value because it enables a rapid prediction of the corrosion resistance of a dissimilar metal couple. Table 4.1 is also a series, often called the *electrochemical* series. It compares the reduction (or oxidation) potentials of the metals, but differs from the galvanic series in a number of ways:

(a) The electrochemical series is an absolute, quantitative series listing electrochemical data for use in precise calculations; the galvanic series is a relative, qualitative series listing an experimental order of nobility (or activity) of metals.

(b) The electrochemical series, of necessity, lists data only for metal elements; the galvanic series contains both pure metals *and* alloys – a considerable practical advantage.

111

Dissimilar metal corrosion

Free corrosion potential, V SCE

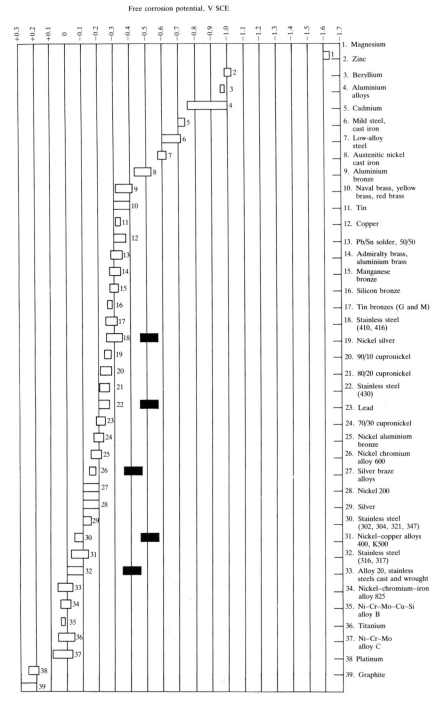

Fig. 5.5 The galvanic series in sea water. Black zones indicate active potential ranges. (After F L LaQue[5])

© Copyright John Wiley and Sons, Inc. 1975. Reprinted by permission.

(c) The electrochemical series is measured under standard conditions and is independent of other species in the environment; the galvanic series is measured under arbitrary (though specified) conditions of temperature, pressure and electrolyte.

Figure 5.5 shows the galvanic series of selected metals at 25 °C in sea water.[5] The potentials listed are actually free corrosion potentials, and, in general, it can be inferred that the bigger the separation of any two metals in the series, the more severe is the corrosion of the more active of them likely to be. Thus the considerable difference in activity between copper and iron or steel underlines the danger of using such a combination, not only in sea water but in any aqueous medium. Very occasionally, such a combination may be justified on the grounds that the corrosion of the steel rather than an expensive copper component may be preferable, especially if the steel is of sufficient dimensions that failure is not possible. (The general rule is that large anodes may be tolerated with small cathodes but small anodes are very dangerous with large cathodes. This is because current density is the factor which determines corrosion rate, not just current.)

The most anodic material listed is magnesium. This is a significant engineering material only when it is alloyed with small amounts of aluminium or zinc to improve the mechanical properties. Its low density makes it extremely useful in the aircraft industry; many early helicopters were made of magnesium alloy, but the severity of attack in marine environments, particularly when coupled to another metal, such as steel, has led to substitution of aluminium alloys for magnesium alloys.

Case 5.5: Figure 5.6 is a magnesium alloy component from a helicopter which crashed into the sea and was immersed for three days before recovery. The component was secured to the main structure by steel bolts which are visible, protruding from the large mass of corrosion product. The danger of using these materials in such an aggressive environment is obvious, even though they are not necessarily immersed!

The problem with graphite, the most noble material, and its coupling to one of the most active metals has already been mentioned. The similar noble position of titanium, which is finding an ever increasing use in aggressive environments, indicates the need for care in the design of components and structures utilising this material.

Surprisingly, titanium is often not a problem because it is a passivating material. This is a property which will be discussed in the next section, and in sections 7.3 and 16.6. The surface condition of these materials which readily form stable oxide films in air gives them an unexpected corrosion resistance which can sometimes lead to erroneous predictions about galvanic corrosion compatibility. For example, aluminium which is a very active material can be used in very many applications with a greatly reduced risk of galvanic corrosion because of such an oxide layer (see section 15.4).

Fig. 5.6 A magnesium alloy helicopter component which suffered immersion in sea water for three days. The corrosion is worst adjacent to the steel bolts which were used to secure it to the main structure.

Many other materials are similarly affected, notably stainless steels and nickel alloys. In some cases, an alloy which passivates only slowly can be used in the active instead of the passive condition and galvanic corrosion can ensue if it forms part of an incompatible couple. This is particularly true of some stainless steels, and this schizophrenic behaviour is clearly shown in Fig. 5.5, where the black zones indicate the active potential ranges.

The coupling of dissimilar metals can sometimes be used to advantage. Cathodic protection by sacrificial anodes is discussed at length in Chapter 16, but sacrificial wasters are also a good method of controlling the corrosion of long pipe runs (see Fig. 12.3). The wasters are anodic to the rest of the pipe and corrode preferentially. By siting them in easily accessible positions, they may be replaced quickly and cheaply.

It should always be borne in mind that in certain circumstances, the order of metals in Fig. 5.5 may be significantly altered by a change in the environment. Some particular occurrences have been described by the term 'cell reversal', the most important example of which is the reversal of zinc and iron in galvanised steel components at elevated temperatures in some potable water systems[6] (see section 14.10). Such instances have only served to emphasise the very complicated nature of metal–environment interactions and the importance of supporting design of engineering components and structures with experimental testing of systems which accurately reproduce field conditions. The next section discusses methods for improving the understanding of galvanic effects and for predicting ways of achieving better corrosion control.

5.2 MIXED POTENTIAL THEORY AND DISSIMILAR METAL CORROSION

Despite the usefulness of the galvanic series in predicting the relative tendencies of metals to corrode, we are yet again faced with the problem of not knowing the rate of corrosion. The true measure of corrosion rate is current or current density and it is therefore necessary to examine corrosion behaviour by means of $E/\lg I$ or $E/\lg i$ plots. Elements of the mixed potential theory, originally devised by Evans and represented by diagrams named after him, have already been described in section 4.10 and Figs. 4.15 and 4.16. From those discussions it is possible to make statements about the effect of coupling two dissimilar metals together:

(a) The galvanic series predicts that the more active metal will become the anode of the couple in a wet corrosion cell (as depicted by Fig. 4.2), while the more noble metal will be the cathode.

(b) The corrosion rate of the more active metal is accelerated, while that of the more noble metal is retarded. Note that the cathode may still corrode, depending upon the magnitude of the cathodic polarisation induced (see below).

Question: *How can I determine an Evans diagram such as this in the laboratory?*

Experiment 5.1

Set up a three-electrode cell, as described in section 4.8, using the apparatus shown in Fig. 5.7. The working and auxiliary electrodes should be suitably mounted specimens of the materials of interest, iron and zinc, say, and of known surface areas. (A first experiment should use equal areas.) Separate measurement of the potentials of anode and cathode is necessary. In a low resistivity solution such as 3.5 per cent sodium chloride, one SCE placed between the two electrodes will suffice, together with a switching device to alternate potential measurement between anode and cathode. The use of a resistance box which can be varied between 10^5 and 0 ohms, enables the measurement of current at different polarisations.

Begin with an open circuit. Measure the E_{corr} values for iron and zinc by changing the switch between positions labelled (1) and (2) in Fig. 5.7. Set the resistance box to 10^5 ohms and repeat the measurements of anode and cathode potential. Change the resistance stepwise by suitable amounts. (A trial run will be needed to determine this; use the whole of the available range.) Read the potential of each specimen for each value of resistance selected. Allow 2 minutes after each resistance step in order for the electrodes to stabilise. The final measurement is of the short-circuit condition,

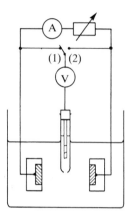

Fig. 5.7 The experimental determination of Evans diagrams.

when the potentials of both anode and cathode should be the same. If they are not then there is a residual resistance in the circuit, either in the electrical connections or in the solution. On the same graph, plot the variation of both anode and cathode potential with lg i.

A typical result for iron and zinc is shown in Fig. 5.8. The immediate impression gained from the results is that the plot is nothing like those which have been drawn from theory. It must be remembered that deviation from linearity will be obtained for most, if not all, of the reasons discussed in section 4.8, that is, at low current densities when i_a and i_c are comparable, and at high current densities when diffusion effects come into play. Hence, the correlation between the expected iron polarisation line (dotted) and the measured line is small. The correlation for zinc is much better, though for different reasons. As can be seen, the polarisation of zinc by coupling to iron is quite small, a factor which, as we shall see in Chapter 16, is of great value in the application to cathodic protection. In this case, it is highly likely that most of the deviation of the iron polarisation line at high current densities is caused by hydrogen evolution from about −0.9 V SCE and below. Notice also that the lines do not meet, even when the electrodes are short-circuited. This is caused by the presence of internal resistances discussed in section 4.10.

Another factor which determines the rate of corrosion is that of the reaction occurring on the cathode. In section 4.1 it was stated that there were two normal cathode reactions: hydrogen evolution and oxygen reduction. The former is the usual reaction in solutions of pH < 7, while the latter occurs in aerated solutions of pH ⩾ 7. The effect of the hydrogen evolution reaction was discussed to a small degree in section 4.10; more complete discussion may be found in the bibliography to Chapter 4. The effect of the oxygen reaction was mentioned briefly above. In a couple where there is no metal reduction reaction at the cathode, the diffusion of oxygen to the

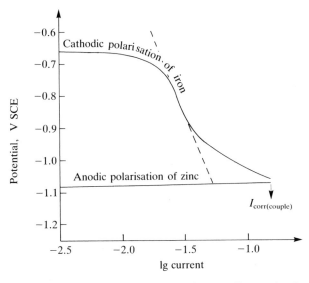

Fig. 5.8 An experimentally determined Evans diagram for the iron/zinc bimetallic couple.

cathode is very often the factor which controls the rate of corrosion. This is because both the concentration of oxygen (0.001 mol dm^{-3} in air saturated solution) and the diffusion coefficient (10^{-5} cm^2 s^{-1}) are relatively small. Equation [4.64] gives an expression for the maximum current, from which it can be calculated that the limiting current density is about 80 μA cm^{-2} in unstirred solution where the thickness of the double layer is about 0.5 mm. In a stirred solution, the double layer is likely[7] to be about 0.01 mm, in which case the current is increased correspondingly to about 4 mA cm^{-2}. Figure 5.9(a) illustrates the effect of cathodic control by oxygen reduction and shows how the corrosion rate is increased in stirred solutions.

In the previous section it was mentioned that titanium is a passivating material. In non-oxidising acids such as sulphuric or hydrochloric acid, its rate of corrosion is governed by the rate of hydrogen evolution (section 4.10). The microstructure of the metal causes it to behave as its own anode and cathode, and the metal corrodes at a potential, $E_{corr(Ti)}$ and a rate, $i_{corr(Ti)}$, as shown in Fig. 5.9(b). The galvanic effect was used to enable the manufacture of titanium alloys with superior corrosion resistance in these media. The effect relies upon the greatly increased rate of hydrogen evolution on noble metals, such as palladium or platinum.[8] The alloying of as little as 0.5 per cent palladium caused spontaneous passivation of the titanium at the potential $E_{corr(alloy)}$ and a correspondingly low corrosion rate, $i_{corr(alloy)}$, in Fig. 5.9(b). Stern and Wissenberg[9] reported that in 10 per cent boiling sulphuric acid and in 10 per cent boiling hydrochloric acid the corrosion rates were 800 to 1000 times slower for Ti/0.5 per cent Pd than for commercial purity titanium. Other alloys containing platinum, rhodium, iridium and gold were similarly resistant to attack.

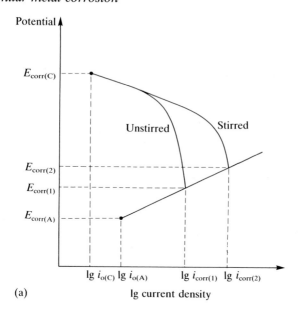

(a)

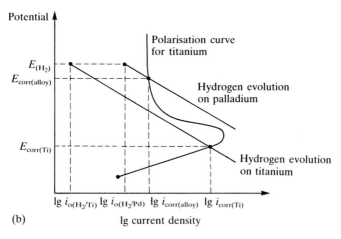

(b)

Fig. 5.9 (a) The effect of stirring on the corrosion rate in a galvanic cell. (b) The effect of noble metal alloying in reducing the rate of corrosion in a passivating metal such as titanium.

A similar effect was obtained with noble metal alloys of chromium, but it would seem to be restricted to chromium and titanium because of the particular way in which the anodic curve intersects the hydrogen evolution line: in other metals, the intersection causes an increased corrosion rate in the alloy.[8]

5.3 REFERENCES

1. Chapin G 1984 Long-term deterioration requires Statue of Liberty rework, *Machine Design* 56(2): 72–73
2. McKenney C R, Banker J G 1971 Explosion-bonded metals for marine structural applications, *Marine Technology* July: 285–92
3. Jefferson T B 1977 Welding aluminium to steel, *Welding Design and Fabrication* May
4. Pask R, Mohammet S 1985 *An evaluation of explosively bonded transition joints in the fabrication of aluminium superstructures for ships*. MESDOC report, Royal Naval Engineering College, Manadon, Plymouth
5. LaQue F L 1975 *Marine corrosion*. John Wiley, p 179
6. Chamberlain J 1985 Cell reversal of the zinc/iron system in sodium bicarbonate and sodium nitrate solutions, *Surface Technology* **25**: 229–57
7. Shreir L L 1979 Corrosion in aqueous systems, in Shreir L L (ed.), *Corrosion* (Vol. 1). Newnes-Butterworths, p 1: 98
8. Fontana M G, Greene N D 1978 *Corrosion engineering*. McGraw-Hill, pp 335 and 341
9. Stern M, Wissenberg H 1959 *Journal of the Electrochem Society* **106**: 759

6　SELECTIVE ATTACK

> The present method of philosophising established by Sir Isaac Newton is to find out the laws of nature by experiments and observations.
>
> (J Rowning: *A Compendious System of Natural Philosophy*, 1738)

The principles of galvanic corrosion which were discussed in the previous chapter are very important in the study of selective attack. Although we normally apply the term 'galvanic corrosion' in cases where dissimilar metals have been coupled together, the corrosion processes which occur, often on a very small scale, within a single piece of metal are also galvanic in origin.

Metals are rarely uniform in composition or structure whether we consider them from a macroscopic or a microscopic viewpoint. Section 2.6 examined the defects which can be found in the crystal structures of metals. (Remember that a defect is considered to be any deviation from the perfect crystal lattice.) The presence of defects can be both beneficial and detrimental to the engineering properties of metals. For example, it is the movement of dislocations which confers upon metals their useful property of ductility, but volume defects, such as cracks cause metals to fail at lower stress levels than they would otherwise (see section 9.3). Metals often contain many undesirable volume defects which result from the production process, yet, even if these heterogeneities could be eliminated by careful quality control, the microscopic structure of the metal is usually non-uniform. One very significant defect type is the grain boundary, which results from the solidification process; other types of defect such as dislocations or point defects, have a finite statistical probability of occurring at any temperature above absolute zero because of the thermodynamic energy of the metal atoms. An atom in the solid state attains its lowest thermodynamic energy only when it occupies a site within a perfect crystal lattice, thus any atom or group of atoms which are in non-perfect lattice sites will, in theory, have a more positive free energy and be more likely to corrosive attack. Since the proportion of atoms which are part of defects is usually small compared to those which are in normal lattice positions, the corrosion processes described under the general heading of selective attack are localised and may be very penetrating. Considerable loss of strength often occurs and a dangerous condition can result, especially in pressurised or other stressed components.

Any corrosion which occurs at preferred sites on a metal surface, for whatever reason, can be described as selective attack. It is therefore possible to include in this category certain forms of corrosion described elsewhere in this book. Pitting corrosion (section 7.2) is a form of selective attack, as also is environment-sensitive cracking, a subject which occupies the whole of Chapter 10, but these have been discussed separately because of their importance, and because there are other aspects of their mechanisms which require special attention. The contents of this chapter are some specific forms of corrosion which are no less damaging and which arise because of metallurgical factors alone.

6.1 GRAIN BOUNDARY CORROSION

Most metals manufactured in bulk for general engineering purposes contain volume defects. Even in the case of a pure metal free from all production defects there can be selective corrosive attack at grain boundaries where, because of mismatch in crystal structure, atoms are less thermodynamically stable than those at perfect lattice sites, and have a greater tendency to corrode. This fact allows the examination of grain size and shape, a vital part of metallographic investigation.

If we wish to study the grain structure of a metal or alloy, the specimen must first be polished to obtain a scratch-free surface. In the polished state it is not possible to see the grain structure: the specimen is flat and, as in a mirror, the illuminating light is reflected evenly from the surface which exhibits no visible topography. If a mildly corrosive liquid is now applied to the surface, attack occurs at the grain boundaries while the bulk material remains largely unaffected. Some skill is required to effect this, or the metal will corrode indiscriminately over its entire surface. The topographical detail which results from careful application of the corrodant corresponds to the grain structure and is readily visible because of the non-uniform reflection of the illuminating light. This process is known in the context of metallography as **etching**, but is mostly grain boundary corrosion.

You may well have seen these effects without realising the corrosion significance. Many zinc coated or brass articles have large grain structures which are very noticeable after being exposed to the atmosphere for a period of time. Brass door handles in particular soon show beautiful and intricate patterns after the touch of hundreds of sweat-moistened hands has caused grain boundary corrosion. Experiment 3.9 examines this further.

Figure 6.1 is a photograph of the microstructure of a sample of mild steel containing 0.15 per cent carbon. The specimen was prepared by the method described above and etched with a solution of 2 per cent nitric acid in ethanol. The specimen, magnified 500 times, shows two distinct regions: one a light-coloured area of ferrite, the solid solution of carbon in body-centred cubic iron; and the other a striped grey material called pearlite, a fine mixture of ferrite and iron carbide. The grain boundaries are clearly visible, as also are manganese sulphide inclusions which appear as dark specks

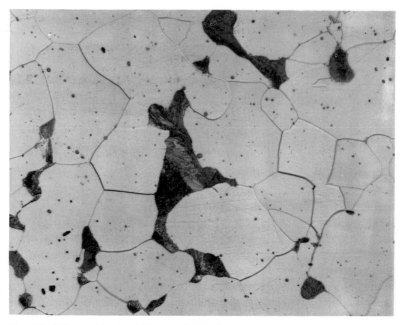

Fig. 6.1 Micrograph of a fully annealed 0.15 per cent carbon steel, etched with 2 per cent nitric acid in ethanol, ×500.

randomly scattered through the material. The corrosion of this material is further discussed in sections 7.2 and 15.1.

It is not the aim of this section to discuss the metallurgical structure of steel, but rather to point out that without grain boundary corrosion the examination of such an important engineering material would not be possible and engineers' understanding of the properties of steels, indeed all metals, would be seriously deficient. Not all corrosion is unwanted!

6.2 INTERGRANULAR CORROSION

Intergranular corrosion occurs when a grain boundary area is attacked because of the presence of precipitates in these regions. Grain boundaries are often the preferred sites for the precipitation and segregation processes observed in many alloys (sections 2.5 and 2.6). Segregates and precipitates are distinguished only by the means by which they were created: from a corrosion point of view they may both be considered as physically distinct from the remainder of the material with their own, different thermodynamic energy. These intruders into the metal structure are of two types:

(a) **Intermetallics** (also known as intermediate constituents) are species formed from metal atoms and having identifiable chemical formulae. They can be either anodic or cathodic to the metal.

(b) **Compounds** formed between metals and the non-metallic elements, hydrogen, carbon, silicon, nitrogen and oxygen. Iron carbide and manganese sulphide, two important constituents of steel, are both cathodic to ferrite (see section 6.1).

In principle, any metal in which intermetallics or compounds are present at grain boundaries will be susceptible to intergranular corrosion. It has been most often reported for austenitic stainless steels, but it can also occur in ferritic and two-phase stainless steels, as well as in nickel base corrosion-resistant alloys. (For a discussion of the composition and properties of these alloys, see Ch. 15.)

Aluminium alloys may suffer severe intergranular corrosion. In the high-strength aluminium alloys used for aircraft, it is the control of precipitates both at the grain boundaries and within the grains that determines the materials' strength. Two common precipitates, $CuAl_2$ and $FeAl_3$, are both cathodic, but Mg_5Al_8 and $MgZn_3$ are anodic to the surrounding metal. The presence of these constituents, whether anodic or cathodic, creates minute localised galvanic cells when an electrolyte is present. Corrosion can then occur such that, if the precipitates are anodic, they dissolve and leave a porous material, or, if they are cathodic, the surrounding metal is attacked. In either case the material may be seriously weakened.

Intergranular corrosion has also been observed in some zinc die-cast alloys, and also in lead, but the most significant problems have been those found in the austenitic group of stainless steels where it has often been called weld decay because of its frequent occurrence in association with the welding of the materials. The remainder of the discussion will be a case study of weld decay, but the principles apply equally well to any alloy system in which precipitation occurs at grain boundaries.

Figure 6.2 shows the simplified diagram of the solid solubility of carbon in an Fe, 18Cr, 8Ni (type 304) alloy. When the carbon content is less than about 0.03 per cent only the γ phase is stable, but for compositions with greater than 0.03 per cent the equilibrium stable phases are γ and a mixed carbide which is believed to have the formula $(FeCr)_{23}C_6$ and is called chromium carbide. The proportions of carbide actually obtained are dependent upon the rate of cooling: fast cooling by water or oil quench from $> 1000\ °C$ suppresses carbide formation. If the material is subsequently reheated, especially within the range 600–850 °C, there is a considerable likelihood that carbide precipitation will occur at the grain boundaries. The material is then said to be sensitised, and is in a dangerous condition from a corrosion point of view. (Below 600 °C the rate of diffusion of chromium is too slow for carbide precipitation to occur.) The presence of chromium (> 12 per cent) in a steel significantly improves the corrosion resistance – hence the 'stainless' steels. However, the precipitation of chromium carbides causes a depletion of chromium to below 12 per cent in the metal adjacent to the precipitates which is thus no longer 'stainless'. Compared to the rest of the grain, the chromium depleted region is very anodic and severe attack occurs adjacent to the grain boundary if the metal comes into contact with

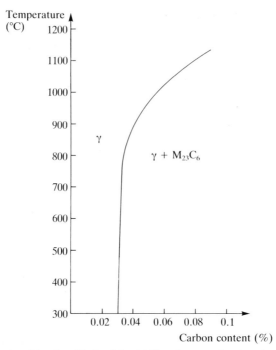

Fig. 6.2 Simplified solid solubility diagram of carbon in an Fe, 18Cr, 8Ni (type 304) alloy.

electrolytes. In extreme cases whole grains become detached from the material which is considerably weakened.

The problems of using such materials are obvious. Even if the alloy is in an unsensitised condition after manufacture with no carbides present, there is always a danger that subsequent use may involve a heating process which will result in sensitisation. The welding of austenitic stainless steels is one example which in the past has led to many serious failures. The problem was so serious that other alloys have now been developed in which there is far less probability of obtaining grain boundary precipitates. Even so, it is reported[1] that type 304 stainless steel is still being specified for use in the construction of boiling water reactors, despite the realisation that the dangers of weld decay make it unsuitable. Furthermore, evidence has been found[1] that sensitisation and the resulting danger of weld decay can occur at temperatures in the range 300–320 °C if chromium carbide nuclei pre-exist in grain boundary regions.

The term 'stabilised stainless steel' is used to denote an alloy which is not susceptible to intergranular corrosion. It is possible to stabilise an austenitic stainless steel, such as the Fe,18Cr,8Ni example used above, by the addition of a small amount of titanium or niobium (known as columbium in the US). These elements in sufficient amounts form carbides in preference to those of chromium and as a result the grain boundary areas are not chromium depleted. It is usual to add about 5–10 times as much titanium or niobium as the amount of carbon present in order to ensure that no chromium carbides are obtained.

Correct training or supervision of welders can assist in guarding against weld decay. For example, it is essential that the work should not be wiped 'clean' with an oily rag prior to welding because this may lead to carbon 'pick-up' by the metal when it is heated.

In principle, then, there are three ways of reducing the susceptibility of a metal such as a 304 stainless steel to intergranular corrosion:

(a) Use a low carbon steel, i.e. less than 0.03 per cent, so that the carbides are not stable (for example, type 304L).

(b) Apply post-weld heat treatment to dissolve the precipitate.

(c) Add titanium or niobium to form carbides preferentially (for example, type 321).

Case 6.1: While attempting to fold the tail pylon of a helicopter it was found that it was not possible to disengage the locking assembly because of a fractured, unstabilised stainless steel pin (Fig. 6.3(a)). Examination showed that the most likely cause of failure was intergranular cracking (Fig. 6.3(b)) initiated by a network of grain boundary precipitates (Fig. 6.3(c)). Subsequent heat treatment of the material redissolved the precipitates and it was recommended that heat treatment be applied to all similar pins in service or in store.

Experiment 6.1

For this experiment you will need specimens of unstabilised AISI type 304 and stabilised type 321 stainless steels. If they can be obtained in the form of tensile test specimens, so much the better. Take two specimens of each material; heat treat in a furnace at 1050 °C for 3 hours and then water quench. This is to ensure that the material contains only one phase. Next reheat the specimens to 750 °C for an hour and furnace cool to about 450 °C. To save time, the specimens may be water quenched when they have cooled to this temperature. The purpose of the second heat treatment is to simulate the effect of a welding process on the alloys.

Prepare a solution of 25 g of hydrated copper(II) sulphate in 200 ml of water. Add **CAREFULLY** 25 g of sulphuric acid and make up to 250 ml. Reflux the four specimens in this solution for 72 hours taking care that they do not lie across each other: wastage because of crevice corrosion (section 7.1) will occur at points of contact between them. After refluxing, use one of each type to examine the microstructures. Section, mount and polish each one, and electrolytically etch using a solution of oxalic acid, 10 g in 100 ml of water. The specimen is made the anode of a corrosion cell by connecting it to the positive terminal of a d.c. power supply, set to deliver 7.5 V. A suitable piece of stainless steel may be used as a cathode. The etching process takes approximately 1 minute. Examination under a microscope should show that the unstabilised specimen is severely attacked, while the stabilised sample is very much less affected.

Now perform a tensile test on the two remaining specimens. You should observe that the unstabilised stainless steel has a much reduced tensile strength compared to the stabilised specimen. Indeed, it is quite common for the unstabilised specimen to break in the hand, in which case you will be able to perform the experiment without a tensile testing machine!

(a)

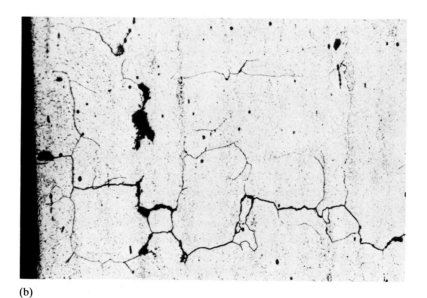

(b)

Fig. 6.3 Intergranular corrosion of an unstabilised stainless steel (case 6.1).
(a) A fractured helicopter tail pylon securing pin.
(b) Micrograph showing intergranular corrosion close to the point of failure. (×200)
(c) Micrograph showing network of cathodic grain boundary carbide precipitates which redissolved when given a suitable heat treatment. (×500)

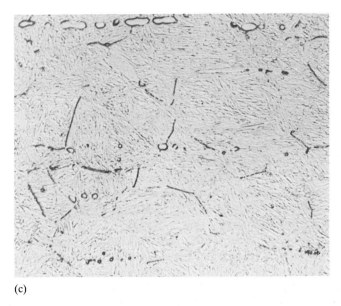

(c)

This experiment is based upon a standard method of testing which was devised by Hatfield[2] to try to predict the susceptibility of alloys to intergranular corrosion. Several variations exist, as well as other different test procedures, and today the methods of testing have been much refined. Hatfield's original idea remains the simplest and is an effective way of demonstrating the phenomenon of intergranular corrosion. It is a severe test and exaggerates the effect observed *in vivo*. The work of Streicher[3] contains a detailed review of the methods of testing.

6.3 SELECTIVE LEACHING

As the term implies, selective leaching is the net removal of one element from an alloy and is thus often referred to as **dealloying** or **demetallification**. The whole of an exposed surface may be attacked leaving the overall geometry unchanged, yet the removal of a large proportion of one of the alloying elements leaves a porous material with virtually no mechanical strength. Sometimes the effect is very localised, in which case perforation may occur. As with the other types of corrosion discussed in this chapter, the prime cause of dealloying is the galvanic effect between the different elements which compose the alloy, although other factors such as differential aeration and temperature are also important.

In the past, many problems have been reported[4] with brasses which corrode by loss of zinc in a process known as **dezincification**. Components designed for use in both sea and fresh water have failed by this form of corrosion in such applications as condensers, valves, taps and pipework, as well as screws, nuts and bolts. The problem has not been confined to brasses, however, and other reports[4] of this form of corrosion have included the loss of nickel, aluminium and tin from copper alloys in processes known respectively as **denickelification**, **dealuminification** and **destannification**. This subject is discussed further in section 15.5, but because the overwhelming majority of reports have concerned dezincification, the discussion of the dealloying mechanism in this section will be restricted to the dezincification of brasses. The principles are the same for dealloying in general.

The compositions of a number of important brasses are listed in Table 15.5. There are two main categories of brass: those with a single phase and those with two-phase structures. Of the former, the most common are the 70/30 brasses, while the second group contains the so-called 60/40 brasses. Because of their poor cold working properties, the use of 60/40 alloys is restricted to castings, while the 70/30 brasses are used where malleability is important, as in tubes, for example.

The problem of dezincification was quickly overcome in the case of the single phase, lower zinc brasses, by the discovery that the addition of very small amounts of arsenic, usually about 0.05 per cent, virtually eliminated this form of corrosion. The 70/30 alloy to which the arsenic addition had been made became known as Admiralty brass (inhibited) and today, most commercial production of 70/30 brass contains arsenic for this purpose. Uninhibited 70/30 brass should never be specified for uses involving immersion in either fresh or sea water.

Unfortunately, the addition of arsenic did not significantly affect the susceptibility of the 60/40 alloys and the problem of dezincification has remained. Efforts have been made to find other elemental additions which might alleviate the corrosion. The addition of 1 per cent tin to Muntz metal, an alloy with 60 per cent Cu, 40 per cent Zn composition, gave another two-phase material known as Naval brass which is widely reported as having greatly reduced susceptibility to dezincification, though there is little evidence to support this. Rogers[5] states that Naval brass does not provide improved resistance and that there is no reliable method of inhibiting the dezincification of two-phase brasses.

Case 6.2: The smaller component in Fig. 6.4 is a 60/40 brass casting used as a connector in a water supply. The dark inner areas are dezincified and the left-hand region has fractured because of the embrittlement caused by the extreme porosity.

Case 6.3: The larger of the two components in Fig. 6.4 is part of a cast aluminium bronze sea-water valve which had originally suffered wastage because of erosion, but which had not dealloyed. The component was recovered by depositing 60/40 brass along the shank and around the closing surface. As can be seen in the figure, the reclaimed areas have suffered dezincification and some has been broken off. (Note the beautiful grain structure exhibited by this specimen.)

Fig. 6.4 Two brass components which failed because of dezincification. The smaller one is described in case 6.2 and the larger in case 6.3.

Alloys formed by the addition of lead (about 4 per cent) to 60/40 brass to improve the machinability are also susceptible to dezincification.[6]

Case 6.4: Figure 6.5 shows a leaded brass bolt, 12 mm in diameter, which suffered dezincification to a depth of 3 mm in several years' exposure to sea water. The fractures which are a direct result of the embrittlement of the material are clearly visible.

The fact that such corrosion can take a long time to develop in practice makes accelerated laboratory simulation a desirable technique. This can be achieved using a method described below.

Experiment 6.2

Set up a three-electrode cell, as described in section 4.8, using a specimen of leaded 60/40 brass as a working electrode, a carbon rod auxiliary electrode, a SCE reference electrode and a solution of 3.5 per cent sodium chloride as electrolyte. Set the potentiostat to apply an anodic potential of -120 mV SCE ($E_{corr} = -240$ mV SCE) and leave the specimen for a convenient length of time. The effect will be observable after one or two days, but several weeks will give better results. After the chosen exposure time, mount the specimen longitudinally to examine the penetration of the dezincified layer. An example of the microstructure you should obtain is shown in Fig. 6.6(a). The features of the specimen prepared in this way are very similar to those of the corroded brass bolt in Fig. 6.5. There are several distinct regions which can be identified and these are illustrated for clarity in Fig. 6.6(b).

In the 60/40 brasses there are two phases, α and β. The α phase contains much less zinc than does the β phase. (In leaded brasses the lead atoms are totally insoluble in the copper–zinc lattices and the lead exists as tiny globules dispersed at random throughout the solid. We use this material in the experiment because it gives good examples of this form of corrosion. The presence of the lead can be ignored for this discussion, although it does influence the corrosion rate.) The E_{corr} values of the α and β phases are respectively -230 mV and -285 mV SCE[7] on the basis of which we would predict from the galvanic effect that the β phase would be more anodic compared to the α phase and that the β phase would be attacked first. The attack of the electrolyte on the β grains is interesting because it seems that the β grains behave as if they were themselves simply galvanic couples of zinc and copper: the zinc is dissolved quickly, while the copper is much less affected. Dezincified specimens produced in expt 6.2 and illustrated in Fig. 6.6 show the following features:

Fig. 6.5 A dezincified leaded brass bolt, 12 mm diameter. The corrosion penetrated to a depth of 3 mm and caused severe porosity and embrittlement (case 6.4).

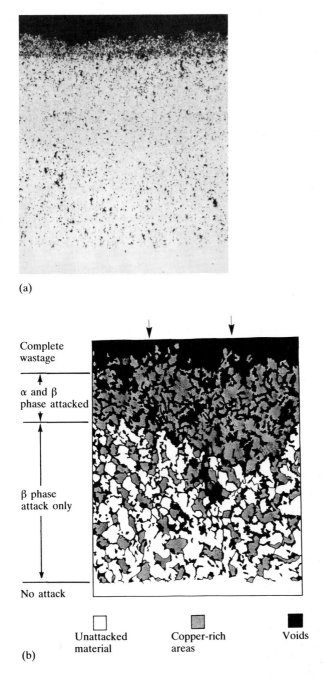

(a)

(b)

Complete wastage

α and β phase attacked

β phase attack only

No attack

☐ Unattacked material   ▨ Copper-rich areas   ■ Voids

Fig. 6.6 (a) Microstructure of a laboratory-prepared dezincified specimen of 60/40 leaded brass (×200, ammonia/peroxide etch).
(b) Schematic illustration of the dezincified layers of the specimen shown in (a).

1. A band which appears pink to the naked eye and which contains a mixture of unattacked α grains and pink, copper-rich grains which were initially β.

2. A band in which the β phase is missing and the α is dezincified.

3. Complete wastage of the material from the outside surface.

One explanation for the observed corrosion is that the following sequence of events takes place. In the early stages, the α is relatively unaffected. The zinc leaching from the β leaves behind an increasing number of pores through which the electrolyte is able to penetrate much deeper into the material. A corrosion front begins to progress into the material. Once the band described in (1) has been established, two more processes occur. The copper in the dezincified β grains begins to corrode and soon the complete grains which were originally β have disappeared. At the same time the α grains are leached of zinc. This yields band (2) above. Eventually the dezincified α is dissolved completely and wastage occurs (3).

There has been much speculation about the exact nature of the mechanism of dezincification. Evidence has been produced that only zinc is leached from brass, while it has also been claimed that both zinc and copper are dissolved. Currently it is proposed[8] that both mechanisms are possible, but that the factor which determines the type of behaviour is the corrosion potential. This would certainly seem to apply to the experiment described above. Measurements[9-11] on the specimens obtained from similar experiments have shown that porosity of up to 40 per cent is normal in the dezincified layers of two-phase brasses. Furthermore, the free corrosion potential of the specimen becomes more anodic when the solution is deaerated by bubbling argon into the flask; indeed, the E_{corr} of Muntz metal changes[11] from -235 to -380 mV SCE when the solution is deoxygenated. It is reasonable to assume that this simulates the differential-aeration cell formed once it has become porous. Such a potential difference between the cathodic areas on the outside of the specimen where oxygen is readily available, and the corroding interface deep within the metal where there can be little oxygen would constitute a major driving force.

6.4 REFERENCES

1. MacDonald D D, Begley J A, Bockris J O'M, Kruger J, Mansfeld F B, Rhodes P R, Staehle R W 1981 Aqueous corrosion problems in energy systems, *Materials Science and Engineering* **50**: 19–42
2. Hatfield W H 1933 *Journal of the Iron and Steel Institute* **127**: 380–83
3. Streicher M A 1978 Theory and application of evaluation tests for detecting susceptibility to intergranular attack in stainless steels and related alloys – problems and opportunities, in Steigerwald R F (ed.), *Intergranular corrosion of stainless alloys*. ASTM STP 656, pp 3–84

4. Heidersbach R H 1968 Clarification of the mechanism of the dealloying phenomenon, *Corrosion*, quarterly report, Feb: 38–43
5. Rogers T H 1968 *Marine corrosion*. Newnes, p 120
6. Pitt C D, Trethewey K R 1982 *Report on a dezincified brass bolt showing an unusual banded structure*. Engineering Materials Section research report REP EMS-827-U-CP, Royal Naval Engineering College, Manadon, Plymouth
7. Tigwell N K 1983 *Aspects of the corrosion of naval brass*. Third year project report, Royal Naval Engineering College, Manadon, Plymouth
8. Verink E D Jr, Heidersbach R H 1972 Evaluation of the tendency for dealloying in metal systems, in *Localized corrosion – cause of metal failure*. ASTM STP 516, pp 303–22
9. Pinwill I, Trethewey K R 1985 *An investigation into the mechanism of penetration of the dezincification of leaded brass*. Engineering Materials research report. Royal Naval Engineering College
10. Trethewey K R, Pinwill I 1987 The dezincification of free-machining brasses in sea water, *Surface and Coatings Technology*, **30**(3): 289–307
11. Trethewey K R 1986 New aspects of the dezincification of brasses, in *Proceedings of UK Corrosion 86*, Institution of Corrosion Science and Technology

6.5   BIBLIOGRAPHY

ASTM 1972 *Localised corrosion – cause of metal failure*. STP 516

Shreir L L 1979 *Corrosion* (Vol.1). Newnes-Butterworths, section 1.6

Steigerwald R F (ed.) 1978 *Intergranular corrosion of stainless alloys*. ASTM STP 656

7 CREVICE AND PITTING CORROSION: CONCENTRATION CELLS

It is the natural habit of a metal to corrode unless prevented by human endeavour.

(T H Rogers: *Marine Corrosion*)

The corrosion cells described in Chapters 5 and 6 are examples in which corrosion results from changes in the nature and composition of metals exposed to electrolytes of *uniform composition*. In contrast, **concentration cells** exist when metals of uniform nature and composition are exposed to aqueous environments in which the components are *heterogeneous*. In Chapter 3, expts 3.6 and 3.7 showed how changes in oxygen or electrolyte concentration give rise to corrosion potentials. The two forms of corrosion described in this chapter result directly from such changes in composition within an electrolyte.

It is commonly believed that stainless steel has excellent resistance to corrosion and that its selection will provide an infallible solution to corrosion problems. While it is certainly true that stainless steels do perform well in most situations, they are a very poor choice for some applications and must be selected with caution (see section 15.2). Experiment 3.15 highlights a common problem. Many failures of stainless steel components have occurred because of corrosion in crevices, or shielded areas where small volumes of stagnant electrolyte can become much more aggressive than they are in bulk solution. Figures 7.1 and 7.2 are good examples in which the designer thought that fabrication from stainless steel (type 304, 18 Cr/8Ni) would provide good service.

Case 7.1: Figure 7.1 shows a thermometer pocket from a condenser. Pitting and serious wastage has occurred along the stem on the downstream side where there was a layer of stagnant water.

Case 7.2: The stainless steel universal joint in Fig. 7.2 is badly pitted in the region over which the rubber grommet fits.

As will become apparent later, there is good reason for considering crevice and pitting corrosion together. Although they may be treated as different forms of corrosion, *per se*, there are many aspects of mechanism and instances of application in which there is little distinction. In the

Fig. 7.1 Stainless steel thermometer pocket from a condenser. A stagnant layer of sea water formed on the downstream side of the stem when the pocket was inserted into the cooling water flowing in the condenser. The difference in oxygen concentration between the stagnant and flowing water initiated a crevice-type attack on the stainless steel. (Case 7.1)

Fig. 7.2 Stainless steel universal joint showing crevice corrosion and pitting in the region where the rubber grommet fitted (case 7.2).

examples quoted above, pitting of stainless steels resulted from crevice conditions. There are other examples in which crevice conditions do not cause pitting, but localised wastage of the surface.

The problems are not restricted to stainless steels: many other alloy systems suffer from the localised corrosion described in this chapter.

Case 7.3: The pitting of copper pipes for fresh-water systems is rare, but it can occur when incorrect fabrication techniques are employed.

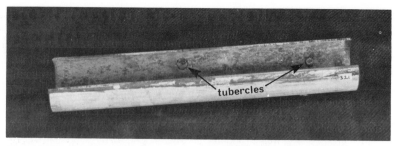

Fig. 7.3 Pits penetrated the wall of a copper water pipe because of the presence of a film of carbon along the bore (case 7.3).

Figure 7.3 shows a section of 25 mm diameter copper piping which failed because a layer of organic lubricant had been left after the pipe was formed. The pipe was annealed and the high temperature formed a carbon film along the inside of the pipe. Any subsequent break in the carbon film resulted in a very active pitting site which penetrated the copper in a short time. The pits are characterised by the formation of tubercles, small scabs of corrosion product over the hole (see section 7.2).

Despite the well-known problems of crevice corrosion, like bimetallic corrosion, engineers are still failing to appreciate the importance of correct design and materials selection in providing adequate corrosion resistance. Many car owners are familiar with the sad story depicted in Fig. 12.5, yet crevice corrosion occurs in successive new models. Such mistakes are perhaps best summed up by the following example.

Case 7.4: In 1981 it was reported[1] that one of the most serious problems in the US nuclear industry was 'denting' (localised attack) of Inconel 600 tubing because of corrosion in crevices between the tubes and carbon steel support plates. With about 60 steam generators affected, the cost of rectifying the known problems was estimated to be $6000 million.

7.1 THE MECHANISM OF CREVICE CORROSION

In the past, the use of the term crevice corrosion has been restricted to the attack upon oxide-passivated alloys by aggressive ions such as chloride in crevices or other shielded areas of a metal surface. Attack in similar circumstances upon non-passivated metals was given the name, **differential-aeration** corrosion. Current practice has tended to neglect such distinctions, and consequently crevice corrosion has been given many other names. Differential-aeration and **concentration cell corrosion** are terms which arise because of aspects of the mechanism of corrosion in crevices. Other, less common, alternative names arise from the specific situations in which crevice corrosion is found, for example, deposit, fissure, gasket, interface, poultice, waterline and wedge corrosion. Thus, a good general definition of crevice corrosion is the attack which occurs because part of a metal surface is in a shielded or restricted environment, compared to the rest of the metal which is exposed to a large volume of electrolyte.

Considerable research has recently been devoted to the electrochemistry within crevices, mainly because of advances in technology which enable accurate measurements to be made within the extremely small confines of a crevice; it is thought that typical widths range from 25 to 100 μm.[2] A unified mechanism which applies to almost all commonly encountered situations has been proposed by Fontana and Greene.[3] As with many mechanisms in science, it is possible to find exceptions, but this mechanism represents a great step forward in the understanding of such processes.

Figure 7.4 is a schematic diagram of the events occurring in a typical metal/electrolyte system in which a crevice is present. The attack by chloride ion upon a passivated metal surface is the example chosen because this combination is the one most commonly encountered in cases of crevice corrosion. The steps involved are as follows:

(a) Initially, the electrolyte is assumed to have uniform composition. Corrosion occurs slowly over the whole of the exposed metal surface, both inside and outside the crevice. The normal anode and cathode processes occur, as described in section 4.2. Under such conditions, the generation of positive metal ions is counterbalanced electrostatically by the creation of negative hydroxyl ions (see the discussion on the principle of electro-neutrality, section 2.2. Some association of positive and negative ions is assumed.) This situation is shown in Fig. 7.4(a).

(b) The consumption of dissolved oxygen results in the diffusion of more oxygen from those electrolyte surfaces which are exposed to the atmosphere. Oxygen is more readily replaced at metal surfaces in the bulk electrolyte than at those within the crevice. Within the crevice this lack of oxygen impedes the cathodic process and generation of negative hydroxyl ions is diminished within the confined space.

(c) The production of excess positive ions in the gap causes negative ions from the bulk electrolyte to diffuse into the crevice to maintain a minimum potential energy situation. In the presence of chlorides, it is likely that complex ions are formed between chloride, metal ions and water molecules. These are thought to undergo hydrolysis (reaction with water), giving the corrosion product, and more importantly, hydrogen ions which reduce the pH. This can be described by the simplified equation:

$$M^+ + H_2O \rightarrow MOH + H^+ \qquad [7.1]$$

The equation describes a general hydrolysis reaction, in which the role of the electrolyte anion (chloride, in this case) is important but too complicated to be described here. Presence of chloride is well known to be conducive

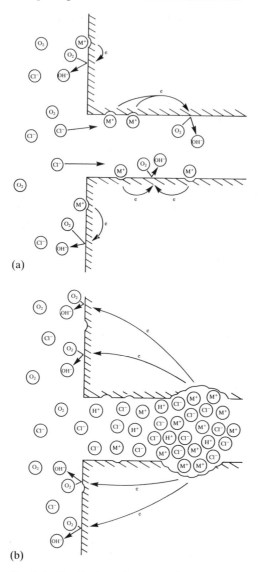

Fig. 7.4 The Fontana–Greene mechanism of crevice corrosion.
(a) Initial conditions: corrosion occurs over the whole of the metal surface.
(b) Final conditions: metal dissolution occurs only inside crevice where acidity increases, concentration of chloride ion increases, and reaction becomes self-sustaining.

to the development of low pHs because of its extremely low tendency to associate with hydrogen ions in water. (Hydrogen chloride, HCl, dissociates completely in water.) Additionally, those metals, such as stainless steels, which rely upon the protection of passive films, are notoriously unstable in chloride environments, and in the active crevice the very species needed to maintain passivity, oxygen, is denied access. (The exception to this statement is that of titanium which has excellent resistance to crevice corrosion because its oxide film is particularly unreactive towards chloride ion.)

(d) The increase of hydrogen ion concentration accelerates the metal dissolution process, which, in turn, exacerbates the problem. So, also, does the concomitant increase of anion (chloride) concentration within the crevice. An important feature of active crevice corrosion cells is that they are **autocatalytic**, that is, once started they are self-sustaining. This is the situation shown in Fig. 7.4(b). The metal within the crevice is corroding rapidly while that outside is cathodically protected.

Interestingly, it is not the corrosion of the iron in stainless steels which is the most damaging process. Evidence has been produced[4] that it is the dissolution and subsequent hydrolysis of chromium which leads to the most significant fall in pH, again, summarised in the simplest form by the equation:

$$Cr^{3+} + 3H_2O \rightarrow Cr(OH)_3 + 3H^+ \qquad [7.2]$$

There is little doubt that the pH of an electrolyte within an active crevice can become extremely acidic: just two reported instances[5] of pH change within a crevice in a titanium alloy are from 8.3 (bulk) to 2.3 (crevice), and from 6 (bulk) to 1 (crevice). Of the two concentration cells which act in the mechanism, namely, oxygen concentration and ion concentration, the latter is thought[6] to have the most significant effect upon the level of attack because of its effect upon local pH.

7.2 PITTING

Pitting is localised corrosion which selectively attacks areas of a metal surface where there is:

(a) a surface scratch or other mechanically induced break in an otherwise protective film;
(b) an emerging dislocation or slip step caused by applied or residual tensile stresses;
(c) a compositional heterogeneity such as an inclusion, segregate or precipitate.

The observation of corrosion pits as a result of crevice corrosion can sometimes cause confusion about the difference between the two forms of corrosion. An important paper on the subject by Wilde[7] proposes considerable similarities in the propagation mechanisms of crevice and pitting corrosion. Once formed, a pit behaves in a very similar manner to a crevice and the processes described in section 7.1 can be deemed to be operative. However, pitting is distinguishable from crevice corrosion in the **initiation** phase. Thus, whereas crevice corrosion is initiated by differential concentration of oxygen or ions in the electrolyte, pitting corrosion is initiated (on plane surfaces) by metallurgical factors alone.

The pitting of iron and steel will be described next because the mechanisms represent significant advances in the understanding of pitting.

It is well known that if a sheet of clean mild steel is exposed to rain, within a few days it will be rusting rapidly with the 'rust' occurring as hard deposits, scabs, or tubercles, in localised areas where water droplets have remained longest. If the 'rust' is subsequently removed with a wire brush the surface will be found to be pitted in the areas previously covered by corrosion products. At this stage, we are using the term 'rust' in quotes because the word, in its commonly understood context, is used to mean *the brown corrosion product formed on corroded iron or steel surfaces*. As we shall see below, this corrosion product is actually a mixture of chemical species and the word 'rust' has a more precise meaning.

The classic description of pit formation beneath a drop of water on an iron surface is credited to Evans.[8] A relevant diagram is shown in Fig. 7.5. The initiation of a pit is preceded by general corrosion over the whole of the wetted surface, probably as a result of simple grain boundary effects. The consumption of oxygen by the normal cathode reaction in neutral solution causes an oxygen concentration gradient within the electrolyte. Obviously, the wetted area adjacent to the air/electrolyte interface receives more oxygen by diffusion than the area at the centre of the drop which is at a greater distance from the oxygen supply. This concentration gradient anodically polarises the central region which actively dissolves:

$$Fe \rightarrow Fe^{2+} + 2e^- \qquad [4.6]$$

The hydroxyl ions generated in the cathode region diffuse inwards and react with the iron ions diffusing outwards, causing the deposition of insoluble corrosion product around the depression, or pit. This further retards the diffusion of oxygen, accelerates the anodic process in the centre of the drop, and causes the reaction to be autocatalytic, Fig. 7.5(b).

At this point, it is worth noting exactly how simplified the expression eqn [4.6] has been made. The actual steps involved are believed[9] to be as follows:

(a) $\quad Fe + H_2O \rightarrow Fe(H_2O)_{ads}$ $\qquad\qquad$ [7.3]

(b) $\quad Fe(H_2O)_{ads} \rightarrow Fe(OH^-)_{ads} + H^+$ $\qquad$ [7.4]

(c) $\quad Fe(OH^-)_{ads} \rightarrow Fe(OH)_{ads} + e^-$ $\qquad$ [7.5]

(d) $\quad Fe(OH)_{ads} \rightarrow Fe(OH)^+ + e^-$ $\qquad\quad$ [7.6]

(e) $\quad Fe(OH)^+ + H^+ \rightarrow Fe^{2+} + H_2O$ $\qquad$ [7.7]

In the above expressions, *ads* represents *adsorbed* and implies that reaction occurs in the solid phase at the solid/liquid interface. The summation of the five equations leads directly to eqn [4.6]. The example is not intended to intimidate students, but is merely a reminder of the complexity of reactions which may otherwise appear very simple, especially in a process as common as rusting.

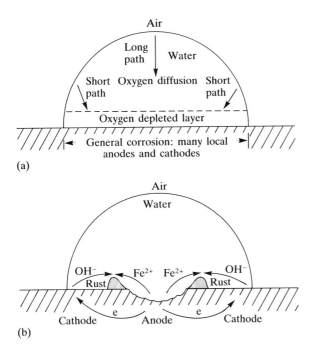

Fig. 7.5 The mechanism of pitting because of differential-aeration beneath a water droplet.
(a) General corrosion over the whole of the wetted metal surface depletes the oxygen levels in the adjacent electrolyte.
(b) Longer diffusion path for oxygen to reach the central area makes this the anode. Metal dissolution occurs in the centre of the droplet and reaction of metal ions with hydroxyl ions formed at the edge generates a ring of rust around the corrosion pit.

It was mentioned in section 2.2 that iron has two valency states, that is, it can lose either two or three electrons. In aqueous solution, numerous iron ions can exist, some oxidised to iron(II) and others to iron(III). The oxidation state can be disguised by reaction with negative hydroxyl ions, and for clarity distinguishing labels will be given in the equations which follow. It is also probable that neutral water molecules are combined with the complex iron ions, but these will not be shown as they have no direct bearing on the chemistry.

The simple explanation described above for the pitting of iron has been extended by Wranglen[10] to explain pit formation on carbon steel. (Further discussion occurs in section 15.1.) The following description by Shreir[11] summarises in simplest form another sequence of complex chemical reactions.

First, a hydrolysis reaction occurs, similar to that in the mechanism of crevice corrosion, in which acidity is increased:

$$Fe^{2+} + H_2O \rightarrow Fe(OH)^+ + H^+ \qquad [7.8]$$
$$\text{iron(II)} \qquad\qquad \text{iron(II)}$$

The formation of iron(III) ions is an oxidation reaction which is facilitated by the presence of oxygen. Even when the iron is combined in the $Fe(OH)^+$ ion, it can still be oxidised to the iron(III) state, as, for example

$$2Fe^{2+} + \tfrac{1}{2}O_2 + 2H^+ \rightarrow 2Fe^{3+} + H_2O \qquad [7.9]$$

or $\quad 2Fe(OH)^+ + \tfrac{1}{2}O_2 + 2H^+ \rightarrow 2Fe(OH)^{2+} + H_2O \qquad [7.10]$
$\qquad$ iron(II) $\qquad\qquad\qquad\qquad$ iron(III)

More hydrolysis reactions are possible, in which the solution is further acidified:

$$Fe(OH)^{2+} + H_2O \rightarrow Fe(OH)_2^+ + H^+ \qquad [7.11]$$

and $\quad Fe^{3+} + H_2O \rightarrow Fe(OH)^{2+} + H^+ \qquad [7.12]$

(All iron ions in eqns [7.11] and [7.12] are iron(III) ions.) The two major corrosion products, magnetite and rust, are respectively denoted by the formulae Fe_3O_4 and $FeO(OH)$ and are formed from the complex ionic species thus:

$$2Fe(OH)^{2+} + Fe^{2+} + 2H_2O \rightarrow Fe_3O_4 + 6H^+ \qquad [7.13]$$

and $\quad Fe(OH)_2^+ + OH^- \rightarrow FeO(OH) + H_2O \qquad [7.14]$
$\qquad\qquad\qquad\qquad$ rust

The corrosion products 'grow' over the pit and its immediate surroundings, forming a scab or tubercle and isolating the environment within the pit from the bulk electrolyte. It is thought that the autocatalytic process is assisted by an increased concentration of chloride ion within the pit.

In section 6.1 it was described how corrosion can be initiated at various inclusions and precipitates. In steels, it is common to find manganese sulphide inclusions which are strongly cathodic to uncombined metal. The areas immediately adjacent to these inclusions make ideal nucleation sites for rusting, especially in cold rolled steels which are produced with a bright surface; most hot rolled steel is produced with a layer of **mill scale** (a layer of oxide) which protects the metal while the scale remains intact, but promotes corrosion at breaks in the film.

The increased acidity described in eqns [7.11], [7.12] and [7.13] above is thought to attack the manganese sulphide, producing S^{2-} and HS^- ions which themselves promote more rapid dissolution of the iron because they decrease the activation polarisation for the dissolution of iron. A further consequence of the acidity increase is the reduction of hydrogen ions by electrons generated in the oxidation reactions. Hydrogen evolution causes occasional breaking of the crust of corrosion product.

The pitting of iron and steel has been described in some detail, not least because they are such common engineering materials, but because, although the detailed mechanism which occurs in the pitting of aluminium or copper alloys, for example, may differ, the features of the events occurring within such micro-environments are broadly similar.

7.3   ELECTROCHEMICAL TECHNIQUES

The most significant difference between pitting and crevice corrosion is in the mechanism by which each process is initiated. The discussion of the previous two sections has emphasised the similarities in the propagation mechanisms. Recent developments in electrochemical techniques have helped to distinguish these two very similar forms of corrosion, while offering methods of assessing the relative performance of alloys under similar conditions.

The most common cause of pitting is the selective attack of a metal where there is a surface scratch or other mechanically induced break in a protective film. Hence, electrochemical techniques[11] which investigate the breakdown of passive films can be used to study pitting propensities. First, it is necessary to discuss the general behaviour of passivating metals. The solid line in Fig. 7.6(a) is a schematic $E/\lg i$ plot for a metal in a mild environment, such as an austenitic stainless steel in dilute sulphuric acid. (The dashed lines should be ignored for the moment.) The metal is not passivated at its free corrosion potential, the point labelled B, but exhibits a change to passive behaviour at applied anodic potentials greater than G. The experimentally measured cathodic polarisation plot appears as the solid line, BA.

At potentials more positive than B, there is an increase in corrosion rate which reaches a maximum at the passivation potential, G. This potential is commonly given the symbol, E_{pp}. The formation of the protective film causes a sudden drop in corrosion current density in the region G to J. From J to P the current density is maintained at a steady, low level, until, at P, sufficient energy is available to cause breakdown of the protective film. It is here that the likelihood of pitting is greatest, and consequently the potential, E_c, often called the **critical pitting** or **breakdown potential**, is a useful parameter in assessing pitting properties of materials.[12] It should be noted that it is not an absolute parameter, and varies according to both metallurgical and environmental conditions. At potentials more positive than P, the current density begins to rise as more and more pits propagate. This region of the curve is known as the **transpassive** region.

Some passivating metals have $E/\lg i$ plots which, at first sight, seem to bear no relation to that shown in Fig. 7.6(a). For example, considerable instability may be noticed in certain potential ranges, as in the plot Fig. 7.6(b). Such an effect is explained by considering the effect of different cathodic processes. Three possibilities are shown as dashed lines, CA, KF, and LN in Fig. 7.6(a). A different type of polarisation plot will be measured, depending upon which cathodic process is operative. (Remember, there are three cathodic processes: hydrogen evolution, oxygen reduction and metal cation reduction. Remember also that the experimental plot is always the sum of the anodic and cathodic currents.) In the plot in Fig. 7.6(a), the observed cathodic line, BA, results from the addition of the

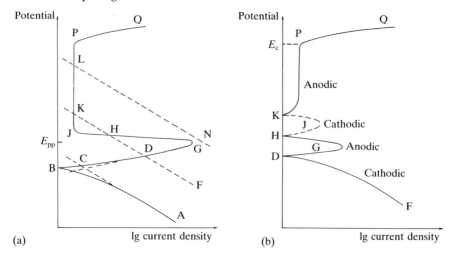

(a)

(b)

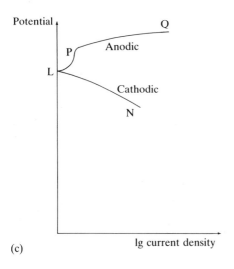

(c)

Fig. 7.6 The schematic potentiodynamic polarisation behaviour of some passivating metals.
(a) A general polarisation plot showing three possible theoretical cathodic processes.
(b) A polarisation plot showing unstable passivity.
(c) A polarisation plot showing stable passivity.

anodic and cathodic currents in that potential range. The cathodic process alone (either metal cation or oxygen reduction) is represented by the line CA.

If this cathodic process is not applicable, but instead, another represented by the dashed line KF (hydrogen evolution in acid conditions) in Fig. 7.6(a) applies, it can be seen that the line intersects the anodic curve in three places labelled D, H and K. Each point of intersection represents equal anodic and cathodic current densities, and hence is an E_{corr} value. The measured polarisation plot is thus that shown in Fig. 7.6(b). Between K and H the cathodic current is greater than the anodic current, while between H and D the reverse is true. Thus, in the potential range KD, the material is unstable and alternates between cathodic and anodic behaviour. The use of a material in such conditions of instability is usually considered to be unwise.

A third type of behaviour is observed when the cathodic process (hydrogen evolution) is represented by the dashed line LN in Fig. 7.6(a); then the measured $E/\lg i$ plot will be as shown in Fig. 7.6(c). E_{corr} may be quite noble and the metal is completely passivated in the range PL. Such a situation is regarded as exhibiting resistance to localised corrosion because the metal is well protected by its passive film.

Further data relating to both crevice and pitting behaviour can be obtained using a method based primarily upon work by Pourbaix.[13] The anodic polarisation scan described in Fig. 7.6(a) is continued to Q, but instead of terminating at this point, the potential is now reduced at the same stepping rate until E_{corr} is reached once more. The graph which results is best described as a *corrosion hysteresis loop*, but is called a **pitting scan**. In general, there are two types of behaviour:

(a) **The metal may be slow to repassivate**. *The pits which have initiated at P continue to propagate, but no new pits are formed.* The corrosion current density remains high until a sufficiently negative potential has been reached, whereupon repassivation occurs once more. Such a situation will lead to a large area in the loop, and usually indicates poor resistance to pitting because if breaks in the passive film occur they do not repair themselves easily. Figure 7.7(a) shows such a pitting scan for type 304 stainless steel in aerated sea water at 25 °C. Although reasonably stable at potentials more positive than E_{corr}, -0.260 V SCE, breakdown occurs at $+0.300$ V SCE. Once the material has passed this potential, attack is severe, even though the potential may be made more negative again. Only if the potential can be brought down to below -0.160 V SCE does the attack cease. If a polarisation scan is carried out in acid conditions with a high concentration of chloride ions, the conditions likely to exist within a crevice, the material is very active at all potentials more positive than E_{corr}. This is shown as the dashed line in Fig. 7.7(a). Thus, the graph illustrates the unsuitability of type 304 stainless steel in a marine environment where there is a danger of crevice conditions.

(b) **The metal repassivates easily**. *The pits which initiated at P and propagated along PQ in Fig. 7.6 are quickly filmed and both initiation and propagation cease.* If this is the case, then the reverse scan will retrace a very similar path, returning to the low level of current density of the passive state. The loop, if there is one, has a small area. Such materials would be expected to show good pitting resistance, since the reforming of surface protection eliminates the local active sites. Figure 7.7(b) shows the pitting scan for Hastelloy C276 in aerated sea water at 25 °C. It shows no hysteresis: the material is efficiently passivated from the moment it is immersed

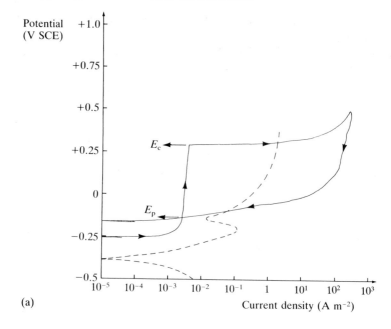

(a)

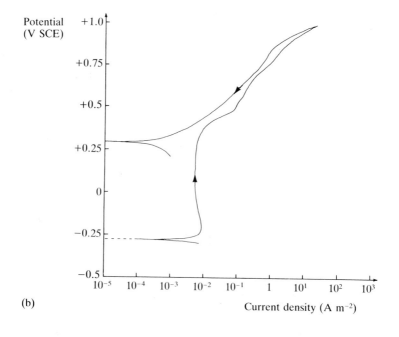

(b)

Fig. 7.7 (a) A pitting 'hysteresis' polarisation scan for stainless steel type 304 in aerated sea water at 25 °C. The dashed line shows a similar plot using aerated sea water + 4.5 M sodium chloride and HCl adjusted to pH 1.2.
(b) A pitting 'hysteresis' polarisation scan for Hastelloy C276 in aerated sea water at 25 °C.

in sea water and, during the reverse scan period, smaller current densities are recorded for the same values of potential. The final E_{corr} at $+0.285$ V SCE is much more noble than the starting E_{corr} at -0.280 V SCE. Such a material is recommended for use in a marine environment.

Pourbaix and his colleagues defined a **protection potential**, E_p, corresponding to the potential at which the reverse scan intersected the forward scan and completed the hysteresis loop. They proposed that, at potentials more negative than this, pit initiation could not occur and existing pits would repassivate. Such a concept has proved useful because it is considered that the area of the hysteresis loop is related to the amount of pit propagation that occurs during the potential sweep. Furthermore, since a growing pit can be considered as a special case of an active crevice, it has been proposed that if a series of alloys is tested under similar conditions, relative susceptibilities to crevice corrosion can be determined by comparison of loop sizes, with those alloys having the smallest loops being the least susceptible. Indeed, to a good approximation, it is often sufficient to measure $E_c - E_p$ instead of loop area. Wilde[7] showed that such an analysis on a series of stainless steels tested in sea water correlated well with similar specimens with simulated crevices, subjected to four years of exposure tests. However, Wilde showed that care should be exercised in the interpretation of such data. He showed that a plane sample of material with no crevice gave very narrow hysteresis loop, and a very small value of $E_c - E_p$; it showed no signs of pitting corrosion. However, when a crevice was introduced into the same material, a large loop was obtained. The specimen showed considerable crevice corrosion but no pitting. It was judged that a growing crevice and the associated anodic micro-environment cathodically protect the rest of the metal surface. The conclusions reached were that E_c measurements relate only to pitting in the absence of crevices and that for stainless steels in sea water, crevice corrosion is more important than pitting, especially since it is difficult to imagine real structures with no crevices whatsoever. Although, once formed, pits and crevices are thought to propagate by similar mechanisms, Wilde concluded that the initiation of pits was primarily a potential-controlled passivity breakdown by adsorption of chloride ions at specific sites. In a crevice, however, initiation occurred when differential-aeration caused passivity breakdown.

7.4 MATERIAL SUSCEPTIBILITY

It has been stated that the most damaging environment giving rise to pitting and crevice corrosion is one of high chloride ion content. In sea water, the resistance of alloys to these forms of corrosion has been ranked[13,14] and is given in Table 7.1. It can be seen that stainless steels fare badly in such an environment, in accordance with the discussion so far. It has been well documented that the addition of chromium, nickel and molybdenum are

Table 7.1 The relative crevice corrosion resistance of metals and alloys in quiet sea water[14]

Metal or alloy	Resistance
Hastelloy C276 Titanium	Inert
Cupronickel (70/30) + 0.5% Fe Cupronickel (90/10) + 1.5% Fe Bronze Brass	Good
Austenitic cast iron Cast iron Carbon steel	Moderate
Incoloy 825 Carpenter 20 Copper	Low
316 stainless steel Ni–Cr alloys 304 stainless steel 400 series stainless steels	Poor – pit initiation at crevices

especially advantageous in combating such corrosion. Thus, for a typical 18 per cent Cr, 8 per cent Ni austenitic stainless steel, it is considered that the addition of 2–3 per cent molybdenum in the 316, or 3–4 per cent in the 317, causes an improvement in the resistance. Only in some of the more highly alloyed materials (not shown in the table) discussed in Chapter 15, does the performance of stainless alloys in crevice conditions reach a satisfactory standard. However, it can be seen from the table that high nickel alloys have been found to give good resistance, together with titanium and cupronickels.

The information given in Table 7.1 should be regarded cautiously. In a comprehensive review of crevice corrosion, France[15] states that the absence of a standard procedure for comparative testing of alloys in environments which lead to crevice and pitting corrosion has led to a situation where there is much conflicting evidence about material susceptibilities. Flow rate, for example, is a factor of considerable importance; titanium and Hastelloy C276 show excellent resistance over a considerable range of flow rate, while stainless steels benefit from a rapid flow rate which allows oxygen to repassivate the material. Indeed, in some non-chloride-containing environments there is evidence to suggest that 316 does not have superior resistance to 304.[16] Thus, the order given in Table 7.1 is likely to be considerably altered in a different environment. France stresses the importance of carrying out evaluation testing in an environment which simulates as closely as possible the real conditions.

7.5 REFERENCES

1. MacDonald D D, Begley J A, Bockris J O'M, Kruger J, Mansfeld F B, Rhodes P R, Staehle R W 1981 Aqueous corrosion problems in energy systems, *Materials Science and Engineering* **50**: 19–42
2. Shreir L L 1976 Localised corrosion, in Shreir L L (ed.), *Corrosion* (Vol. 1). Newnes-Butterworths, p1: 144
3. Fontana M G, Greene N D 1978 *Corrosion engineering*. McGraw-Hill pp 41–4
4. Peterson M H, Lennox T J, Groover R E 1970 *Materials Protection* Jan:23
5. Greiss J C Jr 1968 *Corrosion*, SBIIA **22**: 96
6. Korovin Y M, Ulanovskii I B 1966 *Corrosion*, SBIIA **22**: 16
7. Wilde B E 1974 On pitting and protection potentials: their use and possible misuses for predicting localised corrosion resistance of stainless alloys in halide media, in *Localised corrosion*. NACE, pp 342–52
8. Evans U R 1961 *The corrosion and oxidation of metals*. Edward Arnold, p 127
9. Cohen M 1979 Dissolution of iron, in Brubaker, G R., Phipps, P B (eds.) *Corrosion Chemistry* ACS Symposium Series, No. 89, pp. 126–52.
10. Wranglen G 1974 *Corrosion Science* **14**: 331
11. Shreir L L 1976 Localised corrosion, in Shreir L L (ed.), *Corrosion* (Vol. 1). Newnes Butterworths, pp: 158–62
12. Greene N D 1962 *Corrosion* **18**(4:) 136–42
13. Pourbaix M, Klimzack-Mathieu L, Merterns Ch, Meunier J, Vanleugen-Haghe Cl, de Munck L, Laureys J, Neelemans L, Warzee M 1963 *Corrosion Science* **3**: 239–59
14. Fontana M G, Greene N D 1978 *Corrosion engineering*. McGraw-Hill, p 269
15. France W D 1972 Crevice corrosion of metals, in *Localized corrosion – cause of metal failure*. ASTM-STP 516, pp 164–200
16. Sedriks A J 1979 *Corrosion of stainless steels*. John Wiley, p 93

8 EROSION CORROSION

O ruin'd piece of nature! This great world shall so wear out to nought.
(Shakespeare: *King Lear*)

Erosion corrosion is a poetic and self-explanatory name for a form of corrosion which results when a metal is attacked because of the relative motion between an electrolyte and a metal surface. Although electrochemical processes do occur, many examples of this form of corrosion can be attributed to mechanical effects, such as wear, abrasion and scouring. Soft metals are particularly vulnerable to this form of attack, for example, copper, brass, pure aluminium and lead, but most metals are susceptible to erosion corrosion in particular flow situations.

Three effects are discussed: corrosion associated with laminar flow; damage caused by impingement in turbulent conditions; and cavitation, a special form of erosion corrosion of materials in very rapidly moving environments.

8.1 VELOCITY, TURBULENCE AND IMPINGEMENT

When a fluid flows across a metal surface it can be imagined as a series of parallel layers, each moving at a different velocity. The slowest layer is adjacent to the metal surface where frictional forces and molecular collisions at surface irregularities are greatest, and layer velocity rises to a maximum at some distance into the bulk fluid. This effect is known as laminar flow and can have several consequences, some of which may actually be beneficial.

(a) We saw in section 4.6 that an equilibrium is established at metal surfaces in static systems in which the cathodic and anodic processes are occurring at the same rate. The ionic distribution in the immediate vicinity of the surface is called the double layer. When the ions from the corroded metal are removed from the system by a flowing electrolyte, the equilibrium cannot be established and, in theory, an increased rate of dissolution occurs.

(b) A mitigating factor is the replenishment of oxygen. Differential-aeration cells, which are a very common cause of attack, are minimised and overall corrosion resistance is improved in most cases where the relative motion is not too great. Stainless steels usually have improved corrosion resistance in electrolytes flowing above a given minimum velocity because the replenishment of oxygen maintains the protective oxide films.

(c) A detrimental effect of increased flow rates is the replenishment of aggressive ions such as chloride or sulphide. Conversely, a high flow rate can be an advantage in situations where a steady concentration of added inhibitor is important in controlling a corrosion process (see section 13.3).

(d) If any solid particles are present in the fluid, protective layers may be scoured away and corrosion enhanced. On the other hand, the flow may be sufficient to prevent deposition of silt or dirt which might otherwise cause differential-aeration cells in the crevices beneath.

Such a combination of factors makes the effect of flow rate somewhat unpredictable. However, perhaps the most significant effect which results from high flow rates is the destruction of laminar flow and the onset of turbulence. The molecules of fluid now impinge directly upon the metal and the impact causes mechanical wear. Velocity is just one of the factors which can cause turbulence; the system geometry can have a major effect in determining whether attack will occur or not. Some factors likely to cause impingement corrosion are:

(a) a sudden change in the bore diameter or direction of a pipe;
(b) a badly fitting gasket or joint which introduces a discontinuity in the otherwise smooth metal surface;
(c) a crevice which allows liquid flow outside the main body of fluid;
(d) the presence of a corrosion product or other deposit which may disturb the laminar flow.

Figures 12.10 and 12.12 show typical examples of (a) (b) and (c) above, while section 12.5 further discusses turbulence in tanks and pipe systems.

Case 8.1:   Figure 8.1 illustrates a copper alloy flange suffering from erosion corrosion caused by a badly fitting gasket.

Case 8.2:   Figure 8.2 shows a valve spindle from a hydraulic system which should have been fully closed or fully opened. It was left slightly open and the constriction of flow caused turbulence and erosion corrosion. The damage shown occurred in only 48 hours.

Turbulence is often inevitable because of the necessity for fluid flow from unrestricted regions into small bore or other restricted spaces. Consequently, erosion corrosion is frequently found at the inlet ends of condenser tubes and heat exchangers. Erosion corrosion is easily recognisable because it can create some strange and beautiful effects, often characterised by grooves, rounded holes or gullies.

Fig. 8.1 Copper alloy flange showing erosion by impingement at badly fitting gasket.

Fig. 8.2 Valve spindle showing erosion caused by valve being only half open instead of fully open or closed.

Case 8.3: The upper of the two specimens in Fig. 8.3 is an example of the erosion corrosion found at the end of a brass condenser tube. Wet steam flowed through the tube, but the corrosion was caused in a restricted pocket where turbulent sea water was able to attack the metal.

The peculiar effects produced by erosion corrosion have been attributed to a time dependence of the rate of erosion. On smooth surfaces, the erosion rate is slow, but accelerates as roughening proceeds. When a certain

Fig. 8.3 Upper specimen: Admiralty brass condenser tube exhibiting the characteristic well-rounded erosion pits.
Lower specimen: Admiralty brass condenser tube pierced by impingement of wet steam.

depth of surface roughness has been achieved, a layer of water adheres to the surface or becomes trapped inside the pits to reduce the eroding effects of succeeding liquid flow. Consequently, the erosion rate is observed to decelerate after a maximum rate has been achieved.[1]

Impingement has been a serious problem in several important applications. The destructive effects of high velocity impingement of rain upon the skins of aircraft caused considerable problems when the first supersonic jet fighters were introduced.[2] Considerable testing was necessary to evaluate the erosion resistance of new materials to the high impact velocities of wet steam in high-speed turbine generators.[3] The impingement corrosion which occurs when a liquid is forced to change direction may be quite different from the frequently beautiful effects produced by erosion corrosion.

Case 8.4: The lower specimen in Fig. 8.3 shows an Admiralty brass (see section 6.3) condenser tube which has been pierced by the direct impact of wet steam upon the metal surface.

8.2 CAVITATION

Cavitation is a particular form of erosion corrosion caused by the formation and collapse of bubbles of vapour on metal surfaces. This form of corrosion

Fig. 8.4 Cavitation damage on a copper alloy propeller blade.

tends to be associated more with components which are being driven at high velocity through a fluid, rather than in pipes or tanks where fluid flow occurs across stationary metal surfaces. Thus propellers, impellers and hydraulic turbine gear are the most common instances for encountering corrosion by cavitation.

Case 8.5: Figure 8.4 shows a typical example of cavitation corrosion along the edge of a propeller blade. (The large holes are where samples were removed for analysis.) The characteristic feature is initial surface roughening which subsequently develops into considerable, closely spaced pitting.

When the fluid flow over a metal surface becomes sufficiently great, very localised reductions in hydrodynamic pressure cause the fluid to vaporise and bubbles to nucleate on the metal surface. The same mechanical effect which reduces the pressure also creates pressure increases which cause the bubbles to collapse with considerable force. If these forces exceed the elastic limit of the metal (see section 9.2), deformation of the surface results; protective films are broken and corrosion ensues. The surface roughening, in turn, provides better nucleation sites for new bubble formation and the corrosion process is aggravated.

Cavitation has also been observed on wet liners of diesel engines where vibration resulting from the piston motion is the primary cause of the pressure variations within the flowing coolant.[4] Figure 8.5 is such an example and is the liner described in case 12.16. The cavitation damage is clearly visible on the upper and lower parts of the specimen. In addition,

Fig. 8.5 Wet liner of a diesel engine. Cavitation damage caused by engine vibration is at the top and bottom. Erosion damage is visible in the grooves at the middle of the liner. Crevice corrosion has occurred around the fuel injector seat in the centre hole.

the central part of the liner shows erosion damage where the water was pumped through the specially machined channels. (Crevice corrosion is also evident around the seat of the fuel injector nozzle in the centre of the photograph.)

An important method of controlling corrosion by cavitation is to use very smooth and well-machined components which offer fewer nucleation sites. Sometimes, resistance is obtained by the use of barrier methods, particularly varieties of rubber coatings which absorb the shock waves, though in severe environments they are less effective. It has even been said that cathodic protection is beneficial because evolution of hydrogen can provide a protective 'cushion' against the damaging forces which the metal must otherwise withstand.[5] However, the large currents necessary to produce sufficient hydrogen for protection make this a rather inefficient method of corrosion control, and the danger of hydrogen embrittlement cannot be neglected (section 10.3). Selection of the correct material is equally important in the control of cavitation damage. Stainless steel is widely regarded as the material with the greatest resistance to cavitation, although Stellite, a Co/Cr/W/Fe/C alloy, is reported to have a much longer life than a 304-type stainless steel in a severe environment.[4]

8.3 REFERENCES

1. Heymann F J 1967 On the time dependence of the rate of erosion due to impingement or cavitation, in *Erosion by cavitation or impingement*. ASTM STP-408, pp 70–110
2. Hoff G, Langbein G, Rieger H 1967 Material destruction due to liquid impact, in *Erosion by cavitation or impingement*. ASTM STP-408, pp 42–69
3. Smith A, Kent R P, Armstrong R L 1967 Erosion of steam turbine shield materials, in *Erosion by cavitation or impingement*. ASTM STP-408, pp 125–58
4. Godfrey D J 1976 Cavitation damage, in Shreir L L (ed.), *Corrosion* (Vol. 1). Newnes-Butterworths, pp 8: 124–32
5. Fontana M G, Greene N D 1978 *Corrosion engineering*, McGraw-Hill, p 85

9   ENABLING THEORY FOR ENVIRONMENT-SENSITIVE CRACKING

Ut tensio sic vis.

(Robert Hooke, 1660)

Engineering structures are frequently required to withstand the action of forces which tend to deform them. When forces act upon these structures in an environment which may cause corrosion of the metal, new types of corrosion may be observed which can be broadly grouped under the heading of **environment-sensitive cracking**. Before describing the nature of these corrosion processes it is necessary to achieve an understanding of the effects of forces upon those materials in the absence of a corrosive environment.

9.1   ELASTICITY: STRESS AND STRAIN

Robert Hooke's discovery that elastic properties could be described in the simple relationship **load/extension = constant**, was to have far-reaching consequences for the science of materials subjected to the action of forces. Provided that the load is not too great, most materials will extend and compress uniformly in the direction of the load, in accordance with Hooke's Law. The behaviour is termed elastic if it is both linear and reversible. To examine more precisely the effect of forces upon materials, some fundamental parameters must be defined.

When a force acts upon a material so as to deform it, we use the term **stress**: the stress in the material is the ratio of the force to the area of the material which resists the force, i.e.

$$\text{stress} = \frac{\text{force}}{\text{area}} \tag{9.1}$$

Although the unit of measurement would seem to be N m^{-2}, the SI unit is called the pascal, symbol Pa, such that 1 Pa = 1 N m^{-2}.

The material is said to be in a state of **strain** when it has suffered a deformation as a result of a stress. Strain is defined as follows:

$$\text{strain} = \frac{\text{deformation}}{\text{dimension of material}} \tag{9.2}$$

157

Strain is a dimensionless quantity, but is often expressed as a percentage of the original dimension.

The deformation which results from the application of a stress depends upon the way in which the stress is applied. In Fig. 9.1(a) force P is applied to a material of cross-section area A_0. The material is fixed in space and resists the applied force with a reaction R which is equal, opposite and co-linear. The stress, σ, is thus given by:

$$\sigma = \frac{P}{A_0} \qquad [9.3]$$

This type of stress situation is known as a **tensile**, or **uniaxial** stress because it acts along only one of the principal axes.

The tensile strain, is uniform in the direction of the applied stress and is usually given the symbol ε such that:

$$\varepsilon = \frac{(l - l_0)}{l_0} \qquad [9.4]$$

It is important to realise that although the material extends elastically in length in the direction of the applied force, it also contracts in width and thickness (not shown in Fig. 9.1(a)). As a good approximation it can be said that the volume of material remains constant. The reduction in width and thickness causes a reduction in the area of cross-section and although the load may have remained constant, the *true* stress has increased. It is normal in such situations to call the stress, as defined by eqn [9.3], **engineering stress** in order to distinguish it from the true stress. Similarly, the tensile strain calculated on the basis of eqn [9.4] is better called the **engineering tensile strain**.

Figure 9.1(b) shows a stress situation which is called **shear stress**. The material is fixed in space at the bottom, and the force is applied parallel to the top face. The reaction is equal, opposite, but **NOT** co-linear. The shear strain is non-uniform in the direction of the shear stress, τ. Shear strain is measured in terms of the angle of deformation, θ radians such that for small angles:

$$\theta \text{ (radians)} \simeq \tan \theta = x/y \qquad [9.5]$$

The most general case is depicted in Fig. 9.1(c) which shows an element of material subjected to a **triaxial stress state**. In the figure, σ_x, σ_y, and σ_z are three **principal stresses** which, if they are equal and act so as to compress rather than extend the material, relate to the situation which exists for bodies immersed in fluids. This is termed a **hydrostatic stress**. If one principal stress is zero the stress state is called **plane stress**, a situation found in the loading of thin sheets which do not develop a tensile stress through their thickness.

When a material is subjected to a **uniaxial** tensile stress which increases from zero, the tensile strain increases uniformly and the material initially behaves in an elastic manner. We can write that for most materials there exists a relationship:

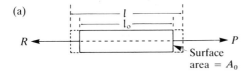

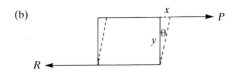

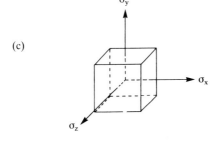

Fig. 9.1 Three simple stress situations: (a) tensile stress (uniaxial); (b) shear stress; (c) triaxial stress.

$$\text{stress/strain} = \text{constant} \qquad\qquad [9.6]$$

For tensile stress and strain, the constant is called **Young's modulus**, given the symbol, E (not to be confused with the symbol for potential). Because strain is a dimensionless quantity, the units of E are the same as for stress. Young's modulus is an important material property used in engineering design.

The linear elastic behaviour of materials ceases when stresses are increased beyond the elastic limit. From this point onward, the behaviour of the material is of even greater importance to the engineer, and this will be discussed next.

9.2   THE TENSILE TEST

A useful way of monitoring the behaviour of materials under the action of tensile stresses is by means of the **tensile test**. A specimen of defined dimensions (it may have either cylindrical or rectangular cross-section) is extended uniformly at a pre-determined rate, between the parting crossheads of a loading machine and the values of the load on the specimen at any given extension are measured. It is usual for this data to be produced on a chart recorder in graphical form as a load/displacement curve. However, it is

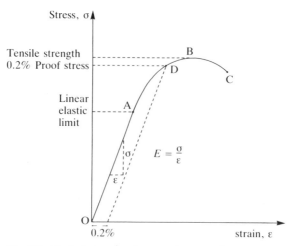

Fig. 9.2 Typical stress/strain curve for a tensile test specimen.

instructive to consider the graph of engineering stress against engineering strain: there is no difference in shape, but the axes are changed. Load is divided by a constant quantity, the original cross-section area, while extension is converted into strain by the relationship, eqn [9.4].

A schematic curve which illustrates some important features of tensile test results is shown in Fig. 9.2. During the initial period, the relationship between the two parameters is linear (eqn [9.6]) and elastic, i.e. if the test were stopped and the stress removed, the material would return to its original dimensions. At point A the material reaches the limit of linear elastic behaviour. As the increments of extension take the specimen beyond A, smaller increments of load are measured. This situation continues to the point represented by B where the material begins to narrow considerably at one point along its length, rather than along the whole of its length, an effect known as **necking**. The stress on the specimen at the point of necking, B, is called the **tensile strength**.

Further strain leads to actual decreases in engineering stress until, at C, the specimen breaks. The two broken pieces can be matched together to determine the plastic elongation of the specimen at the point of fracture. The greater this elongation, the greater is the property known as **ductility**. If the fracture surfaces of a ductile material are examined, they will exhibit much tearing of the material. Such a specimen is said to have undergone a **ductile fracture**. The specimen on the right of Fig. 9.3 shows a typical example.

In some cases, there is little or no permanent deformation of a material before fracture occurs. The surfaces are typically bright and shiny with almost none of the macroscopic tearing associated with ductile fracture. The specimen on the left of Fig. 9.3 is said to have fractured in a **brittle** manner. Brittle failure is the scourge of engineers, because components and struc-

Fig. 9.3 Fracture surfaces exhibiting brittle fracture (left) and ductile fracture (right).

tures give no warning of impending failure, as they do if the metal deforms plastically.

It is often difficult to determine the point at which *plastic* deformation (dislocation movement, which occurs to a very small degree from the first application of a load) begins to predominate over *elastic* behaviour (bond stretching). Consequently, engineers have defined another parameter known as the **proof stress**. For example, a 0.2 per cent proof stress is that stress which will result in a permanent deformation of 0.2 per cent. If, in Fig. 9.2, the specimen were strained from O to D and then the stress removed, the specimen would contract to release the elastic part of the extension, but the permanent 0.2 per cent elongation would remain. For the sake of clarity, the curve shown in the figure has been exaggerated: real curves may look rather different from this, with, for example, D much closer to A and a much steeper slope, E.

9.3 STRESS CONCENTRATION

Let us now consider the effects of a tensile stress upon the bonding between the atoms of a metal. In a perfect crystal structure, the stress is absorbed uniformly between all the bonds concerned, but when a defect in the structure is present, the stress distribution cannot be uniform. Some bonds will therefore be under greater stress than others. Figure 9.4(a) illustrates this. The atomic structure is shown as a simple cubic arrangement for simplicity

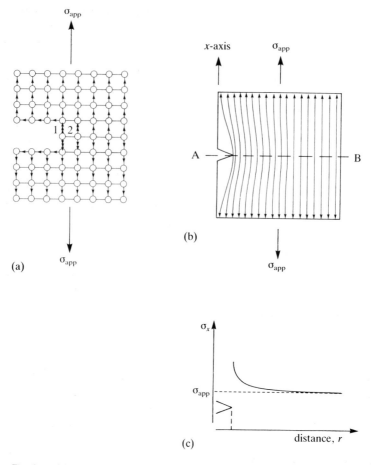

Fig. 9.4 (a) The effect of an applied stress, σ_{app}, upon the atomic bonds in a crystal lattice containing a crack.
(b) Stress concentration at a crack tip.
(c) The variation of effective stress in the x-direction, σ_x, with distance r ahead of the crack tip.

and arrows indicate how, when a tensile stress σ_{app} is applied to the metal, the stress is concentrated in the bonds which are adjacent to the defect. Thus bonds 1 and 2 are subjected to greater stresses than that applied externally. As distance is increased away from the defect and into the material, the stress diminishes until an average level is reached which is the same as the external stress. For small values of σ_{app} the bonds are merely stretched, but for large σ_{app} the stress upon bond 1 may be larger than the bond strength, in which case the bond will break. Not only is the full effect of the stress concentration now transferred to bond 2, but the concentration is greater than before and the material is very likely to undergo total failure by a 'domino' effect of bond rupture through the material.

The action of a stress on the atomic bonds within a material may be

162

likened to lines of magnetic force (except that stress lines act only through solids) and the diagram of Fig. 9.4(a) is sometimes drawn as in Fig. 9.4(b) to show how the concentration of lines of stress is greatest at the crack tip. Consider the line AB through the material, and define the *x*-axis as the direction of the applied stress. It has been shown that, to a fair approximation, the effective stress in the *x*-direction at points along AB is similar to that shown in Fig. 9.4(c).

Any of the defects described in section 2.6 which locally reduce either the number or the strength of bonds has the potential to concentrate stress in this manner. Not surprisingly, they are referred to as **stress raisers**. They may be contained within the body of the material, in which case they may have been caused as a result of the manufacturing process. Alternatively, they may be on the exterior and may have been caused either by mechanical or corrosive means.

9.4 CONCEPTS OF LINEAR ELASTIC FRACTURE MECHANICS

In earlier times, the most obvious criterion by which engineering structures could be designed was the tensile strength, the greatest load which could be applied to a structure without causing immediate failure (see Fig. 9.2). As engineering knowledge increased, this idea was revised and the elastic limit substituted instead. Even so, it was not good enough to design structures so that they deformed: a safety factor was divided into the chosen proof stress to produce a design stress and the magnitude of the safety factor was largely dependent upon the scale of disaster which would result from a failure. For example, a component upon which human lives were dependent might have a safety factor of 10 or more, so that the maximum stress which the designer anticipated being applied to the component was still only one-tenth that which **MIGHT** lead to failure. Even so, it was found by experience that catastrophic failure occurred unexpectedly. A better method of predicting failure was necessary and as a result the subject of **linear elastic fracture mechanics**, often simply called fracture mechanics, was devised.

The failure of a metal component occurs as a result of the presence of defects, whether they be present in the original material or introduced during the lifetime of the component. The likelihood of catastrophic failure increases with increasing defect size. The aim of fracture mechanics is to predict whether a given defect size will lead to catastrophic failure or not. In Chapter 2, a number of different defects were described, any one of which may contribute to the failure process. Therefore, for simplicity, we shall refer to defects as cracks. In any failure, there are two distinct phases to be considered. First, there is an **initiation** phase. Cracks may either be present in the original material as a result of the manufacturing process, or they can be created by mechanical or corrosive mechanisms. Once initiated, cracks may **propagate** either by mechanical or corrosive means. Obviously, this book is concerned with the corrosive aspects but consideration of the mechanical mechanism is also essential.

A very important material property which results from considerations of fracture mechanics is **fracture toughness**, a measure of the *resistance to crack growth*. If a material is able to sustain the presence of a crack and perhaps allow the growth of the crack to occur slowly, then it will probably have desirable engineering properties. Brittle materials, on the other hand have little resistance to crack growth once a critical tensile stress is applied. Above this stress, a crack propagates very rapidly to give a catastrophic failure. Some brittle materials such as concrete are used only in compression because their resistance to tensile stress is so low.

It is an unfortunate fact that most materials become more brittle as their strength is increased, a property which works against the interests of the designer and enforces compromise. The engineer who seeks to use the strongest materials must have a very good predictive capability for typical crack growth rates under different loading conditions. Such information can be gleaned from a study of fracture mechanics. A determination of fracture toughness, for example, can be made so that it is a conservative parameter and represents the worst possible case which a material must withstand. This can be done in the following way.

Figure 9.5 shows three modes by which crack growth can occur. A crack in a material is simulated by a sharp-edged notch, cut across the whole of the thickness of the specimen.

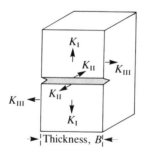

Fig. 9.5 The three modes of crack growth.
I: Crack opening mode.
II: In-plane sliding mode.
III: Anti-plane shear mode.

Mode I is considered to be the most potentially damaging and is therefore the one most frequently described because it leads to the most conservative value of fracture toughness. It is the **crack opening mode**, in which crack surfaces move directly apart. Therefore, mode I relates precisely to a material under the action of a tensile stress.

Mode II is called the **in-plane sliding mode** and corresponds to crack surfaces sliding over one another in a direction perpendicular to the leading edge of the crack.

Mode III is the **anti-plane shearing mode**, a tearing mode in which crack surfaces slide over one another in a direction parallel to the leading edge of the crack.

Modes II and III are less damaging situations and will not be discussed further. **Henceforth, all terms with the subscript I relate to mode I crack opening.**

Consider a specimen similar to that shown in Fig. 9.5, and subjected to a constant applied mode I tensile stress, σ_{app}. We wish to study the effect of increasing the size of a material defect (represented by the machined notch) upon the stress immediately ahead of the crack tip. Initially, the stress concentration for small cracks is not sufficient to cause significant growth. As crack length increases, so also the stress concentration at the crack tip increases, until the effective stress becomes so great that the crack begins to grow at a very fast rate and failure ensues. It is possible to determine a **critical crack length** below which the material can be considered stable. Thus, just as with the simple tensile test it is possible to find the tensile strength of a material, so also can fracture toughness be determined for the failure of the material under a mode I crack opening regime.

In Fig. 9.5, the thickness of the material is represented by B, and the crack extends throughout the whole of the thickness. This is called a **through-thickness crack**. Using the schematic graph of Fig. 9.4(c) to represent the variation of effective stress with distance in an infinite body containing a through-thickness crack of length $2a$, it can be shown that the effective stress at a distance r ahead of the crack tip, when r is very small, is given by:

$$\sigma = \frac{K}{\sqrt{(2\pi r)}} \qquad [9.7]$$

K is the **stress intensity factor** and has the value:

$$K = \sigma_{app}\sqrt{(\pi a)} \qquad [9.8]$$

For mode I crack opening, K is replaced by K_I. When the crack is of critical size such that failure results, K_I is minimised and becomes the useful material property, fracture toughness, denoted K_{Ic}.

$$K_{Ic} = \sigma_{app}\sqrt{(\pi a_{crit})} \qquad [9.9]$$

It would appear that the determination of fracture toughness for engineering materials is a straightforward exercise. Unfortunately, the situation is not quite so simple, for it is found that different values are obtained with specimens of the same material, but of different dimensions. This is mostly because stress concentration is strongly dependent upon the dimensions of the specimen, and in particular, the thickness. If a *thin* specimen is used (B is small) the material will be found to be quite tough. Such a result might

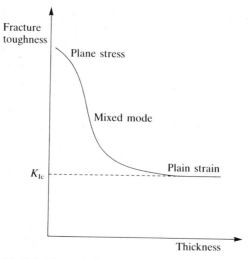

Fig. 9.6 The variation of fracture toughness with thickness.

be thought to be favourable; however, if a *thicker* specimen were to be tested a much lower critical stress intensity factor, K_c, would be measured. The engineer must know the least possible value of K_c for the material he plans to use in order to avoid the possibility of catastrophic failure. Figure 9.6 is a graph showing the variation of critical stress intensity factor with thickness of a test material. It is the lowest value of K_c, which is a major design parameter in modern engineering. This is the fracture toughness, K_{Ic}.

If the fracture surface of a broken specimen exhibits **shear lips**, as in the case of aluminium, shown at the bottom in Fig. 9.7, such a fracture is usually associated with ductility and relatively high toughness value. The specimen is not thick enough for a valid measurement of K_{Ic} to be made. In specimens of the correct thickness for valid K_{Ic} measurement, it is found that the shear lips almost disappear, as in the titanium alloy specimen shown at the top in Fig. 9.7. In the case of thin specimens which exhibit shear lips, the stress condition is known as **plane stress**, while the condition causing brittle fracture in thicker specimens is called **plane strain**. Inter-mediate cases which show elements of both types of character are known as **mixed mode fractures**.

The procedures for performing mechanical tests are laid down in the references[1,2] and must be carefully adhered to if meaningful data are to be obtained. Experimentally, K_{Ic} is evaluated by a series of careful, time-consuming experiments. A number of identical notched specimens are stressed under mode I conditions and measurements of crack growth rates made. When the variation of stress intensity factor is plotted against the crack growth rate, a graph similar to that shown in Fig. 9.8 is obtained. For convenience, log scales are used for crack growth rates. Units of stress intensity factor (and fracture toughness) are $MPa\sqrt{m}$. Such graphs are very important in the analysis of corrosion mechanisms of environment-sensitive cracking.

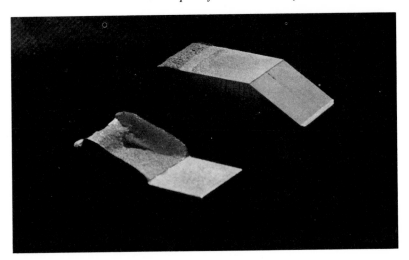

Fig. 9.7 Fracture surfaces showing:
Top: a titanium alloy specimen showing almost no evidence of shear lips (fracture in plane strain — specimen thick enough for valid test).
Bottom: an aluminium specimen exhibiting shear lips (fracture in plane stress — specimen too thin for valid K_{Ic} test)

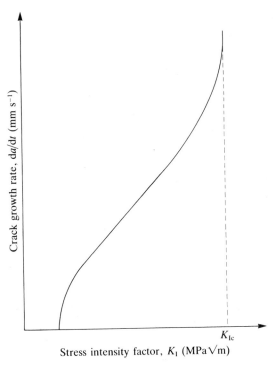

Fig. 9.8 The variation of crack growth rate with stress intensity factor.

9.5   FATIGUE

The discussions so far have assumed that applied stress is constant in magnitude and direction. In reality, this may not be so. Engineering structures and components may be subjected to stresses which continuously rise and fall in magnitude, and may also change direction. If a metal component fails as a result of such a condition, it is referred to as a **fatigue failure**. The potential problems caused by fatigue were known over a century ago when the British engineer, Sir William Fairbairn, found that a steel girder would support a static load of 12 tons indefinitely, yet it failed under a cyclic load of only 3 tonf applied 3 million times. It was only after a series of Comet airliner crashes in the 1950s that the severity and the unpredicability of fatigue failure were brought to the attention of the public.

Fatigue is produced by stress *cycling*, a process which introduces many new variables not present in static stress situations. Figure 9.9(a)–(c) illustrates three different load cycles to which a material may be subjected. Figure 9.9(a) represents a sinusoidal load variation which oscillates with equal magnitude about a zero mean. Figure 9.9(b) shows a similar load cycle which is never negative in sense, but increases from and returns to zero, while the cycle in Fig. 9.9(c) shows the load always positive. It is possible to describe each situation by considering the ratio of the minimum and maximum applied loads. P_{min}/P_{max} is a parameter denoted by R. Thus, in the first case, $R = -1$, while in the second, $R = 0$ and in the last, $R > 0$. As well as the amplitude of the stress cycle, the frequency (measured in cycles per second, Hz) is also important. It should be emphasised that in real engineering situations, loading patterns may be most irregular and quite unlike the idealised cycles used for laboratory simulations. In such situations the mean stress, indicated in Fig. 9.9, is a useful parameter. The parameters described above must be carefully defined for all fatigue tests.

One common type of fatigue test consists of the application of a particular stress cycle to a material and observation of the number of applied cycles necessary to cause failure. Data are plotted in the form, stress amplitude, S, against the logarithm (base 10) of the cycles to failure, N. Such graphs are called S–N curves, a typical example of which is shown in Fig. 9.10.

The upper curve demonstrates the existence of a stress level below which the material will never fail by fatigue. Such behaviour is typical of iron and ferrous alloys, but many non-ferrous alloys do not possess such a property, as evidenced by the lower curve. Structures made of these metals must be designed with fatigue firmly in mind.

The development of methods of analysing and predicting failure using fracture mechanics has proved to be a major advance in the study of fatigue. The theory described in sections 9.3 and 9.4 applies equally well to fatigue and modification of the experimental technique to include stress cycling is straightforward. Data obtained from fatigue testing can be plotted on a graph similar to that shown in Fig. 9.8, except that stress intensity factor

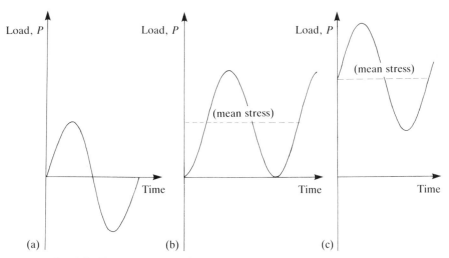

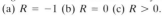

Fig. 9.9 Three common load cycles:
(a) $R = -1$ (b) $R = 0$ (c) $R > 0$.

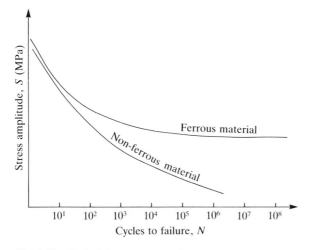

Fig. 9.10 Typical S–N curves for fatigue test in air:
(a) A ferrous alloy.
(b) A non-ferrous alloy.

is replaced by the difference between the maximum and minimum stress intensities encountered in each cycle, i.e.

$$\Delta K = K_{\text{max}} - K_{\text{min}} \qquad [9.10]$$

ΔK is plotted against the crack growth rates measured in terms of the increment of crack growth per cycle, $\text{d}a/\text{d}N$.

Observations of fatigue crack growth rates in the intermediate range of stress amplitudes are often found to behave in accordance with the equation:

$$\mathrm{d}a/\mathrm{d}N = C\Delta K^m \qquad [9.11]$$

where C is an experimental constant. Over a wide range of materials, it has been found that the average value of m is between 2 and 3. Equation [9.11] is frequently referred to as the Paris Law.

9.6 REFERENCES

1. British Standards Institution 1977 *Methods of test for plane strain fracture toughness K_{Ic} of metallic materials*. BS 5447
2. ASTM 1965 *Fracture toughness testing and its applications*. STP 381

9.7 BIBLIOGRAPHY

Hertzberg R W 1976 *Deformation and fracture mechanics of engineering materials*. John Wiley

Knott J F 1973 *Fundamentals of fracture mechanics*. Butterworths

Knott J F 1979 *Worked examples in fracture mechanics*. The Institution of Metallurgists

10 ENVIRONMENT-SENSITIVE CRACKING

The image of stress corrosion I see
Is that of a huge unwanted tree,
Against whose trunk we chop and chop,
But which outgrows the chips that drop;

And from each gash made in its bark
A new branch grows to make more dark
The shade of ignorance around its base,
Where scientists toil with puzzled face.

(From 'On Stress Corrosion' by S P Rideout 1967)*

In the late nineteenth century, British rule in India was at its strongest. Infantry, artillery and cavalry were all present in strength and well supported with arms and ammunition. During the wettest months of the monsoon season, however, military activity was diminished and ammunition was stored in the stables until the dry weather returned. Then a problem with cartridges became apparent; many were found to be useless because of considerable cracking of the brass. It was not until 1921 that the problem was satisfactorily explained, when, in a classic investigative study, Moore, Beckinsale and Mallinson[1] found that residual stresses caused cracking in 70/30 cartridge brass in the presence of specific chemical agents. It was concluded that residual stresses from the manufacturing process, combined with ammonia from horse urine had been the agents responsible for the cracking of the cartridges in India. A new form of corrosion had been identified, which, because of the circumstances in which it was found, was first called **season cracking**, and later, **stress-corrosion cracking**.

Case 10.1: Figure 10.1(a) shows a cold-drawn 70 Cu, 30 Zn brass tube (essentially the same material as that of the cartridges) which had been left in a metal rack in a chemical laboratory. The periodic use of ammonia solutions some 15 m away from the rack caused cracking along the 300 mm length in six months.

Case 10.2: Figure 10.1(b) shows a 70/30 brass fitting to a high pressure air bottle. When withdrawn from stores after a long period on the shelf,

* Published in *Proceedings of the Conference on Fundamental Aspects of Stress Corrosion Cracking* (1969). Reprinted with permission from the National Association of Corrosion Engineers, Houston, Texas.

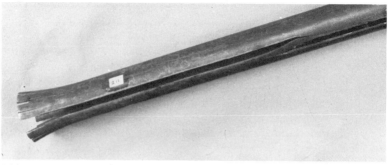

(a)

(b)

Fig. 10.1 Stress-corrosion cracking of 70/30 brass:
(a) A cold-drawn tube left in metal rack in chemical laboratory for six months.
Ammonia in the working environment coupled with residual stresses in the tube to
cause cracking.
(b) A fitting for a high pressure air bottle which had cracked after a period in
store in an environment containing an unidentified aggressive species. Application
of a stress-relieving heat treatment to other uncracked specimens resulted in no
further failures.

the fitting was found to be cracked from end to end. Inspection of
the remainder of the stock showed that 80 per cent were suffering
from the same problem. The source of the aggressive species
responsible for the corrosion was never identified, but was probably
airborn ammoniacal vapours. The remaining 20 per cent were given
a stress-relieving heat treatment and did not crack thereafter.

During the long period which intervened between the observation and explanation of season cracking, a new and seemingly unrelated problem was regularly occurring in riveted steam boilers. The age of steam power had arrived but the materials technology was still deficient in important areas. Boiler explosions were common, many of them a direct result of corrosion; indeed, in the final 20 years of the nineteenth century, 600 lives were lost in the UK alone. Although new methods were being introduced to control the corrosion of boiler tubes, it was found that the presence of caustics (alkalis), usually concentrated in crevices around rivets, combined with the considerable fabrication stresses around rivet holes to cause cracking of the steel boiler shells and tube plates with consequent catastrophic failure.[2]

Case 10.3: Figure 10.2 is a sample of riveted mild steel boiler plate suffering from caustic embrittlement. Caustic deposits collecting under the rivet head created an environment of pH 11–12, which, together with residual stresses around the drilled hole, provided the conditions necessary for caustic cracking.

The caustic embrittlement of boilers and the season cracking of brass are linked by an insidious combination: a *particular metal/environment pair* in the presence of *tensile stress*. They were the forerunners of a problem which has blighted the technological society for decades, for, with the development of each new alloy, has come the seemingly unpredictable risk of catastrophic failure. In the early 1900s, mild steel was found to crack in nitrates as well

cracks

Fig. 10.2 A sample of riveted mild steel boiler plate showing caustic embrittlement. The combination of residual stress around the drilled hole, together with caustic deposits collecting under rivet head, created ideal conditions for cracking.

as caustics, while aluminium alloys were found to crack simply in moist atmospheres. In the 1930s, magnesium alloys were also found to be susceptible to cracking in moist atmosphere, and stainless steels suffered badly from cracking in aerated chloride-containing environments.

Case 10.4: Figure 10.3 shows part of an austenitic stainless steel component for sieving molasses liquor. A combination of the stress which resulted from the drilling of the holes, and chloride ions in the liquor, caused stress-corrosion. Cracks which emanate from the holes are visible in the figure. The explosive failure of the component killed a man.

In the 1950s and 1960s martensitic steels for aerospace applications were discovered to be vulnerable to stress-corrosion cracking, and even the very corrosion-resistant titanium alloys could be cracked in environments containing methanol. The problem seemed worse with every new application, and the poem at the beginning of this chapter captures well the cloud of despair which, even today, occasionally descends upon those who work in the field of stress-corrosion cracking. The efforts made by both pure and applied researchers to solve the problems have been considerable. Those stress-corrosion failures which result in fatalities or injuries usually receive detailed investigation, yet many others are still not identified as such. When, in 1975, a detailed survey of corrosion failures in the chemical industry was carried out, it was discovered that more than one-third were caused by environment-sensitive cracking.[3]

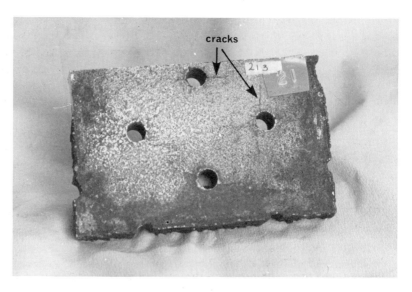

Fig. 10.3 A piece of austenitic stainless steel (18Cr, 8Ni) component for sieving molasses liquor. The combination of residual stress around the drilled holes and chloride ions in the liquor caused stress-corrosion cracking and a fatal explosion.

This chapter uses the term environment-sensitive cracking to describe all failures in which stress combines with particular metal/electrolyte pairs to cause cracking. There is an important subdivision in which *cyclic* or *periodic* stress is differentiated from *static* stress, the former being classified as **corrosion fatigue**. Traditional texts have further separated cases of **hydrogen embrittlement**, but most current workers believe that this is so closely allied to stress-corrosion as to be an integral part of mechanistic discussions.

10.1 STRESS-CORROSION CRACKING (SCC)

Stress-corrosion cracking is a term given to the intergranular or transgranular cracking of a metal by the conjoint action of a static tensile stress and a specific environment. This form of corrosion is very common throughout industry, and despite decades of intensive research, after which workers are only now beginning to understand the processes involved, measures adopted to control it are still frequently unsuccessful.

In the technology of the boiling water reactor, the intergranular SCC of stainless steel (type 304) piping is said to be the major corrosion problem, while in pressurised water reactors the same material has been found cracked in such applications as boric acid lines and spent fuel liners. Stress-corrosion failure of turbine blades made from stainless steel (type 403) was said to be at a level of 4 per cent per annum.[4] In the chemical industry, SCC of stainless steel owing to chloride leaching from thermal insulation continues to be a problem, despite the fact that its cause is well known. In 1973, one instance alone of the failure of a stainless steel component caused the loss of $1 m.[5] Similar problems continually plague the oil industry where pipes in deep, high-pressure wells require the use of high-strength steels which are known to be susceptible to SCC, particularly in the presence of hydrogen sulphide. An inhibitor has been used consistently in an attempt to alleviate corrosion problems in such situations, yet failures in the presence of the inhibitor were still being reported 10 years after the inhibitor had been proved ineffective.[6]

One feature of stress-corrosion cracking which occurs repeatedly is the unexpectedness of such cases. Often, a material, chosen for its corrosion resistance in a given environment, is found to fail at a stress level well below its normal fracture stress. Rarely is there any obvious evidence of an impending failure, and because it can occur in components which are apparently unstressed, it is even more of a surprise when components are found defective.

Case 10.5: Stainless steel pipe was stored close to the sea while waiting to be used in a construction programme in the Middle East. The high day temperature and low night temperature, together with the salt environment, caused a massive salt build-up and resulted in SCC of the pipe, even before it was installed in the plant.[7]

Problems with pipes and tubes are very common because of the hoop stresses which result from the fabrication processes. Stress-relieving heat treatments are thus a vital part of the control of SCC. In Chapter 3, expt 3.16 shows, in minutes, how the residual stresses in a cold-drawn brass tube combine with the aggressive mercury(I) ion to destroy the integrity of the tube. An identical tube which has been annealed is unaffected. Such a test has frequently been used to show the presence of residual stress in metal components: it is **NOT** a test for susceptibility to stress-corrosion cracking, though as a quick and simple demonstration it is ideal. As will be seen later, the testing for SCC susceptibility is rather more involved.

It is now generally agreed that there is no single mechanism for stress-corrosion cracking.[8] Any combination of a number of significant factors may contribute to a given SCC failure. Rather than list a large number of specific instances, the discussion in this chapter will focus on the main elements involved in general types of environment-sensitive cracking. In particular, it is a good idea to refer frequently to the *stress-corrosion spectrum* which has been compiled by Parkins[9] and which lists the important alloy/environment combinations, as well as the essential features of the mechanism involved. This important information source is shown as Table 10.1. Information about specific cases listed, will be found in the bibliography.

The main features of stress-corrosion cracking have been listed by Brown[10]:

(a) The requirement for tensile stress is beyond doubt. Environment-sensitive cracking is the synergism of stress and corrosion: the absence of either eliminates the problem. The stress can be applied directly during the

Table 10.1 The stress-corrosion spectrum (after Parkins)[9]

	↑ INTERGRANULAR CORROSION	
Corrosion dominated (solution requirements highly specific)	Carbon-steels in NO_3^- solns	Intergranular fracture along pre-existing paths
	Al–Zn–Mg alloys in Cl^- solns	
	Cu–Zn alloys in NH_3 solns	
	Fe–Cr–Ni steels in Cl^- solns	Transgranular fracture along strain-generated paths
	Mg–Al alloys in $CrO_4^{2-} + Cl^-$ solns	
	Cu-Zn alloys in NH_3 solns	Mixed crack paths by adsorption, decohesion or fracture of brittle phase
Stress dominated (solution requirements less specific)	Ti alloys in methanol	
	High strength steels in Cl^- solns	
	↓ BRITTLE FRACTURE	

operational lifetime; alternatively, it can be present in the component as a result of the fabrication or installation process.

(b) In general, it is found that alloys are more susceptible than pure metals, though there are some well-documented exceptions, notably copper.

(c) Cracking of a particular metal is observed only for relatively few chemical species in the environment, and these need not be present in large concentrations.

(d) In the absence of stress, the alloy is usually inert to the same species in the environment which would otherwise lead to cracking.

(e) Even when a material is particularly ductile, stress-corrosion cracks have the appearance of a brittle fracture.

(f) It is usually possible to determine a threshold stress below which SCC does not occur.

In addition to the features listed by Brown, it is now clear that there are certain potential ranges within which SCC is likely or unlikely, and that corrosion potential is an important consideration.

10.2 MECHANISMS OF ENVIRONMENT-SENSITIVE CRACKING: INITIATION

The failure of metal components can be conveniently divided into the **initiation** phase during which a stress raiser is formed, and a second phase of crack **propagation** leading to failure. Because stress-corrosion cracking is observed in metal/environment pairs where, in the absence of stress, corrosion would not be a serious problem, the first question which must be answered is '*How does stress initiate SCC?*'

In the first instance, some form of attack upon very localised anodes on the metal surface must occur, the result of which is best described as a pit. The action of tensile stresses upon materials can have many effects, some of which were described in Chapter 9. The most fundamental possibility is that the application of a tensile stress to a crystal lattice, which is otherwise in equilibrium, results in a raising of the thermodynamic energy of the atomic bonds. If this effect is localised at the surface, anodes will be formed, even though the material is being stressed within its elastic limit. Such arguments can be used only in the cases of SCC which occur when the stresses are well below the yield strength and there is no evidence of significant structural defects in the original material. As we shall see later, this is unlikely.

Once the stress exceeds the yield stress of the material, plastic deformation is obtained, that is, the crystal structure suffers bond breaking and reforming so that its shape is permanently altered. The mechanism for this is well documented in the metallurgical literature and can be thought of most simply as the creation and motion of defects, usually dislocations, through the crystal structure. The movement of dislocations will be halted when they reach either the surface of the metal or a grain boundary. Dislocation movement can be prevented in a number of other ways, but these are the most significant in stress-corrosion mechanisms. The 'pile-up' of dislocations at grain boundaries results in the anodic polarisation of these regions because of the increased irregularities in the crystal structure. This has no effect upon the initiation phase if it is within the material, but is most important in the propagation stage. At the surface a local blemish occurs on an otherwise 'smooth' surface: this is known as a slip step and it is at this site that the material is most vulnerable to initial corrosive attack.

Alloys which rely on thin films of oxide or other material for their corrosion protection are especially vulnerable because the slip step uncovers a microscopic quantity of bare metal which is highly anodic compared to the surrounding surfaces. Such a process is shown in Fig. 10.4.

If the metal is able to repassivate quickly, then little danger ensues, but if the passivation time is long enough to allow corrosion of the exposed area to occur and a pit to form, then the criteria for commencement of SCC have been fulfilled.

Even in metals which are not passivated, the formation of slip steps on the surface represents a corrosion problem for the discontinuity of crystal

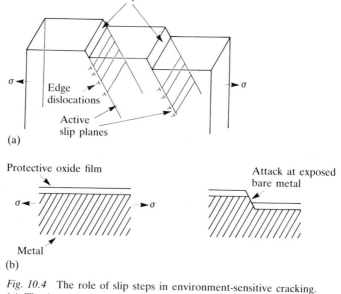

Fig. 10.4 The role of slip steps in environment-sensitive cracking.
(a) The formation of slip steps on the surface of a metal by movement of dislocations along active slip planes under the action of tensile stress.
(b) A slip step on the surface of a passivated metal creates an active site for the initiation of a stress-concentrating pit.

structure causes local anodes. Brown[10] places great importance on pit formation in which a porous cap of corrosion product creates a tiny concentration cell. The aggressive species are concentrated within the confines of the cell to form an even more aggressive local environment. The diffusion process operates up a concentration gradient because the energy change which results from the dissolution process at the exposed bare metal is so favourable. Much has already been said about the mechanism of pit formation in Chapter 7, all of which applies to this situation.

Considerable evidence has been accumulated that the local environment inside pits and cracks becomes very acidic, sometimes as low as pH 2, a factor which not only accelerates the dissolution processes, but adds considerably to the weight of evidence for the involvement of hydrogen in most, if not all, SCC mechanisms. As pH falls, the thermodynamic tendency for hydrogen evolution becomes greater at constant potential. The potential is not constant, however. The creation of an oxygen concentration differential causes an anodic polarisation, which further increases the hydrogen evolution tendency.

The formation of a pit, then, is the usual precursor to stress-corrosion cracking, and represents the initiation phase. It is the most unpredictable aspect of this form of corrosion: it may take anything from days to many years to occur. Even then, a pit may not be of the size and shape necessary to promote a crack in the pit bottom. However, once a pit has been produced with the right conditions, crack growth begins and the propagation phase takes over. This phase of the process is much more quantifiable using concepts developed from fracture mechanics, an aspect which will be discussed later.

Many workers place little importance upon initiation processes, and take the view that in normal engineering materials there are almost always enough material defects and surface irregularities present as a result of the fabrication process to provide ample initiation sites. In steels, the typical size of defect in a well-made weld is $100–500$ μm,[11] yet the calculated critical crack sizes for SCC of mild steel in caustic solution is only 1 μm; for stainless steel SCC in chloride solution, the critical crack size has been quoted as 180 μm.[12] In such cases a treatment of environment-sensitive cracking may deal only with propagation and treat initiation time as effectively zero.

There are many instances of SCC not being initiated at pits, for example, mild steels in nitrates, hydroxides and bicarbonates. The SCC initiators in these cases are localised cracks at the grain boundaries. In all of these environments, the grain boundaries are attacked because of chemical heterogeneity, even in the absence of stress.

Acidification at the SCC initiation site does not invariably occur. Thus, stress-corrosion cracks in mild steels exposed to hydroxides or bicarbonates contain environments not essentially different from those outside cracks, because those solutions are effectively buffered and the solubility of iron is very low. Acidification will only occur when hydrolysis can take place, as with chlorides and steels.

10.3 MECHANISMS OF ENVIRONMENT-SENSITIVE CRACKING: PROPAGATION

Many mechanisms have been proposed for the propagation of cracks in environment-sensitive cracking, but three are generally applicable[8] and will be described here. They are:

(a) pre-existing active paths;
(b) strain-generated active paths;
(c) adsorption-related mechanisms.

It is worth mentioning again that no single mechanism can account for all the experimental observations concerning instances of stress-corrosion cracking. As is usually the case where more than one mechanism can be proved, it is likely that each is applicable under differing service conditions.

Pre-existing active paths mechanism

In this mechanism, propagation is believed to occur preferentially along active grain boundary regions. The mechanism is essentially the same as for intergranular corrosion, already described in section 6.2. Grain boundaries may be anodically polarised for a variety of metallurgical reasons, such as the segregation or denudation of alloying elements. It is quite possible that dislocation pile-up can produce the same effect, although this may be less likely where SCC occurs at low stress levels, in which case the role of tensile stress may simply be to keep the crack open, and to allow electrolyte access to the crack tip region. The mechanism can be viewed as predominating in cases where the SCC is governed by electrochemical or metallurgical considerations, rather than the influence of stress. Systems which seem to fit well into this category are shown at the top of Table 10.1

Evidence in favour of such a mechanism is extensive. Most alloy systems in which grain boundary precipitates are present fail by intergranular cracking. The existence of an active path in unstressed mild steel has been shown by its destruction in boiling nitrate solution when anodic currents are applied. Similar confirmatory evidence correlating metallurgical structure in the grain boundaries with cracking propensity has been produced for aluminium/copper and aluminium/magnesium alloys with appropriate heat treatments.[13]

Strain-generated active paths mechanism

In contrast to the cases of cracking which are dominated by the influence of corrosion, there are many examples, some listed at the bottom of Table 10.1, in which strain is the controlling influence. Such instances have led to the development of the strain-generated active path mechanism. One feature of SCC described in section 10.1 was that in the absence of stress

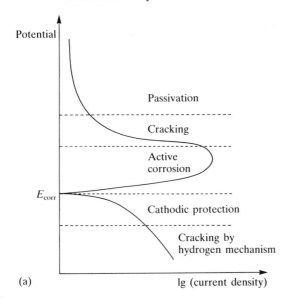

(a)

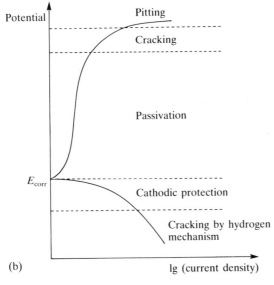

(b)

Fig. 10.5 The relationship of cracking susceptibility to the potentiodynamic polarisation curves for:
(a) a weakly passivating metal;
(b) a strongly passivating metal.

alone, the alloy is usually unreactive to the environment responsible for the cracking, normally because of the existence of a protective surface film. If crack propagation by dissolution occurs then the growth rate must be greatest at the crack tip where anodic dissolution occurs, rather than, say, at the sides of the crack which have been passivated because they have been exposed to the environment for a longer time. The mechanism is thus tightly linked to active–passive behaviour, which, in turn, has strong electro-chemical connections.

The strain-generated active path mechanism is based upon the idea of a strain-induced rupture of the film, followed by metal dissolution at the rupture. The propagation rate is governed by three criteria:

1. **Rate of film rupture**. This is determined by the applied strain rate, or, in the case of static loading, by the creep rate.

2. **Solution renewal and removal rates at the crack tip**. This diffusion-controlled process is also governed by the accessibility of the crack tip to the aggressive species.

3. **Passivation rate**. This is a vital consideration, for if repassivation is very slow, excessive metal dissolution can occur at both the crack tip and sides. The crack widens considerably and blunts, and the usual result is that the crack growth is arrested. Thus, in poorly passivating alloys, general corrosion rather than cracking is to be expected. Conversely, very rapid repassivation leads to slow propagation rates; it is at moderate repassivation rates that the greatest damage is caused. This concept suggests that the conditions which are most likely to result in cracking are those close to active/passive conditions, found from a potentiodynamic polarisation curve. Figure 10.5 shows two such curves, one for a weakly passivating and one for a more strongly passivating material. At potentials more negative than the free corrosion potential, both types of metals are, in the first instance, not susceptible to cracking. This corresponds with the well-known method of control of SCC by cathodic protection. However, at strongly negative potentials cracking which is usually attributed to hydrogen evolution may be observed. This is particularly dangerous in high-strength materials, which are frequently used in situations of high stress, and sounds a loud warning against overprotection by cathodic methods described in sections 16.4 and 16.5.

Adsorption-related mechanisms

When a weakly passivating metal is at potentials more positive than E_{corr}, general corrosion is observed as far as the passivation potential. It is the region immediately more positive than the passivation potential which represents the intermediate passivating rates most likely to result in cracking. A strongly passivating metal, however, shows rather different behaviour. Passivation occurs at potentials immediately more positive than E_{corr}, and it is not until potential is sufficiently anodic for the film to be destabilised (the transpassive region usually associated with pitting corrosion) that the cracking regime occurs. There is good experimental agreement with such a theory which offers good predictive ability.

In the past it was thought that hydrogen embrittlement could be distinguished from 'pure' SCC because the hydrogen effect could only occur if the specimen was cathodically polarised. Recently, evidence has been found that in the confines of a crack, the solution composition may bear little resemblance to that existing in the bulk. Thus even though a bulk material may seem to be outside the potential range for the evolution of hydrogen, the combination of pH and potential existing at the crack tip may allow such a cathode reaction. Thus the role attributed to hydrogen in the mechanism of SCC has increased greatly in latter years and care should be exercised in the interpretation of results.

Adsorption-related mechanisms propose that active species in the electrolyte degrade the mechanical integrity of the crack tip region, thus facilitating fracture of bonds at energies much lower than would be expected. In one mechanism, the aggressive ion which is specific for the case in question is thought to reduce the bond strength between metal atoms at the crack tip by an adsorption process which results in the formation of metal–species bonds. The energy used in binding the aggressor to the metal atoms reduces the metal–metal bond energy allowing mechanical separation to occur more easily. It is possible that the specific ion (which is normally unreactive towards the metal) is more reactive because of the increase of thermodynamic energy which occurs in the metal–metal bond as a result of the tensile stress. Figure 10.6(a) illustrates, in schematic form, a possible mechanism.

A second adsorption-related mechanism is based upon the formation of hydrogen atoms by the reduction of hydrogen ions within the crack. The hydrogen atoms are adsorbed by the metal and are thought to cause weakening, or embrittlement, of the metal–metal bonds just beneath the surface at the crack tip. A number of possibilities exist by which this can occur. Three of these are shown in Fig. 10.6(b). One presupposes the formation of metal hydrides, discrete chemical species which are well known for their brittle nature. In the past, hydride formation has been probably the most favoured mechanism for the general phenomenon of hydrogen embrittlement, though other mechanisms have been proposed. For example, it is also possible that decrease of the bond strength occurs by an adsorption process, similar to that described for the specific ion.

A third possibility is that hydrogen gas is formed in tiny amounts: the thermodynamic tendency for this to occur is very great. It has been shown that atomic hydrogen permeates steel, but combines to form hydrogen gas in cavities. The hydrogen molecule is unable to diffuse through the metal lattice and therefore the pressure in the cavities rises. The extreme pressures which can build up, given sufficient time, can rupture most materials. Steels are able to resist between 3000 and 20,000 atmospheres (0.3 to 2 GPa).[14] The pressure of hydrogen in a defect can exceed this, but crack growth would be expected to occur before such a pressure was reached. Any increase in pressure caused by hydrogen gas in a locally dilated area would augment the existing tensile stress and assist in crack propagation. Current

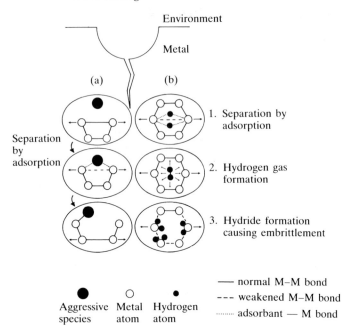

Fig. 10.6 Mechanisms for the propagation of cracks:
(a) The aggressive species is adsorbed at the crack tip and causes a reduction in the metal–metal bond strength.
(b) (1) Separation by adsorption of hydrogen in locally dilated areas immediately ahead of the crack tip.
(2) Hydrogen gas formed in locally dilated areas or along slip planes. Pressure of gas assists in rupture of metal–metal bonds.
(3) Formation of metal hydrides causes reduction in metal–metal bond strength and embrittlement of region ahead of crack tip.

thinking places the separation by adsorption mechanism as more likely than brittle failure caused by the hydrostatic pressures of build-up of hydrogen gas. The exact role of hydrogen in the embrittlement of metals at crack tips is still subject to much speculation and is likely to remain so for some time.

10.4 PRACTICAL ASPECTS OF ENVIRONMENT-SENSITIVE CRACKING

Stress-corrosion cracking is a complex phenomenon. The testing for material susceptibility to it requires a considerable understanding of all the factors involved. Methods currently available are numerous, and a full discussion appears elsewhere.[15] Only two important methods will be discussed because they have a bearing upon the discussion of the remainder of Chapter 10.

Mechanical tests for stress-corrosion cracking have evolved both for plain and precracked specimens. Methods developed by Parkins[15,16] have involved

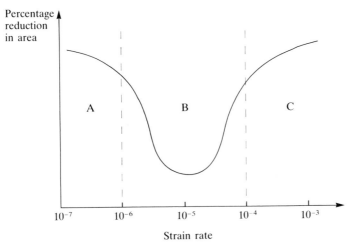

Fig. 10.7 The evaluation of susceptibility to stress-corrosion cracking at different strain rates. Measurements of the ratio of reduction in cross-sectional area for tensile specimens statically loaded in air and an aggressive environment show a marked change in region B. This corresponds to a reduction in ductility caused by SCC at the appropriate strain rates.

the slow straining of smooth specimens at a constant, pre-determined rate. Tensile test specimens are subjected to different constant strain rates while immersed in a test electrolyte and, after failure, the percentage reduction in area of cross-section is determined. Since a specimen which fails with a small reduction in area is considered to be less ductile than one failing with a large reduction in area, the parameter is a fair measure of cracking susceptibility. A typical set of specimens might produce a graph similar to that shown in the schematic Fig. 10.7.

At low strain rates, repassivation is fast enough and straining slow enough to have little effect upon the specimens. At high strain rates, the mechanical effect predominates over the corrosion processes: the environment does not have enough contact time with the specimen and the failure occurs by mechanical means only. At intermediate strain rates, usually about 10^{-4} to 10^{-6} s^{-1}, the environmental effect is greatest.

The slow strain rate test method is useful because it always produces a fracture surface which can be examined by traditional fractography: other tests on smooth specimens which do not result in fracture are often less effective. Although the time involved for the test is usually quite short, its major shortcoming is that it is largely qualitative.

Modern quantitative test methods utilise the recent developments in fracture mechanics to quantify crack growth rates of notched specimens. In Chapter 9, it was shown how a graph of stress intensity against crack growth rate could be used to determine the material property known as fracture toughness (Fig. 9.8). If a series of crack growth measurements are carried out both in the presence and absence of an environment which gives rise to SCC then data will be obtained similar to those shown in Fig. 10.8.

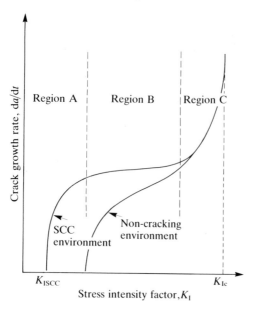

Fig. 10.8 Testing of susceptibility to SCC by crack growth rate measurements. Crack growth rates are plotted against stress intensity factor in air and corrosive environments.
Region A: At low values of K_I, rate is strongly dependent upon K_I. Extrapolation indicates a threshold level, K_{ISCC}.
Region B: Crack growth almost independent of stress intensity factor and controlled by corrosion processes.
Region C: Crack growth occurs by mechanical mechanism and approaches the fracture toughness, K_{Ic}.

The data obtained from the experiments in the presence of the corrosive environment are usually divided into three regions:

Region A. Crack growth is strongly related to stress intensity but drops very rapidly, nominally to zero. Extrapolation indicates the existence of a threshold stress intensity, below which crack growth does not occur.

Region B. In this region there is little, if any, dependence upon stress intensity: crack growth occurs at almost a constant rate which is faster than that in the control environment.

Region C. Here the mechanical duress is so great that there is little effect of the environment. Fracture is strongly dependent upon stress intensity and follows closely the behaviour of specimens tested in the control environment.

The difference in behaviour which results from the presence of the specific environment is clear from Fig. 10.8: for a susceptible material, crack growth rates show a general shift upwards and to the left of the data obtained in an inert environment. However, as the value of stress intensity factor increases, the deviation of the two curves becomes less and approaches a behaviour consistent with purely mechanical fracture.

The discovery of the existence of a threshold stress value is an important advance since it indicates that the idea that stress must be completely eliminated before control of SCC can be achieved is not quite true: it is sufficient to reduce the stress levels below the threshold. This significant material property (which only relates to the environment specified) is denoted by the symbol K_{ISCC}, and the threshold stress which relates to it is given the symbol, σ_{TH}.

A major advantage in the determination of a threshold stress is that it allows the calculation of the maximum flaw size which can be tolerated in the material/environment pair. Fracture mechanics allows the calculation of a tensile strength for any given flaw size, a. Brown[10] has shown that

$$a = 0.2 \left(\frac{K_{ISCC}}{\sigma_y} \right)^2 \qquad [10.1]$$

Figure 10.9 shows the relationship of tensile strength and critical crack size in schematic form. On the basis of the threshold stress, σ_{TH}, a maximum design stress, σ_{max}, will be evaluted, from which a value of critical crack size is found. If the smallest defect which can be measured by non-destructive examination is a_{nde}, it is vital that $a_{max} > a_{nde}$, to ensure that routine examination will always find cracks that can lead to failure and allow remedial action to be taken before failure occurs.

Experiments on steels in sea water have shown that K_{ISCC} decreases as yield stress increases, from which it can be deduced that SCC is much more of a problem in high-strength steels.

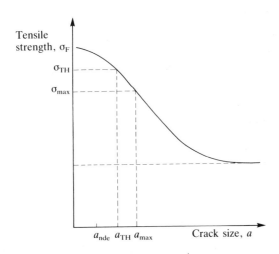

Fig. 10.9 The effect of critical crack size upon tensile strength.

10.5 CORROSION FATIGUE

There are many similarities between corrosion fatigue and SCC, but the most significant difference is that corrosion fatigue is extremely non-specific. Mechanical fatigue affects all metals, causing them to fail at stress levels well below those at which static stress leads to failure. In aqueous environments, it is frequently found that a metal's fatigue resistance is reduced. This makes corrosion fatigue a common, and dangerous, form of corrosion.

The stages in the development of fatigue cracks are thought to be as follows[11]:

(a) The formation of slip bands leading to intrusions or extrusions of material;
(b) The nucleation of an embryo crack approximately 10 μm long;
(c) Extension of the embryo crack along favourable directions;
(d) Macroscopic (0.1 to 1 mm) crack propagation in a direction perpendicular to the maximum principal stress, leading to failure.

Furthermore, examples of corrosion fatigue may be considered as taking place in one of three different categories:

1. **Active:** freely corroding, such as a carbon steel in sea water;
2. **Immune:** in which the metal is protected either cathodically (Ch. 16) or with a coating (Ch. 14);
3. **Passive:** in which the metal is proteced by a corrosion-generated surface film, usually an oxide.

The discussion which follows will refer to the conditions of corrosion and crack growth defined above.

Figure 10.8 showed that under SCC conditions, crack growth rates are enhanced at lower stress levels by a greater amount than at values approaching K_{Ic}. Under corrosion fatigue conditions the tolerable stress levels leading to similar crack growth rates are lower still. Figure 10.10 shows the fatigue and corrosion fatigue characteristics of a typical low alloy steel in both inert and aqueous sodium chloride environments. In the inert environment the behaviour is as discussed in section 9.5, but when the aqueous environment is introduced, it is apparent that the effect is greater at lower stress levels; at high stress levels the behaviour is more akin to a mechanism of crack growth by purely mechanical means. The corrosion fatigue curve can be conveniently divided into three regions, as was done for the SCC crack growth curve in Fig. 10.8, i.e. A: initiation, B: propagation and C: failure.

An indication of the threshold value for SCC is shown in Fig. 10.10, from which it can be judged that corrosion fatigue can occur at stress levels much lower than those for SCC. Whereas SCC crack growth rates in regime B are usually independent of stress intensity factor (parallel to the *x*-axis), this is not so for true corrosion fatigue: behaviour is usually in accordance with

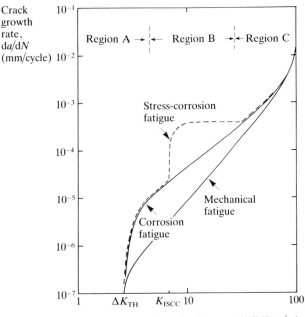

Fig. 10.10 The general characteristics of corrosion fatigue curves.

the Paris Law, as stated in eqn [9.11], except when SCC behaviour may be superimposed upon that of corrosion fatigue. In this case the graph would appear similar to the dashed line in Fig. 10.10. The onset of SCC occurs at stress levels corresponding to the threshold level and causes a further increase in the rate of crack growth. Observation of a 'plateau' on the curve is typical of such a dual mechanism, which is frequently referred to as stress-corrosion fatigue, even if the effect may be attributable to a hydrogen embrittlement mechanism.

An example of such behaviour is found in the corrosion fatigue behaviour of structural steels used for offshore platforms.[11] The use of cathodic protection on such structures has been reported as giving favourable corrosion fatigue properties. The metal is in the immune condition because of the applied potential, but additionally is protected by the formation of calcareous deposits (Ch. 16) which tend to plug incipient cracks. No evidence was found for the lowering of the crack growth threshold, but superimposed upon the crack growth by corrosion fatigue was a plateau which corresponded to the onset of increased propagation rates by hydrogen embrittlement. Although the overall impression of designers is that offshore platforms in service in 1985 have good corrosion fatigue resistance, the reliance upon protection in conditions where the complex stressing may lead to break-up of the calcareous film, or unexpected overprotection and increased hydrogen evolution, is considered to be unwise.[11]

Environment-sensitive cracking

The observation of threshold levels for corrosion fatigue cannot be relied upon. In some systems, it is possible to define such a threshold which can be used as a design parameter. In many structural metals, it can be as low as 2 MPa√m, even in air.[11] The factor which most affects the threshold level is whether the environment causes crack initiation or not. In welded steels it is thought that sufficient defects are always present for initiation processes to have already occurred and for crack propagation to dominate the corrosion fatigue behaviour.

In smooth specimens, the behaviour is quite different because it is necessary for the environment to act conjointly with stress to carry out phases (a), (b) and (c) defined above. Good evidence of this is provided by the differences observed in the behaviour of smooth and notched specimens in corrosion fatigue endurance (*S–N* curves) tests. When smooth specimens of a low alloy steel were tested in air and sodium chloride solutions there was a severe reduction in the fatigue resistance in the aqueous, compared to the air, environment, Fig. 10.11(a). When notched specimens were used, the relative effect of the two environments was much reduced, with the notched specimens behaving more like the smooth specimens in salt water, Fig. 10.11(b). From this it was concluded that the main effect of the environment in plain specimens is to introduce stress concentrations with virtually the same effect as a machined notch. Such results highlight the fact that preservation of a good surface finish by effective corrosion protection measures is an effective way of reducing the susceptibility of materials to corrosion fatigue.

The complexity introduced into corrosion fatigue by stress cycling makes the practical investigation of corrosion fatigue behaviour even more difficult than analysis of SCC resistance. In general, it is possible to examine two contributions to the corrosion, one made by the frequency of the cycling, and the other made by the mean stress.

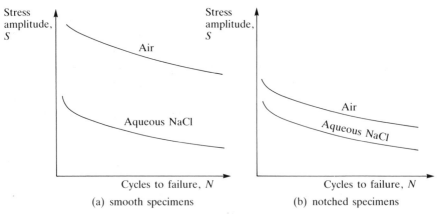

Fig. 10.11 Corrosion fatigue endurance curves for a typical low-alloy steel in air and aqueous sodium chloride.
(a) Smooth specimens.
(b) Notched specimens.

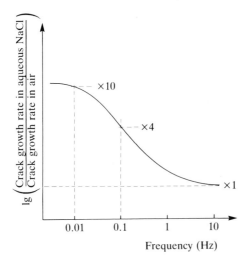

Fig. 10.12 The dependence of crack growth rate upon frequency.

The effect of frequency on the crack growth rate in a structural steel in sea water is shown in Fig. 10.12. At 10 Hz there is no significant environmental effect, and the crack growth is by mechanical, rather than corrosion, fatigue. For the same stress intensity, the crack growth maximises at about 0.01 Hz, at a rate 10 times greater than in air. This is easily explained because the corrosive environment is allowed a greater interaction time at low frequencies. The importance of carrying out tests at the correct frequency is thus obvious: the frequency with which a marine structure is likely to be stressed by wave action will be in the range 0.1 to 0.01 Hz and not 10 Hz.

It is thus apparent that in order for an environment to have a significant effect upon a fatigue process, sufficient time must be given for corrosion processes to occur. The dangers of this are apparent when the following example is studied. A simple fatigue experiment at 25 Hz which runs to a million cycles (a normal number for such experiments) can be completed in 11 hours; at 0.01 Hz the same experiment takes over three years. The use of accelerated testing and extrapolation of laboratory data obtained over short time periods to the tens of years for most design lifetimes requires the utmost caution.

The mean stress is a most important variable because different values can be used for constant ΔK. As has been shown, tensile mean stresses are detrimental to corrosion fatigue resistance if the frequency is in the range of maximum effect. As the mean stress is increased, for the same ΔK (i.e. R increases to more positive values), the crack growth rate increases also. Compressive mean stresses have been shown to be beneficial in carbon steels.[17] Corrosion fatigue resistance was substantially increased in both air and hydrochloric acid by using a compressive mean stress at low frequency.

This lends support to the previously well-known beneficial effect of shot-peening, a process applied to the surface of metal components which has been shown to result in compressive stresses in a region of the material 50–75 μm below the surface. Other methods of introducing such static compressive stresses into structures or components will probably be an important means of protecting against corrosion fatigue. In an environment which may cause pitting, however, this will not help.

It should not be assumed that the traditional *S–N* type testing has been made obsolete by the introduction of more modern fracture mechanics techniques. Corrosion fatigue endurance testing continues to play an important role in lifetime determinations. This is because there are still many occasions when methods of fracture mechanics are not sufficiently accurate to describe the behaviour. In a definitive description of current corrosion fatigue theory, Scott[11] has described the usefulness of considering a combination of crack growth rate data and endurance testing. By assuming crack growth rates in accordance with the Paris Law,

$$da/dN = C\Delta K^m \qquad [9.11]$$

where *m* and *C* are found experimentally, fatigue endurance limits can be calculated for different initial and final crack lengths, a_i and a_f respectively, based upon a typical low value threshold stress intensity for crack propagation of 2 MPa $\sqrt{m}$. By means of the equation:

$$N = \frac{1}{C} \int_{a_i}^{a_f} \frac{da}{\Delta K^m} \qquad [10.2]$$

$$\text{where } \Delta K = \Delta \sigma \sqrt{(\pi a)} \cdot f(a/w) \qquad [10.3]$$

endurance values are calculated and plotted on the practical *S–N* curve to give a lower bound for corrosion fatigue behaviour, Fig. 10.13. By such analysis it is possible to estimate the errors which may be present in the fracture mechanical approach because of the adoption of an inaccurate compliance function ($f(a/w)$, eqn [10.3]) or a failure to describe the growth of short cracks. A design curve which incorporates a safety factor can be defined for a particular case.[18] For example, for a nuclear pressure vessel the safety factors used are 2 for stress and 20 for cyclic life, below the mean of the experimental data. Scott has shown that the worst corrosion can do is to reduce fatigue life to that determined by the assumption that lifetime is controlled by macroscopic crack growth.

Despite the advances made in the understanding of corrosion fatigue, it remains one of the most complex and least understood of all forms of corrosion. It has been stated[4] that more research into corrosion fatigue in the oil, gas, nuclear, geothermal and wind power industries is of a high to critical importance if vital resources are not to be wasted.

A useful schematic diagram which summarises the effect of stress upon crack size is shown in Fig. 10.14. While it should be remembered that there may be considerable overlap of the corrosion fatigue and SCC regimes, the diagram is a good representation of the effects of environment-sensitive cracking.

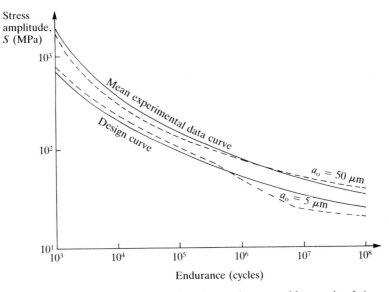

Fig. 10.13 Comparison of integrated crack growth curves with corrosion fatigue endurance curves.

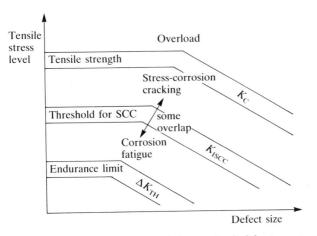

Fig. 10.14 Schematic diagram relating mechanical fracture, stress-corrosion cracking and corrosion fatigue with increasing defect size.

10.6 REFERENCES

1. Moore H, Beckinsale S, Mallinson C E 1921 The season cracking of brass and other copper alloys, *Journal of the Institute of Metals* **25**(1): 35–152
2. Parr S W, Straub F J 1927 The cause and prevention of the embrittlement of boiler plate, *The Engineer* **117**: 496

3. Spahn H, Wagner G H 1976 Corrosion fatigue of steels, in *Bruchuntersuchungen und schadenklarung, probleme bei eisenwerkstoffen*. Allianz Versicherungs-AG, Munich, pp 59–74 (in English)

4. MacDonald D D, Begley J A, Bockris J O'M, Kruger J, Mansfeld F B, Rhodes P R, Staehle R W 1981 Aqueous corrosion problems in energy systems *Materials Science and Engineering* **50**: 19–42

5. Collins J A, Monack M L 1973, *Materials Protection* May: 22

6. McReynolds R F, Vennett R M 1975. *Materials Performance* **14**(1): 23

7. Seibert O 1983 Classic blunders in corrosion protection revisited, *Materials Performance* **22**(10): 9–12

8. Ford F P 1984 Current understanding of the mechanisms of stress corrosion and corrosion fatigue, in Dean S W, Pugh E N, Ugiansky G M (eds), *Environment-sensitive fracture*. ASTM-STP 821, pp 32–51

9. Parkins R N 1976 Mechanisms of stress-corrosion cracking, in Shreir L L (ed.), *Corrosion* (2nd edn, Vol. 1). Newnes-Butterworths, p 8:24

10. Brown B F 1972 A preface to the problem of stress corrosion cracking, in Craig H L Jr (ed.) *Stress-corrosion cracking of metals – a state of the art*. ASTM: STP 518, pp 3–15

11. Scott P M 1983 Design and inspection related applications of corrosion fatigue data, in *Memoires et etudes scientifiques, Revue de metallurgie*, pp 651–60

12. West J M 1980 *Basic corrosion and oxidation*. Ellis Horwood p 158

13. Doig P, Edington J W 1974 *British Corrosion Journal* **9**: 88

14. West J M 1980 *Basic corrosion and oxidation*. Ellis Horwood, p 150

15. Parkins R N 1976 Mechanisms of stress-corrosion cracking, in Shreir L L (ed.), *Corrosion* (2nd edn, Vol. 1). Newnes-Butterworths, pp 8: 139–65

16. Parkins R N 1979 Development of strain rate testing and its implications, in Ugiansky G M, Payer J H (eds), *Stress corrosion cracking: the slow strain rate technique*. ASTM-STP 665, pp 5–25

17. Endo K, Komai K 1973 Metalloberflache, *Angewandte Elektrochemie* **27**: 378–82

18. Scott P M, Tomkins B, Foreman A J E 1983 Development of engineering codes of practice for corrosion fatigue, *Journal of Pressure Vessel Technology* **105**, Aug: 155–262

10.7 BIBLIOGRAPHY

ASTM 1972 *Hydrogen embrittlement testing*. ASTM-STP 543

Craig H L Jr (ed) 1972 *Stress-corrosion cracking of metals – a state of the art*. ASTM-STP 518

Craig H L Jr (ed.) 1976 *Stress-corrosion – new approaches*. ASTM-STP 610

Crooker T W, Leis B N (eds) 1983 *Corrosion fatigue: mechanics, metallurgy, electrochemistry and engineering*. ASTM-STP 801

Dean S W, Pugh E N, Ugiansky G M (eds) 1984 *Environment-sensitive fracture: evaluation and comparison of test methods*. ASTM-STP 821

Karpenko G V, Vasilenko I I 1979 *Stress-corrosion cracking of steels* (2nd edn). Tel Aviv: Freund Publishing House

Logan H L 1966, *The stress-corrosion of metals*. New York: John Wiley

Scully J C (ed.) 1971 *The theory of stress-corrosion cracking in alloys*. Brussels: NATO

Shreir L L (ed) 1976 *Corrosion* (2nd edn. Vol. 1). Newnes-Butterworths, section 8: Effect of mechanical factors on corrosion

Staehle R W 1969 *Comments on the history, engineering and science of stress-corrosion cracking*, in *Proceedings of conference on fundamental aspects of stress-corrosion cracking*. NACE, p 34

Ugiansky G M, Payer J H (eds) 1979 *Stress-corrosion cracking – the slow strain rate technique*. ASTM-STP 665

Yahalom J, Aladjem A (eds) 1980 *Stress-corrosion cracking*. Tel Aviv: Freund Publishing House

11 BASIC PRINCIPLES OF CORROSION CONTROL

Treppenwitz – loose translation: those things or statements which we wished we had done or said beforehand, especially if something later went wrong.

(Oliver W Siebert: *Materials Performance*, April 1978)

It is easy to be wise after the event. Engineers seem to be prepared to spend long hours on stress calculations and the layout and styling of a structure, but not to give the same attention to the selection of materials, or the elimination of design features which promote corrosion. Later they are amazed, and even offended, by the corrosion problems generated amidst the bad design features and arrays of incompatible materials they have specified.

In Chapter 1 we saw numerous examples of how poor corrosion control leads to considerable extra costs in replacement materials, or post-production remedial action to try to prevent further deterioration of the structure.

Fact: In the United States it was estimated that in 1975 about 40 per cent of the steel production was used to replace parts which had rusted and 92 million gallons of maintenance paint for use on steel was sold.

Corrosion has already been defined as the degradation of a metal by electrochemical reaction with its environment. In most practical situations this attack cannot be prevented, it can only be controlled so that a useful life is obtained from the structure.

There are *three* significant stages in the lifetime of a component: **design, manufacture** and **use**. Corrosion control plays an important role in each of these stages, summarised in Fig. 11.1. The failure of *any* of these aspects of corrosion control will probably result in premature failure of the component.

Control may be exercised in a number of ways but the most important are:

(a) modification to the design;
(b) modification to the environment;
(c) application of barrier coats;
(d) selection of materials;
(e) cathodic or anodic protection.

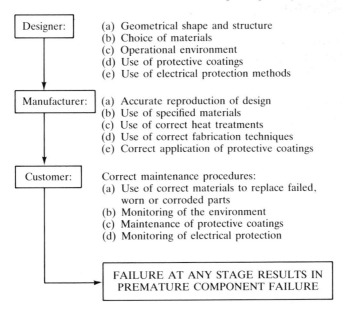

Fig. 11.1 Considerations of corrosion control in the design, fabrication and service life of a component.

The control method should always be considered as part of the overall design concept for the structure, taking its place alongside other design parameters such as stress calculations, creep and fatigue, and fabrication techniques. Its effectiveness should be monitored throughout the life of the structure, including periods spent in storage or transit, as well as in service.

The reliability of the corrosion control method selected can influence other design parameters by:

(a) ensuring that a structure has adequate safety margins to fulfil its function for the design life of the structure;
(b) allowing overdesign to be minimised, resulting in thinner sections, lower weight and reduced capital cost.

In transport systems, for example, the running costs will also be reduced because of the lighter structures.

A typical example of increased capital cost and reduced profitability follows.

Case 11.1: Corrosion occurred in a buried oil pipeline 360 km long, which had a 200 mm external diameter. A wall thickness of 6.3 mm was adequate to cope with the stress levels imposed on the pipe, but to allow for corrosion this was increased to 8.3 mm. This corrosion allowance used an extra 3000 tonnes of steel and caused a loss of 4 per cent carrying capacity in the line.

The type of corrosion attack and the rate of metal loss are strongly dependent upon the environment. Small changes in the environment to such factors as pH, temperature, dissolved oxygen levels, flow rates and pollutants can radically alter the nature and severity of corrosion. In selecting a control method particular attention must be paid to any changes in the environment which may be expected during the life of the structure, or any periods of exposure to particularly hostile environments which may occur during fabrication, erection or maintenance.

Copper-based alloys, which are known to give excellent service in seawater systems on ships, have often failed during the first few months of service. The culprit is a thin film of sulphide which develops on the metal surface during exposure to polluted estuarine water in the final stages of fitting out the vessel. The sulphide is cathodic to the copper alloy and produces severe pitting corrosion at breaks in the film. Once the normal passive film has developed after exposure to fresh sea water over a period of time, the alloys become immune to this form of attack. It can be controlled during the fitting out stage by adding 0.01 per cent of an inhibitor, sodium dimethyldithiocarbamate, to the polluted sea water.

Corrosion damage may only become apparent some months, or even years, after the event which initiated the problem. In Chapter 7 we saw the example of pitting corrosion in the walls of copper tubes in a domestic hot water system. Prior to dispatch from the factory the tubes had been annealed. Lubricating fluid residues in the bores turned to carbon, which caused pitting at breaks in the cathodic film.

Case 11.2: A series of stainless steel heat exchangers, cooled with recirculating treated water, had given over ten years problem-free service. However, in an emergency, untreated river water was used to cool the units for 48 hours. Several weeks later, five units failed because of stress-corrosion cracking – a classic example of the 'It's only for a short period' syndrome.

Careful consideration of the whole structure at the design stage will predict many of the areas of the system which are likely to corrode. The designer can then make provision for easy access for inspection, maintenance or replacement of corroded sections at these points. Indeed, the designer, acknowledging that corrosion is inevitable, may deliberately cause certain sites to become the anodes in the total system, thus giving sacrificial protection to those areas which are less accessible. Having introduced a system of corrosion control, rigorous checks must then be applied during manufacture, storage and in-service use to ensure that subsequent modifications do not upset the design philosophy by replacing these 'persistent areas of corrosion failure' in a more noble metal, thus driving the corrosion to another part of the system where access may not be so easy.

Case 11.3: A fitter, tired of replacing the mild steel shaft in a sea-water pump every two years, installed a stainless steel one. This suffered crevice corrosion beneath the impeller and resulted in failure of the shaft in six months. The remainder of the pump was badly pitted having lost the sacrificial protection afforded by the mild steel one.

In some cases a corrosion protection system may be superfluous owing to the nature of the task to be undertaken by the equipment. There is little point in providing an externally applied corrosion control system on the blades of earth-moving machines where abrasion damage will be much faster than corrosion losses. However, such units still require protection during storage, rest period and transit. Such protection is often neglected, particularly when working in remote geographical regions, or under adverse weather conditions.

The cost of a corrosion control system must be balanced against the required life of the structure, the capital and maintenance costs and the scrap value of the materials. While the engineer will aim to achieve maximum cost effectiveness, other factors may influence the decision over the choice of a system. These range from aesthetic appeal (rust, even on non-essential components in a unit, will create sales resistance) to the overriding need for reliability of performance in safety equipment or weapons systems.

The true cost of failure may be much greater than that of replacing the damaged parts. The loss of confidence by the public in a motor car manufacturer's product once a reputation for poor corrosion resistance is established is catastrophic for the company. The incidental cost of failure can far outweigh the direct cost. For example, when corrosion caused the malfunction of a signal unit at Clapham Junction on the railway system serving London, a repair was effected at a very small direct cost, but the disruption to the railway commuter service was enormous and entailed thousands of lost working hours.

Corrosion control seeks to regulate the rate of corrosion, keeping it within acceptable, or at least predictable, limits for the life of the structure. The chapters which follow examine five different aspects of corrosion control. Though they are separated for logical discussion, the topics are inter-related as the case histories are aimed to show.

12 CORROSION CONTROL BY DESIGN

Now in the building of chaises, I tell you what,
There is always, somewhere, a weakest spot.

(Oliver Wendell Holmes, *The Deacon's Masterpiece*)

In far too many engineering structures the 'weakest spot' is the lack of consideration given to corrosion control during the design stages. To an onlooker with even an elementary knowledge of corrosion principles, it seems that corrosion is designed into, rather than out of, many structures.

The principles of good design which minimise corrosion problems have been known for many years. The introductory paragraphs of Chapter 1 showed how the most basic errors leading to bimetallic corrosion are still being made more than 200 years after the knowledge to prevent it became available. It is a sad comment on the education system, and on the ability of engineers to communicate with each other, that the principles are continually flaunted with predictable, disastrous results.

In this chapter we will discuss a number of principles and design features. They are not listed in any order of merit, and their relative importance will vary from project to project. As section 15.9 describes, there are many aspects to the overall design of engineering components and structures, of which corrosion control is a small but essential part. This chapter aims to give the main features which should lead to a satisfactory design life.

12.1 HOW LONG DO I EXPECT IT TO LAST?

Having established the life to be expected from a component or structure, this must be compared with the life of the corrosion control system which is to be used. If the control system life is shorter than that of the structure then the method of renewal should be considered at the design stage, and provision made for any special access which may be required for inspection, maintenance and replacement. Should the control system of part of the structure fail for any unforeseen reason before the life required for the whole structure, the question must be asked, 'Will the difficulty or cost of replacing the part become the life-limiting factor for the whole?'

The application of a coating may be a routine process in a factory where the temperature and relative humidity can easily be controlled, and the standard of surface preparation and the quality of the coating also can be closely monitored. Renewal of the coating in an exposed situation under adverse weather conditions could be a very different proposition.

Case 12.1: The 500 mm square box section of a 3 m-long steel strengthening beam was successfully coated in a factory by dipping. It was a very different matter to renew the internal coating after the beam had been installed in a lattice girder bridge spanning an estuary.

If it is not possible to renew the protection system then the original application must be capable of withstanding all the hazards of fabrication and installation which occur after it has been applied, as well as lasting the lifetime of the structure.

Case 12.2: Figure 12.1 shows a set of galvanised railings installed at a car park located about 500 m from the sea. These should have given a maintenance-free life of 12–15 years. However, rusting had occurred around each of the welds only weeks after installation. No post-welding protection had been applied to compensate for the galvanising burnt off during during fabrication. The sound coating on the adjacent surfaces was sacrificially corroding to protect the bare steel, welded areas.

When the time for replacement arrives, the performance of the existing structure and materials should be considered before a change to more expensive or more corrosion resistant materials is made.

Fig. 12.1 Case 12.2: corrosion at welding damage on galvanised steel railings.

Case 12.3: A lead-lined wooden tank used to mix sulphuric acid and zinc chloride gave satisfactory service for over a hundred years. When it was replaced, a tank made from the same materials was unacceptable to the owners despite the long service of the original tank.[1]

It is uneconomic for the life of some elements in a plant to be much greater than that of the whole plant if they will be scrapped with it. Fitting a stainless steel exhaust system to a motor car may not be cost effective if the life of the exhaust is much longer than that of the car. Conversely, the fitting of mild steel exhausts with a life of two to three years is clearly not satisfactory.

12.2 THE EVER-CHANGING ENVIRONMENT

Components will be exposed to a wide variety of environments during the different stages of manufacture, transit and storage, as well as the daily and seasonal variations experienced in service. If the structure is mobile the changes in exposure conditions will be even greater.

The rate of corrosion or deterioration of protective coatings applied to the metal will be affected by changes in factors such as:

(a) relative humidity;
(b) temperature;
(c) pH;
(d) oxygen concentration;
(e) solid or dissolved pollutants;
(f) concentration;
(g) electrolyte velocity.

These variations in the environment must be identified, as far as possible, at the design stage. Clearly, steps will be taken to control the in-service corrosion, but special protection may be required during a particularly hazardous, if temporary, phase in production, transit or storage.

Coal tar enamels and bitumen coatings are often used to control the corrosion of steel pipes which will be buried in the earth. However, if the pipes are stored or transported in bright sunlight prior to burial, the coating can degrade, leading to a shorter in-service life than predicted. To guard against this deterioration, the coating may be overpainted with high-density emulsion paint, which is superfluous once the pipe is buried.[2]

Corrosion damage during storage is a common problem. It can be caused by the microclimate generated inside packing cases and store rooms, in systems such as the drained water cooling circuits of stored petrol and diesel engines, or in boilers and condensers at power stations during shutdown. It may also occur in components stacked in warehouses when the free circulation of air is impeded. General climatic changes also generate damage in unheated warehouses which do not have humidity controllers.

The volatile constituents of wood pose particular hazards for metals stored in wooden containers. While this is dealt with more fully in section 13.2, it should be noted that all closed wooden crates must have a sealed lining to prevent deposition of acid films on stored components.

Case 12.4: An example of the generation of local climatic conditions, the significance of which was not appreciated until considerable damage had been caused, involved cold drawn brass tube. The tube was stored in a shed adjacent to a workshop. Birds roosted on the rafters inside the shed and the ammoniacal content of their droppings caused severe stress-corrosion cracking of the tubes.

A factor which can affect in-service performance is the interaction between adjacent installations, or between apparently unrelated parts of a complete system. Units which would give an entirely satisfactory corrosion performance when considered as separate entities, have failed catastrophically owing to environmental effects caused by their relative positions on a site or in a plant. When two or more design offices are responsible for the individual parts of a system, such risks are exacerbated. Care is needed to ensure that there is sufficient liaison and understanding between them, so that the parts are mutually compatible with each other and with installations which already exist on the site.

Case 12.5: A copper lightning conductor earthing a chimney was separated from an aluminium roof by a vertical gap of 6 metres. When it rained, or mist precipitated on to the conductor, a small amount of copper dissolved in the water as it ran down over the metal. When the water dripped on to the roof panels, ion exchange plated out the copper on to the aluminium and severe corrosion of the roof resulted. The problem was eliminated after consultation between the electrical department and the plant design office: the conductor was moved to the other side of the chimney.

Case 12.6: An aluminium aerial support was quite adequate for the designed task until it was placed just behind the funnel of a ship. Acid produced from the burnt fuel oil removed the paint from the support, while carbon deposits from the smoke coated it with a very effective cathode. Corrosion soon rendered the assembly unsafe.

In our modern society, rules and regulations proliferate. Considered in isolation, one set of safety regulations may be consistent and beneficial, but severe hazards can be produced when they interact with those of another organisation.

Case 12.7: Health regulations require farm milking machines to be washed regularly, while electrical safety regulations require the milking machines to be earthed. On a Devon farm, thorough washing of a machine caused corrosion between the copper earthing strap and the stainless steel body of the machine. The electrical resistance of the joint between the two became so high that the earth failed to function when a fault developed in the machine. A cow being milked at the time was electrocuted.[3]

12.3 AVOID ALL UNNECESSARY BIMETALLIC CORROSION CELLS

It is not true to say that all dissimilar metal couples must prove disastrous. There are countless examples of bimetallic couplings which have given many years of useful service although the metals involved are widely separated in the galvanic series. Corrosion of the coupling will occur only if a galvanic cell is formed. You will recall that a basic wet corrosion cell consists of four parts. If one part of the cell is eliminated, corrosion ceases. Although two dissimilar metals provide the possible anode and cathode in a bimetallic coupling, corrosion control is achieved by preventing access of an electrolyte to the joint, or alternatively the joint may be insulated by breaking the coupling to stop the flow of electrons between the two metals.

Case 12.8:   The aluminium valve body shown in Fig. 12.2 has bronze inserts in the top face to take the spindle and securing bolts for the valve cover. Although this would appear to be most unsatisfactory, no corrosion has occurred around the bimetallic couples where only three parts of the cell are present – the gasket and valve cover effectively prevent any electrolyte from reaching the metals. At the flange, however, where brass bolts were used to secure the two halves of the valve, the free access of water completed the cell and caused severe galvanic corrosion of the aluminium.

Wherever two different metals are in contact, bimetallic corrosion will always remain a possibility should the remaining parts of the cell develop in service.

Case 12.9:   Cold liquid flowing in a steel pipe which passed through a heated room caused condensation on the outside of the pipe. This gave rise to general corrosion of the exterior of the pipe as the paint film deteriorated. The pipe was adjacent to and in contact with a brass valve and also suffered severe bimetallic corrosion on the external surface. Lagging the pipe to prevent condensation of the electrolyte cured the problem.

Sometimes the cell develops through unexpected routes, either over long distances, where the conducting path is hidden and escapes attention, or through alternative routes which bypass the careful arrangements of the designer. Tunnel vision which concentrates the designer's attention in one area only is dangerous.

Fig. 12.2 Case 12.8: aluminium valve body with bronze inserts. There is no corrosion around the bronze inserts on the top face where the gasket prevented the ingress of any electrolyte. On the flange there is heavy corrosion where there was free access of electrolyte to the brass securing bolts.

Case 12.10: A dry dock was constructed from steel-reinforced concrete with the steel buried well inside the concrete. In the bottom of the dock a number of iron supports were mounted to assist in the location of ships entering the dock. Around the top of the inside of the dock a 90/10 cupro-nickel pipe system was installed. The pipe and the supports were in contact with the steel reinforcing bars inside the concrete. When the dock was flooded, the sea water covered the pipe and completed the circuit to the iron which corroded so vigorously that it could be heard fizzing!

Case 12.11: In a plumbing system a mild steel pipe had to be coupled to a copper one. A ceramic insulator was used to insulate the pipes from each other. However, the carefully engineered assembly was attached by metal brackets to a wall panel made of sheet metal. The panel bypassed the insulated joint and allowed electrons to flow from the steel to the copper. The steel pipe leaked in a very short time.[4]

In practical engineering it will be impossible to eliminate every bimetallic cell. Where it is necessary to accept a cell in a design, steps can be taken to minimise the corrosion damage:

(a) Anodes should always be kept as large as is practical in the particular component or location to reduce the current density.

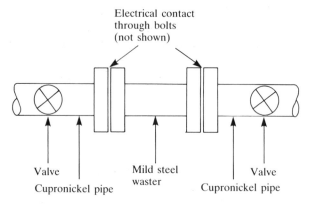

Electrical contact
through bolts
(not shown)

Valve
Cupronickel pipe

Mild steel
waster

Valve
Cupronickel pipe

Fig. 12.3 Sacrificial waster inserted in pipe system. The waster is made from a metal which is anodic to the remainder of the system.

(b) If the electrolyte flows through the system, the anodes should be kept upstream from the cathodes to prevent ion exchange plating out local cathodes on to the anode. Local cathodes cause pitting of the anode (see Ch. 7 and case 12.5). Indeed, in a flowing system short pieces of pipe which are anodic to all the metals in the remainder of the system may be introduced, Fig. 12.3. These act as *sacrificial wasters* and corrode to protect the other metals. It is common practice to put a valve on each side of the waster and fit it with flanged joints so that it can be removed without draining the whole system. Wasters should always be located in positions where they are accessible and there is room to replace them.

(c) The electrolyte can be modified to make it less aggressive. This will be discussed in section 13.3.

(d) In all cases where dissimilar metal joints are inevitable the first aim should be to isolate both sides of the joint from the electrolyte by applying a barrier coating. Alternatively, the metals should be insulated from each other to prevent the flow of electrons across the joint.

12.4 AVOID DIFFERENTIAL-AERATION CELLS

While bimetallic corrosion is destructive, cells caused by oxygen differentials in the electrolyte are more widespread and cause greater damage. In section 7.2 we discussed the development of the differential-aeration cell. One zone of the electrolyte which is low in oxygen produces an anode on the metal surface, while that which is richer in oxygen produces a cathode, even though the materials are identical. This effect was demonstrated in expt 3.11. A differential in oxygen level can develop in any situation where water is in contact with a surface. Therefore, every effort should be made to prevent the ingress of moisture or free-standing water to areas where it will be held for long periods on the metal surface. Potential moisture traps should be sealed, fitted with drainage holes, or ventilated to dry off the water. Surfaces in free contact with water should be protected with barrier coats or cathodic protection systems.

Examples of differential-aeration cells are legion. Some of those which cause the most damage are described below.

Crevices

Any point at which two metal surfaces are separated by a narrow gap is a possible cell. Moisture enters the gap, often drawn in by capillary action. Where the liquid is in contact with air, the oxygen is replenished but the centre of the water film becomes impoverished in oxygen and corrosion occurs at that point. Crevices are formed behind spot-welded overlays or butt joints, under the rim of sheet metal which has been folded to give a smooth outer edge, at bolted or riveted joints and at shingled or overlayed plates. Crevice cells have also been formed beneath loose, inert fastenings, string or wire wrapped round metal supports (expt 3.15), and behind tallies and labels improperly secured to metal surfaces. Rubber gromets have produced severe crevice corrosion on stainless steel shafts and couplings. Examples of some aspects of good and bad design are shown in Fig. 12.4.

Crevice-type attack is found in many areas of the modern motor car and is the factor which often limits the life of the vehicle. It occurs in seams, box sections, the bottom of doors, behind overlay trims and fasteners, and under the folded edges of boot lids and doors. Additional cells are created by owners who embellish the vehicle with fittings such as spotlights clamped to bumpers, advertising stickers behind which water accumulates, retro-fitted spoilers and other individual trims. The site of the corrosion is usually inside the structure of the vehicle and penetrates through the metal to the external surface. As a result it is often called **inside-out corrosion**. Figure 12.5 illustrates the typical inside-out corrosion found along the bottom of a car door.

Debris traps

Debris which will absorb or hold water, such as mud, loose corrosion product, leaves, old rags and paper, set up differential-aeration cells. The corrosion occurs out of sight beneath the debris which forms a 'poultice' on the metal surface. If inside-out corrosion is not to occur, box sections should be designed so that debris cannot accumulate in the section. Rounding internal corners and edges, and providing access so that they may be easily swept and cleaned minimises the risk of damage, as does the provision of access to drainage covers and filters so that they can be cleaned. An untidy, dirty system is more likely to suffer corrosion than one which is well maintained and clean (see Fig. 12.4(f)).

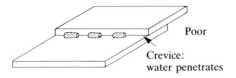

Poor

Crevice:
water penetrates

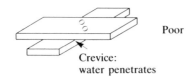

Poor

Crevice:
water penetrates

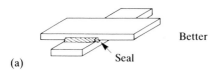

Better

Seal

(a)

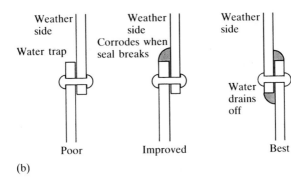

Weather
side

Water trap

Weather
side
Corrodes when
seal breaks

Weather
side

Water
drains
off

Poor

Improved

Best

(b)

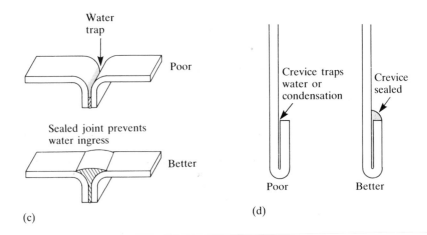

Water
trap

Poor

Sealed joint prevents
water ingress

Better

Crevice traps
water or
condensation

Crevice
sealed

Poor

Better

(c)

(d)

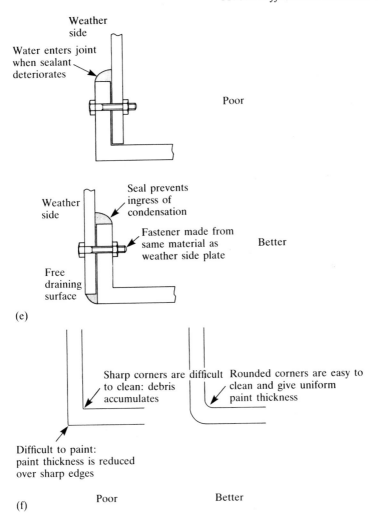

Fig. 12.4 Typical sites at which crevice-type corrosion will be found and methods for reducing the corrosion.

(a) Tack welds or other methods for localised fixing of metal plates leave crevices which water can penetrate. Continuous welds exclude water by eliminating crevices.

(b) The riveting of vertical plates can create crevices unless the geometrical arrangement is correct so that rainwater runs off the joint, rather than into it. A sealant should also be used to prevent ingress of water by capillary action.

(c) Potential water traps occur at rounded-edge joints. Careful, continuous sealing is necessary.

(d) Folded sheet steel provides ideal sites for crevice corrosion, especially on motor cars. Once again, these should be carefully sealed.

(e) Correct geometrical arrangement is essential for the elimination of crevices in structures which are exposed to the weather. Note the possibility of dissimilar metal corrosion between plate and fastener.

(f) Rounded edges and corners are always preferable to sharp ones, especially when a paint coating is to be used for corrosion protection.

209

Fig. 12.5 Inside-out corrosion on a car. Most of this damage originates from differential-aeration cells where crevices or mud poultices have restricted the oxygen supply to the metal surface.

Case 12.12: A mud poultice was trapped in the rear of an open channel section which supported the shock absorber on the front wheel of a sports car. Corrosion penetrated the bottom of the channel which distorted when the vehicle gently hit a grass verge. The distortion caused loss of steering control and a much more serious crash ensued, Fig. 12.6. It should be emphasised that the vehicle was 10 years old at the time and the corrosion should have been found during routine inspection before the accident. The vehicle manufacturers have for some time galvanised the underframe of this model of car, including the shock absorber supports, Fig. 12.7.

Case 12.13: The painted, open-ended mild steel pillar sections of a bus body were failing within three to seven years service because of inside-out corrosion. In some cases complete penetration had severely weakened the structure. Rectification involved the replacement of the pillars and entailed the removal of many body panels. The differential-aeration cell started beneath a 15 cm mud poultice which held water and road de-icing salts on the internal surface. The manufacturer solved the problem by swaging the bottom, open end of the pillar until it was almost closed (leaving an opening to provide drainage) and orientating the closure so that spray and mud were deflected from the remaining opening. The modified pillars were constructed from zinc coated steel and stoved after painting. The additional cost to the manufacturer was only 1 per cent of the cost to the user in repairing the damage, the safety improvement being incalculable.[5]

Fig. 12.6 Case 12.12: corrosion to a metal strut supporting the front suspension of a car. Increased damage resulted from a relatively minor accident.

Fig. 12.7 Case 12.12: galvanising is now used to reduce the risk of the corrosion which contributed to the extensive damage to the vehicle shown in Fig. 12.6. (Reproduced by permission of Lotus Cars Ltd.)

Inadequate drainage and ventilation

If light rain or spray falls on a bare steel surface, rings of rust will be found after the water has evaporated. Each droplet acts as a differential-aeration cell and the rust ring develops where the iron(II) ions from the anode meet the hydroxyl ions generated on the cathode (see expt 3.11 and Fig. 7.5). If the surface is free-draining, or there is adequate ventilation to dry the water droplets rapidly, the corrosion damage will be limited. Even on painted surfaces, there will be damage if the droplets persist for long periods. The risk of paint failure beneath the droplet will rise, followed by pitting corrosion at the centre of the droplet.

Increased damage is often found on the bottom surfaces of a structure where the ventilation is likely to be less efficient, because the area is protected from prevailing air currents. Water droplets will remain on the surface until they grow sufficiently large to fall under gravity. Severe corrosion has been found on the bottom surfaces of the outside of tanks, on the undersides of overhanging units on steel bridges and towers, as well as on the lower surfaces of girders, beams and crane jibs. In such cases, adjacent vertical and top surfaces which are exposed to prevailing air currents have remained in reasonably sound condition, Fig. 12.8.

All channels and box sections should be free-draining so that water will not sit in the section. The internal surfaces and the edges of any drainage holes should be covered with long-lasting barrier coats. Ventilation, either along or through the section, will also help by keeping the surfaces dry. If

Fig. 12.8 Poor ventilation. Increased corrosion on the bottom of oil storage tanks owing to poor ventilation which left water droplets suspended from the surface for prolonged periods. The free flow of air on the sides of the tank kept the surfaces drier and reduced corrosion.

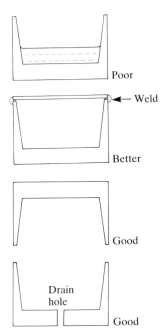

Fig. 12.9 Improved drainage and ventilation reduce the risk of corrosion in box sections and channels.

a channel is closed by welding after the coating has been applied, any damage to the coating, internal as well as external, must be made good, Fig. 12.9.

Water-absorbent soundproofing and insulation

Porous materials are often used to absorb sound and to reduce noise emission levels. They may also act as sponges and absorb water. Oxygen differentials in the water held within the 'sponge' then generate differential-aeration cells on metal surfaces. Pads of foam rubber are frequently placed in the ends of internal box sections in cars to absorb the drumming sound created in the section when the car is moving. Condensation and spray from wet roads soak the rubber and start inside-out corrosion of the section wall.

Case 12.14: The doors of an expensive saloon car were made from thin sheet steel. Naturally, when the door was closed it produced a cheap-sounding 'clang'. In order to produce a more satisfying luxurious 'clunk', which was much more in keeping with the car, the doors were lined with sound absorbent foam. It was a common occurrence to find rows of blisters in the external paintwork of the door after only two or three years' service. The blisters were produced by the corrosion from the differential-aeration cells, formed beneath the water-absorbent foam on the inside of the door skin which had penetrated the metal.

213

Lagging on pipes can produce a similar problem, as the following example shows.

Case 12.15:   Severe corrosion was found on the external surface of a lagged steel pipe which fed a factory heating system. The corrosion occurred inside a building where it was impossible for water to have dripped on to the outside of the lagging, nor was there any evidence of leakage from the pipe. It was found that water was dripping on to the lagging at a point several metres away. The lagging acted as a wick and conducted the water along the pipe. However, the heat from the pipe prevented the water from penetrating to the metal surface at the point of ingress, but as the pipe cooled along its length the water penetrated further through the lagging and wetted the metal surface at the point where the corrosion was found.

Water-absorbent materials should either be eliminated from all positions where the relative humidity will exceed 60 per cent (see section 13.2) or the outer surface should be fully sealed to prevent the ingress of water. Sealing should include all joints and repairs to the lagging.

12.5   TANKS AND PIPE SYSTEMS

There are several factors to be considered when designing tank and pipe systems for the storage or transport of electrolytes. Clearly, the galvanic and differential-aeration cells, discussed in sections 12.3 and 12. 4, may develop because of the various constructional materials used, or at crevices in joints or behind gaskets.

Additional problems arise when electrolyte is trapped in the system by badly sited drain taps or poorly designed bends and junctions in the pipes. Some examples of these are shown in Fig. 12.10 and 12.11. Drain taps must always be positioned so that the system empties completely, and all bends should be rounded with smooth transitions through 'T' joints and into tank bottoms. In most cases, trapped electrolyte will evaporate, exacerbating the corrosion problems as the concentration of dissolved, aggressive ions and the electrical conductivity of the solution increases. However, some electrolytes, notably sulphuric acid, may absorb water from the atmosphere. The rise in the level of the liquid increases the area of potential corrosion. If the more dilute electrolyte is not capable of sustaining any passive film which may be protecting the metal surface, pitting corrosion will occur, followed by general wastage of the metal wall.

Any obstruction in a pipe system may cause turbulence which will produce erosion corrosion by cavitation or impingement (see Ch. 8). Turbulence can be caused by poorly fitted gaskets, burrs, weld metal or solder protruding into the pipe bore at joints, bends which have too tight

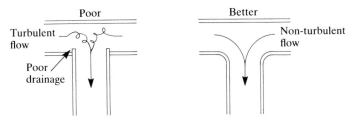

Fig. 12.10 Poorly designed joints introduce water traps and turbulence in pipe systems.

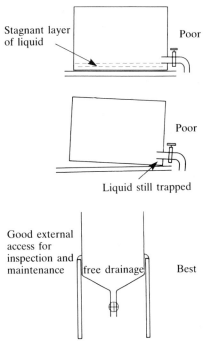

Fig. 12.11 Improved drainage enables tanks to be emptied completely and reduces the risk of internal corrosion. The external arrangements are also improved.

a radius, or any other disturbance to flow. Figure 12.12 shows the effect upon flow of a badly fitting gasket.

Cavitation damage has been found on the bores of pipes in which the liquid appears to be stationary. Vibrations in the tank or pipe wall produced by adjacent machinery, pumps, or even the flow of liquid in other sections of the system, set up transverse pressure oscillation in the layer of liquid adjacent to the wall. The pressure changes produce cavitation and pitting attack on the wall. Pipe runs, valves and pumps must be well clamped and damped to minimise vibration along pipes.

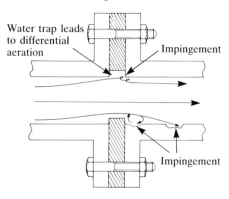

Fig. 12.12 Badly fitted gaskets produce water traps, turbulence and cavitation damage in pipe systems.

Case 12.16: Severe pitting corrosion was found on the outside surface of a diesel engine wet cylinder liner. The flow of coolant over the liner had been carefully smoothed and regulated to avoid cavitation damage, but the vibration in the liner wall caused by the force of the gas ignition and the passage of the piston up and down the liner, caused very large pressure changes in the liquid boundary layer. These pressure changes led to pitting damage by cavitation which penetrated to one eighth of the thickness of the wall (see Fig. 8.5).

Many materials used to manufacture valves, pumps and pipes have limiting velocities for the flow of electrolyte over the metal surface. If this is exceeded protective, passive films may be removed and the metal corrodes and erodes rapidly. While this will be covered more fully in Chapter 15, it should be noted that liquid velocities in pumps, through valves and around bends can be much higher than in the straight runs of pipe. Although most metals have *maximum* limiting velocities, some have *minimum* values which must be maintained in the system. An example of *minimum* velocity is found in the stainless steel/sea water combination. Sea water is very agressive to the chromium oxide film on the stainless steel and pitting corrosion usually results in static systems. However, the material performs better in *flowing* sea water. If the metal is specified for condenser tubes, a minimum velocity of 1.5 m s^{-1} of well-aerated sea water is required to maintain the film. It is normal practice to select stainless steels containing up to 25 per cent chromium and 6 per cent molybdenum for use in such situations: the molybdenum addition reinforces the surface film.

Where aggressive electrolyte must flow through a system it is essential that provision be made for the system to be thoroughly flushed through after draining down so that the aggressive ions are removed or neutralised. If the system is to be left unused for a prolonged period it may be advisable to treat it with a film former, which coats the surface with an inert barrier layer, or an inhibitor. The manufacturer's instructions and limitations on use for both these products must be followed.

There have been many instances of stainless steel giving excellent service over prolonged periods while oxygenated, fast-flowing sea water passed through the system. However, massive crevice corrosion failure has occurred immediately after shutdown, even though the system has been drained. Chloride in the water which was trapped in crevices, inside pumps, below impellers, around shaft bearings, valve stems and under gaskets has rapidly destroyed the protective chromium oxide film on the metal surface leading to failure.

12.6 STRAY CURRENT CORROSION

Metal structures often carry unsuspected electric currents. These currents may be induced from neighbouring electrical conductors, say, in cast iron water mains by adjacent power cables when the two are buried close together to supply services in urban areas. They may arise whenever current finds a path of least resistance less than the one envisaged by the designer, for example, from current leakage when buried metal structures are used as earth conductors to ground electrical distribution systems, or during welding in industrial fabrication and repair work. In very long systems, such as pipelines running over many tens of kilometres, they have even been attributed to electromagnetic induction produced by the Earth's rotation. Collectively, these are known as **stray-currents**.[6]

Should these currents pass from the structure into an electrolyte, severe corrosion damage can occur at the point where the conventional electric current leaves the structure. In Chapter 4 we saw that anodes are sites on a metal surface where electrons are released to flow through the metal, while ions conduct the current through the electrolyte. By definition, *the conventional current passes in the opposite direction to the electron flow.* Therefore, where the conventional current leaves the metal, the electron flow is *into* the metal and that site becomes the anode, as shown in Fig. 12.13a. Unlike normal corrosion, in which an energy difference between the anode and cathode generates a potential and gives rise to the current, here the current is produced by an external source and the corrosion rate is dependent on that imposed current, not on differences between metal surfaces. This gives rise to two important points:

(a) The corrosion rate is not dependent upon the rate of oxygen supply to the cathode. This is often rate-determining in normal corrosion processes.

(b) Although coatings reduce the area which corrodes, the current density and the corrosion rate increase at any breaks in the coating.

Because of the large separation which is usually found between the anode and cathode in stray current corrosion systems, the cathode products do not form films on the anode to stifle the corrosion. However, chlorides, nitrates and other anions in the soil often migrate to the anode site to give corrosion

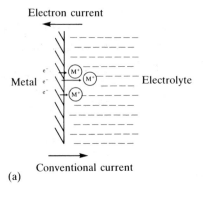

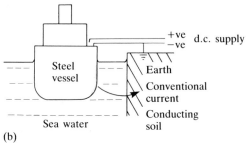

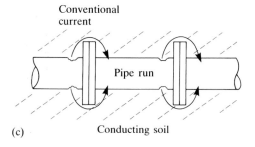

Fig. 12.13 Stray current corrosion:
(a) The development of an anode by an imposed current.
(b) Corrosion caused by incorrect wiring of a welding unit.
(c) Conventional current flow around joints with high electrical resistance in a pipe system.

products not normally found in natural corrosion cells. Where the conventional current enters the metal from an electrolyte, cathodic reactions occur and the resulting alkali may damage some metal surfaces or barrier coats.

The electrolyte may be free water, such as a river or the sea, or water contained in the soil. For buried structures, the magnitude and site of the corrosion may vary with the climatic conditions. Corrosion may cease in dry weather, only to recommence after the next fall of rain. Changes in soil conductivity because of dissolved ions will also alter the severity of the attack.

In the days when trams were a common sight in city streets, very large electrical currents were often measured in metal systems carrying water and other services. In Brooklyn in the early 1900s, 70 A was measured in a cast iron water main.[7] Today, it is more normal to find currents limited to a few amps, although momentarily large surges can occur during lightning strikes.

Case 12.17:   A 65 km multiple pipeline route crossed under four sets of railway lines which were electrified on the 750 V d.c. third rail system. The pipelines were protected by impressed current cathodic protection at four substations. After the final pipeline was installed, the cathodic protection failed to control the corrosion resulting from the stray currents which flowed from the pipeline to the rail negative conductor, via the soil in which the pipe was buried. To prevent corrosion damage the pipelines and the negative rail were electrically bonded together so that the current did not have to leave the metal at any point. This in turn produced two other interesting effects:

(a) At times, the conventional current flowed from the railway to the pipe down the bonded joint. This caused corrosion at other sites. A diode rated at 150 A was included in the rail bond to ensure that current only flowed from the pipe to the rail. (Note the rating of the diode!)

(b) By placing suitable resistors in the individual bonds to each separate run of pipe, the potential of the pipe was limited to -0.85 to -2.5 V CSE, thus providing its own cathodic protection system for the whole 65 km of pipe. The cathodic protection substations were disconnected.[6]

Case 12.18:   A flanged and bolted, underground, stainless steel pipeline was being replaced with an all-welded system because of joint leaks. During the night it rained, and submerged the completed portion of the pipeline already laid in the trench, which had not been backfilled. The next day, stray currents from the welding machine 'drilled' thousands of holes through the wall of the submerged pipe as the current found an easier return through the water than by following the welder's clamped-on earth connection.[1]

Similar problems have affected ships undergoing weld repairs when d.c. electrical supplies for welding units have been taken from a shore-based generator. If the current finds an easier return from the hull, through the water in which it floats, and the earth to the grounded terminal of the generator, thinning of the hull results (see Fig. 12.13(b)). In serious cases, hulls have perforated and ships sunk at the quayside.

Increased resistance across flanged and bolted joints in pipelines has caused stray currents to bypass the joints, flowing either through the soil or into the liquid in the pipe if it is a conductor. The current often returns to the pipe on the other side of the joint. Engineers have been surprised to find a repeated pattern of corrosion at each joint along a pipe run, the same side badly corroded in each case, the other side free from corrosion. This effect is shown in Fig. 12.13(c).

Methods of controlling stray current corrosion include:

(a) Electrical bonding of the structure so the current does not leave the metal at any point.

(b) Using impressed current cathodic protection.

(c) By attaching sacrificial metal pieces to the buried structure at the points where the conventional current leaves the system. These should be orientated in the direction of the current flow in the surrounding medium, carrying the current well away from the structure. The sacrificial pieces should be inspected regularly and replaced as required.

12.7 REFERENCES

1. Seibert O W 1978 Classic blunders in corrosion protection, *Materials Performance* Apr: 33–37
2. *The handling and storage of coated and wrapped steel pipes.* Department of Industry and Central Office of Information, 1981
3. Johnstone K 1972 Private communication. South Western Electricity Board, Plymouth, UK
4. Seibert O W 1983 Classic blunders in corrosion protection revisited, *Materials Performance* Oct: 9–12
5. *Controlling corrosion: case studies.* Department of Industry Committee on Corrosion, 1979, pamphlet No. 5
6. Allen M D, Ames D W 1982 Interaction and stray current effects on buried pipelines. Six case histories, in *Proceedings of cathodic protection theory and practice – the present status.* Institution of Corrosion Science and Technology, Science Division
7. Evans U R 1948 *Metallic corrosion, passivity and protection.* Edward Arnold, p 37

12.8 BIBLIOGRAPHY

Pludek V R 1977 *Design and corrosion control.* MacMillan Press
Shreir L L (ed.) 1979 *Corrosion* (Vol. 2). Newnes-Butterworths, section 10

13 CORROSION CONTROL BY ENVIRONMENTAL CHANGE

What fool hath added water to the sea . . .

(William Shakespeare: *Titus Andronicus*, III. i. 68)

O! When mine eyes did see Olivia first, methought she purged the air of pestilence.

(William Shakespeare: *Twelfth Night*, I. i. 20)

Corrosion is the reaction between a metal and its environment, therefore any modification to the environment which makes it less aggressive will be beneficial in limiting the attack upon the metal. We shall consider three situations.

(a) The bulk of the environment is gaseous. Normally, this will be air in the temperature range $-10\,°C$ to $+30\,°C$. Some of the methods used to reduce corrosion rates in the atmosphere are:

(i) Lower the relative humidity;
(ii) Eliminate volatile components given off by surrounding materials;
(iii) Change the temperature (but see section 13.1);
(iv) Remove contaminants (including solid abrasive particles), deposits which will form cathodes (soot, for example), and agressive ions.

(b) The material is immersed in free water which contains sufficient ions to become an electrolyte. Modifications to electrolytes include:

(i) Lowering the ionic conductivity;
(ii) Altering the pH;
(iii) Uniformly reducing the oxygen content;
(iv) Changing the temperature.

(c) The metal is buried in soil, dissolved minerals forming the electrolyte. Control is normally by cathodic protection or surface coating, but the environment can be made less aggressive by using backfills to improve drainage, control pH and change the conductivity.

Both atmospheric and wet corrosion processes can be controlled by the use of special chemicals called inhibitors. When these are added to the environment, the rate of attack is reduced. Although Shakespeare talks of diluting the sea and removing pollutants from the air, in practice such modifications are only economically viable when the environment to be treated is confined to a limited space. It may be the atmosphere in a warehouse, storeroom, packing case, or a box section in a structure. It may be

221

the electrolyte in a tank, or in a closed loop system such as a cooling circuit for an engine, or a processing plant.

However, reducing the general level of pollutants released by processing plants into the atmosphere or a river system is likely to lower the corrosion rates for exposed structures in the vicinity.

13.1 ATMOSPHERIC CORROSION

Before describing the steps which can be taken to control atmospheric corrosion, it will be profitable to consider some of the factors which cause it.

The most important factor is the presence of water which may arise from rain, mist or condensation owing to a high relative humidity. These will not all have the same effects. Heavy rain can be beneficial by washing away the pollutants which have collected on a metal surface. Care should always be exercised in the design of a structure to ensure that rain water drains away freely and there is sufficient ventilation to dry all surfaces (see Ch. 12).

Mist and condensation constitute the corrosion hazard from the atmosphere, wetting all surfaces including internal ones. The thin films of water they produce may not drain away but remain on the surface until dried by air currents or increases in temperature. A thin, invisible film of water is all that is necessary for the attack to begin. Many metals, iron, steel, nickel, copper and zinc, corrode if the relative humidity exceeds 60 per cent. If it exceeds 80 per cent the rust on iron and steel becomes hygroscopic (water absorbent) and the rate of attack is further increased.

The thin film of moisture which forms from mist or high relative humidities is easily saturated with oxygen from the atmosphere, therefore the cathodic reaction, either oxygen reduction or hydrogen evolution, is not the rate-controlling step in the corrosion process. The rate and severity of the attack is usually determined by the conductivity of the electrolyte, which depends upon the level of dissolved contaminants. These vary from carbon dioxide (giving slightly acidic solutions) in rural areas, to sulphur dioxide, sulphur trioxide, nitrous compounds, hydrogen sulphide and ammonium ions in industrial areas, and chloride ions in marine locations. Exposure tests have shown a corrosion rate for steel of 5 μm per year at Nkpoku, a rural area in Nigeria, rising to 90 μm per year in Sheffield, an industrial area in the UK. In marine environments, especially near a shore line, it can be considerably higher.

Temperature affects atmospheric corrosion in two ways. First, there is the increase in reaction rate which normally follows an increase in temperature. In general, the rate of a reaction approximately doubles for every 10 °C rise in temperature. However, at higher temperatures, the solubility of oxygen is reduced and it has been suggested that this lowers cathodic reaction rates, limiting the corrosion. In thin films with a good oxygen supply from the atmosphere the limiting effect will be small.

Second, changes in temperature affect the relative humidity and can cause dew point condensation. If the temperature falls below the dew point, the air becomes saturated with water vapour and free water droplets condense on any exposed surfaces. The condensation forms on all surfaces which are cool enough, internal as well as external. The droplets can collect in water traps to produce pools of free electrolyte in sheltered areas, causing problems inside a structure where corrosion might not normally be expected.

Dew point condensation is responsible for much of the damage to exhausts on cars and flues carrying burnt fuel gases. If the gas stream temperature falls below its dew point before it is released to the atmosphere, condensation forms. The fuel gases usually contain sulphur oxides and nitrogen compounds. These form aggressive electrolytes leading to rapid corrosion failure which originates on internal surfaces.

Sulphur trioxide produces sulphuric acid, a particularly aggressive electrolyte. Increased sulphur trioxide also raises the dew point of the gas, causing the condensate to form at higher temperatures and thus exposing greater lengths of the exhaust or flue to the electrolyte. Another common fuel impurity, vanadium, acts as a catalyst for the conversion of sulphur dioxide to trioxide in a combustion chamber; again this increases the possibility of early failure in the exhaust system. Flue gas with no sulphur trioxide present has a dew point in the range 38 to 46 °C; 5 ppm of sulphur trioxide raises this to 100 °C, while 40 ppm will give a dew point of 168 °C.

Solid particles carried in air currents can be abrasive to paint finishes and protective films on metal surfaces. The damaged areas are likely to corrode first when an electrolyte forms on the surface.

Aircraft are prone to corrosion damage resulting from abrasive particles. Near to the ground damage to paint films is caused by sand and dust swept up in the turbulence caused by the aircraft movement. Very fine ice particles, and to a lesser extent rain, cause damage at altitude. Corrosion ensues over the damaged areas when the aircraft enters moister climates. Aircraft-finish, epoxy-based paints are more vulnerable than polyurethane ones, particularly to fine ice particles which puncture tiny holes through paint.

Abrasive particles ingested into gas turbine engines can remove the high-temperature coatings applied to blades, resulting in hot or high-temperature corrosion (Ch. 17).

When the relative humidity is high, miniature differential-aeration cells can be generated beneath dust or grit particles which adhere to metal surfaces. The result is a surface, pock-marked with tiny rust spots over the corrosion pits. Stainless steel, used for decorative façades on buildings in busy towns and cities, has often suffered extensively in this way and required overcoating with transparent oils, lacquers or varnishes to stop the formation of the cells. Some particles, such as soot, may act as active cathodes, generating local cells to pit the metal surface.

13.2 CONTROL OF ATMOSPHERIC CORROSION

The most effective way to minimise atmospheric corrosion is to remove the atmosphere under vacuum and seal the components in impervious envelopes. The protection is only effective while the integrity of the covering is maintained. Should it be punctured with even a small hole, corrosion will commence. It is prudent to include drying agents inside evacuated sealed covers as an additional protection.

In warehouses and storerooms air can be heated to lower the relative humidity below the 60 per cent figure at which corrosion begins on many common metals. This does not remove water vapour from the air and condensation will still form on any surface which cools the atmosphere below the dew point. Anyone who wears spectacles will have experienced this phenomenon as their lenses mist over when walking into a warm room from a colder one. It is the metal temperature which must be considered in heated storerooms. Light corrosion often affects metal objects freshly introduced into temperature-controlled warehouses, before the metal attains the air temperature of the room. Serious problems arise if water collects on poorly ventilated internal surfaces where drying is slow. The same effect will happen if goods are packed and hermetically sealed in warm processing areas and then taken into colder environments for storage or transportation. Condensation forms and is trapped inside the sealing.

Water vapour can be removed from the air before it is circulated in a warehouse. Chilling the air by passing it over surfaces cooled well below the working temperature of the warehouse precipitates sufficient water to keep the relative humidity below the critical level at the operating temperature of the store area. **Freeze-drying** can be used to produce very low residual water contents if required. To maintain an average relative humidity below 60 per cent it will be necessary to keep the general level much lower, perhaps at 30 to 40 per cent, to allow for the exchange of air through open doors and by natural ventilation. The optimum relative humidity for comfortable air conditions for people is in the range 40 to 65 per cent[1]. The slightly lower levels for warehouses are not thought to be harmful to health, although staff may experience some dryness in the throat over prolonged periods.

In packing cases or stores where the expense of chill drying is not justified, **desiccants** are used to dry the air. Again, for maximum effect, the atmosphere must be contained by sealing components and the desiccant in impervious envelopes, or by reducing the rate of air exchange in the room to a minimum. The desiccants should be non-corrosive to the metals involved, and preferably they should be cheap and easy to handle. In practice, the commonest ones are silica gel and activated alumina with molecular sieves being used to obtain very low humidities. Colour changes indicate the deterioration of the drying capacity of the desiccant, which can be reactivated by heating in an oven.

The packaging for a desiccant requires careful consideration. Air should be free to circulate over the drying agent while spillage is prevented. In storerooms and other large stable spaces, open trays of desiccant are often used, but in packing cases and other containers that are liable to be inverted, closed perforated metal boxes or porous bags are used.

Contaminants can be removed by scrubbing the air with fine water sprays before it is introduced into the system. The large surface area of the fine spray enables gaseous impurities to dissolve quickly while solid particles are swept from the air stream. Clearly, the scrubbing process should take place before the air is dried.

An alternative method of protecting steel components during transport and storage is to use **vapour phase inhibitors** (**VPIs**). As the name implies, these components are volatile. They spread through the free space in a container and precipitate a water-repellent (hydrophobic) film on to exposed surfaces. While VPIs are beneficial towards ferrous metals, they can increase the rate of attack on other materials; if non-ferrous metals, paints or plastics form part of the structure, the use of these inhibitors should be treated with great caution.

The mechanism of protection afforded by VPIs is still not resolved, but some points can be made. The compounds consist of a volatile organic cation, usually an amine, and an anion which acts as an inhibitor. On the metal surface, the cation produces a thin adsorbed film which has two important functions: it is hydrophobic, and it controls the pH of any moisture layer which forms on its surface. Carbonate and nitrite are the anions in two typical VPIs. These are carried with the volatile cation to deposit on the metal surface. Should free water form on the surface, or the component become immersed so that the cation film breaks down, the anion acts as a normal inhibitor (section 13.3) to control the rate of corrosion by polarising the electrode reactions. Immersed in water, the anion usually gives the same level of protection as the sodium salt of that anion would give.

For steels and aluminium, two common VPIs are dicyclohexylamine nitrite (DCHN) and cyclohexylamine carbonate (CHC). DCHN has the lower vapour pressure and takes longer to produce an effective surface film, but maintains protection for a much longer period. It has a vapour pressure of 0.027 Pa at 25 °C, giving a pH of 6.8 in water. One gram saturates 550 m^3 of air and renders it relatively non-corrosive to steel. In secure packaging at room temperature it will inhibit corrosion for several years. However, it can cause increased attack on some non-ferrous metals, plastics, paints and dyes. Zinc, magnesium and cadmium are particularly affected. CHC has a high vapour pressure, 21.3 Pa at 25 °C. It produces a pH of 10.2 in water. The higher vapour pressure yields a protective coating much more quickly than does DCHN, but for a given weight of inhibitor the system has a much shorter life. CHC is most useful in containers or stores which are opened periodically; the VPI can be renewed regularly and the protection is rapidly restored once the unit is closed. It has no inhibiting effect on the corrosion of cadmium, but increases the attack rate on copper, brass and magnesium.

Plastics, paints and dyes may also deteriorate in a CHC atmosphere.

DCHN-impregnated paper is widely used in precision engineering to wrap tools, measuring equipment and other close tolerance components. The slightly greasy paper gives protection for many years, providing the packaging is not disturbed. CHC is normally inserted in open trays or porous containers.

Mixtures of the two VPIs are available. The CHC rapidly builds a protective film, while the DCHN contributes the longer term control. Other VPIs have been produced at various times to protect different metals, borates for zinc and chromates for copper and its alloys, but these do not appear to be widely used.

Vacuum sealing or desiccants are used to protect electronic and optical equipment, the greasy film produced by VPIs prohibits their use for this purpose.

Many organic materials, wood, plastic and paints give off aggressive vapours which can initiate and contribute to corrosion damage on adjacent metal surfaces. Maleic acid, glycol and styrene have been desorbed from glass reinforced plastics, while wood gives acetic acid vapours. The latter product poses a serious problem when wooden cases are used to store or transport metal objects.

Cellulose is the principal constituent of wood. It consists of long chains of sugar molecules which contain basic hydroxyl groups, some of which are combined with acetic acid in the form of esters. These groups can react with water to release free acetic acid, thus:

$$CH_3COOR \ + \ HOH \ \rightleftharpoons ROH \ \ + \ CH_3COOH \qquad [13.1]$$

ester   water   alcohol   acid

The equilibrium for the reaction always means that the moisture in wood is slightly acidic. However, the main corrosion hazard arises from the volatile nature of the acetic acid which escapes from the wood to fill the surrounding atmosphere with acid vapour. An excellent electrolyte is formed when this vapour condenses on to a metal surface.[2]

Open wooden crates which allow free circulation of the air can reduce the degree of damage. In confined spaces where the microclimate can become heavily charged with the acid vapours, or when it is necessary for objects to be entirely enclosed within the crate, desiccants or VPIs are used. All wooden crates should be lined to keep the vapour away from the metal. Polyethylene, tar paper, bitumen-coated kraft paper and even zinc sheeting have been used as liners. The desiccant or VPI is placed inside the lining material which should be carefully sealed along all joints. The most severe attack occurs on mild steel, zinc, cadmium and magnesium alloys, while titanium, tin and austenitic stainless steels suffer least damage.

Corrosion from acidic vapours is always a possibility when wood and metals are in close proximity and there is insufficient ventilation to remove the volatilised acid.

Case 13.1: Steel wire cables wrapped on wooden drums had been so badly corroded after a period in store that they were condemned as unsafe for use as securing strops.

Preservatives and fire retardants applied to wood can also increase the risk of corrosion damage. Diammonium phosphate, ammonium sulphate and borax mixtures form the basis of many proprietary fire retardants; copper chrome arsenic is a common fungicide. These compounds can release ammonia, soluble sulphate and copper; all present corrosion hazards to metals. Ammonia is particularly harmful to copper and some copper alloys because it leads to general wastage and the possibility of stress-corrosion cracking. Sulphate ions produce very aggressive electrolytes and copper can form effective cathodes after plating out by ion exchange on base metals.

Case 13.2: An unusual case of vapour corrosion happened on the inside of a copper belfry roof in Denmark. The roof was supported on wooden beams which had been treated with an ammonium-based fire retardant. Condensation on the inside of the copper roof leached the ammonium salts from the wood and then dripped on to a concrete floor. The alkali in the concrete released ammonia vapour from the condensate, which rose into the roof space and attacked the copper roof. The initial cuprammino corrosion product decomposed to sulphate/carbonate and released ammonia to continue the attack. A thickness of 0.8 mm was penetrated in three years.[2]

13.3 MODIFICATION OF THE ELECTROLYTE

You will recall that there are four basic processes involved in the corrosion of a metal:

(a) the anode reaction;
(b) the cathode reaction;
(c) ionic conduction through the electrolyte;
(d) electron conduction through the metal.

In this chapter we are concerned with the efficiency of the first three processes; the electron conduction through the metal is not normally affected by the electrolyte.

The rate at which the metal corrodes is controlled by the slowest of the cell processes. The metal cannot corrode and produce ions faster than the cathode can utilise the resulting electrons, or the electrolyte can transport the current by ionic conduction.

The properties of an electrolyte, which can be varied to limit its aggressiveness to a metal surface, were given in the introduction to this chapter. Dissolved ions will affect the corrosion rate by:

(a) changing the conductivity of the electrolyte;
(b) attacking or strengthening passive films on the metal surface;
(c) changing the pH.

A thin film of protective oxide forms on the surface of most metals exposed to air at room temperature. When the metal is subsequently placed in an electrolyte, Brasher[3] has shown that the anion concentration plays a significant role in the behaviour of the electrolyte towards this protective film. On mild steel, the anions are aggressive at low concentrations and attack the film, while at higher concentrations they can become inhibitive. An inhibitive anion is adsorbed on to the metal at weak points in the film. It then suppresses anodic dissolution of the metal and allows oxide formation to take place, strengthening the film. The inhibitive role of some anions is very weak and they can be considered aggressive towards oxide films. The most powerful of the aggressive ions towards mild steel surfaces are sulphates, thiosulphates, sulphites and thiocyanates. Chlorides are also very aggressive, while nitrates are only mildly so.

Dissolved cations, which are more noble than the metal exposed to the electrolyte, can plate on to the metal surface to set up local galvanic cells which often result in pitting attack. Thus, reducing the concentration of ions in an electrolyte will reduce the efficiency of the ionic current transport and change the anodic processes on the metal surface. A change from sea water to distilled water, for example, not only reduces the rate of attack upon the anode, but also the conductivity of the electrolyte, and reduces the maximum distance which can separate the anode and cathode in an active cell.

Although many investigators have worked in the field, the inhibitive mechanism by which some ions retard the rate of corrosion is not fully determined. For the purposes of this study, we can deal with inhibitors qualitatively, dividing them into three groups: anodic, cathodic and adsorption inhibitors.

(a) **Anodic inhibitors** increase the polarisation of the anode by reaction with the ions of the corroding metals to produce either thin passive films, or salt layers of limited solubility which coat the anode, Fig. 13.1. For iron and steel, two types of inhibitor are important: those which require dissolved oxygen to be effective, such as molybdates, silicates, phosphates and borates, and those which are themselves oxidising agents such as chromates, nitrites and nitrates.

(b) **Cathodic inhibitors** affect both the usual cathodic reactions. In the first:

$$2H_2O + O_2 + 4e^- \rightarrow 4OH^- \qquad [4.21]$$

the inhibitor reacts with the hydroxyl ion to precipitate insoluble compounds on the cathode site, thus blanketing the cathode from the electrolyte and preventing access of oxygen to the site. The most widely used inhibitors of

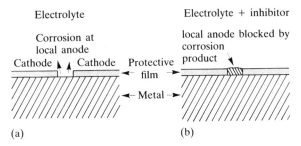

Fig. 13.1 The effect of concentration of anodic inhibitor on corrosion rate.
(a) Corrosion at a local break in the oxide film on the metal surface. The surrounding film is the cathode, the bare metal is the anode.
(b) The anion in the anodic inhibitor reacts with the metal ions in solution to seal off the anodic site.

this type are salts of zinc and magnesium, which form insoluble hydroxides, calcium which produces insoluble carbonate if the water chemistry is adjusted (see below), and polyphosphates.[4]

In the second cathodic reaction:

$$2H^+ + 2e \rightarrow 2H \rightarrow H_2 \qquad [4.19]$$

the evolution of hydrogen is controlled by increasing the overvoltage (polarisation) of the system, as shown schematically in Fig. 13.2. The salts of

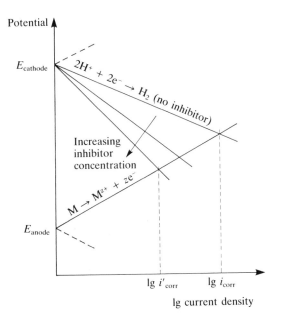

Fig. 13.2 Cathodic inhibitor controlling the hydrogen evolution reaction. As the concentration of inhibitor increases, the cathode reaction is polarised. The value of E_{corr} for the cell tends to the anode value, and i_{corr} (the corrosion rate) falls to i'_{corr}.

Table 13.1   Some common inhibitors and their uses

Inhibitor	Use	Metals protected	Typical* concentration
Sodium nitrite	Cooling water	Steel	0.05%
	Brines	Steel	Up to 5%
	Sea water	Steel	0.5%
	Engine cooling	Steel	Up to 1%
Sodium nitrate	Guard against caustic cracking	Steel	
Sodium hydrogen phosphate	Cooling water	Steel	1%
	Boilers	Steel, copper, zinc	10 ppm
	Sea water (with sodium nitrite)	Steel	10 ppm
Borax	Engine cooling	Steel	1%
	Glycol cooling systems	Steel	1%
Sodium silicates	Potable water	Steel, copper, zinc	10 to 20 ppm
	Oilfield brines	Steel	0.1%
	Seawater	Zinc	10 ppm
Arsenate ion	Most conc. acids	Steel	0.5%
Organic amines	Vapour condensate in boilers	Steel	Variable
	Acids	Steel	
	Oilfield brines	Steel	
Hydrazine	Oxygen scavenger (high temp.)	Steel	As required
Sodium sulphite	Oxygen scavenger (low temp.)	Steel	As required

* 10 ppm = 0.001%

metals such as arsenic, bismuth and antimony are added for this purpose, to form a layer of adsorbed hydrogen on the surface of the cathode. Owing to the toxicity of these metals, organic compounds which achieve the same end have been developed. Inhibitors which control the evolution of hydrogen gas from the surface may allow hydrogen atoms to diffuse into steel and cause hydrogen embrittlement (Ch. 10).

(c) **Adsorption inhibitors** are long organic molecules with side chains which are adsorbed and desorbed from the metal surface. The bulky molecules can limit the diffusion of oxygen to the surface, or trap the metal ions on the surface, stabilise the double layer and reduce the rate of dissolution. It is claimed that as little as 0.2 per cent agar-agar in distilled water will reduce the rate of corrosion to only 2.7 per cent of the rate without agar-agar.

A list of some common inhibitors and their uses is given in Table 13.1.

Cathodic inhibitors are said to be *safe*. Even if too little inhibitor to elim-inate all the cathodic sites is added, the active cathode area, and hence the corrosion rate, is reduced, as shown schematically in Fig. 13.3(a). On the other hand, anodic inhibitors are said to be *dangerous* because the addition of too little inhibitor fails to eliminate all the anode sites and actually increases the corrosion rate on those which are left. The most vigorous attack occurs just before complete inhibition is achieved, see Fig. 13.3(b). Thus, an anodically protected system changes from a condition of no attack, to one of deep localised pitting attack, as a result of a slight dilution of the inhibitor concentration. In summary, local increases in anodic attack are caused when:

(a) insufficient inhibitor is added to the electrolyte;
(b) the electrolyte is diluted after the inhibitor is added;
(c) large concentrations of depolariser ions such as sulphate or chloride are introduced to reduce the effectiveness of the inhibitor present in the solution;
(d) the inhibitor fails to penetrate 'dead legs' in the system.

In general, anodic inhibitors are more efficient than cathodic ones. Modern inhibitor systems reduce the risk from the dangerous, anodic, type, while profiting from their greater efficiency by using a combination of both anodic and cathodic types. The cathodic type retards the overall corrosion rate and allows the anodic one to seal off the anode sites at lower concen-trations than would be needed if it acted alone. Such synergistic inhibitors are typified by the chromate/polyphosphate/zinc system.

The concentration of all inhibitors should be monitored regularly and additions made to maintain effective control.

The effects of temperature, pH and dissolved oxygen content are inter-related. Figure 13.4 shows the effect of pH and oxygen content on the corrosion rate of steel at 25 °C. In section 13.1 it was stated that the rate of reaction approximately doubled for each 10 °C rise in temperature. The same is true for totally immersed conditions. However, as the temperature is increased the solubility of oxygen decreases. In contact with air, the solu-bility of oxygen at 25 °C is approximately 8.5 ppm in tap water and 6.5 ppm in sea water.[6] At 60 °C, the solubility in tap water falls to 5.6 ppm, while at 100 °C it is all removed.

In large volumes of electrolyte, the diffusion path from the free air/liquid interface to the cathode will be much longer than in a moisture film formed on a metal surface. Hence the level of available oxygen at the cathode will fall, and limit the corrosion rate.

If the pH $\geq$ 7, the cathodic reaction is oxygen reduction, eqn [4.21], but oxygen reduction can occur if the pH < 7:

$$O_2 + 4H^+ + 4e^- \rightarrow 2H_2O \tag{13.2}$$

231

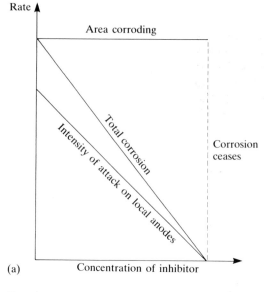

(a)

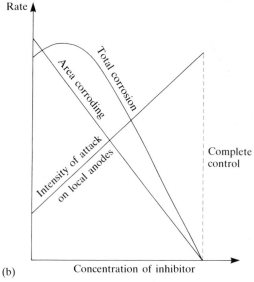

(b)

Fig. 13.3 Safe and dangerous inhibitors.

(a) Cathodic inhibitors are 'safe'. An increase in the concentration of inhibitor, although below the critical amount, reduces the cathodic area and the current demand from the anodes. Hence, while the area corroding remains unaltered, additions of the inhibitor below the critical concentration will reduce the corrosion rate. (After Evans[5])

(b) Anodic inhibitors are 'dangerous'. An increase in the concentration of inhibitor, while below the critical amount, reduces the anodic area. The cathodic area is unaffected, therefore the current demand from the anodes which are still active increases to reach a maximum value just below the critical concentration. The corrosion rate rises on the active anodes and produces severe pitting attack. (After Evans[5])

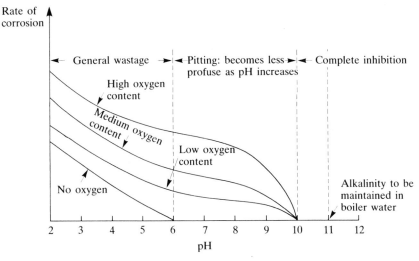

Fig. 13.4 The relationship between the corrosion rate, pH and the oxygen content of water at 25 °C and 1 atmosphere pressure.

At 50 °C in tap water, the corrosion penetration rate for low carbon steel falls from 7.5 mmpy at 6 ppm oxygen, to 1 mmpy at 1 ppm oxygen;[7] although as little as 0.1 ppm oxygen can cause corrosion rates to increase in a flowing system.

Problems arise when the level of dissolved oxygen varies in different parts of the system; differential-aeration corrosion may then occur.

Reducing the level of dissolved oxygen uniformly to very low levels can reduce the corrosion rate to negligible proportions if the pH is neutral or alkaline. This is seen in domestic central heating systems which use copper pipes and steel radiators, and in car engine cooling circuits where cast iron blocks are linked to copper radiators. The bimetallic corrosion in both systems is high just after they have been filled or refilled with water. However, the cathode reaction quickly consumes the dissolved oxygen, a thin layer of black magnetite forms on the steel surface, eqn [13.3], and the corrosion rate falls.

$$3Fe(OH)_2 \rightarrow Fe_3O_4 \quad + H_2 + 2H_2O \qquad [13.3]$$
iron (II)   magnetite
hydroxide

Should the system leak, the replenishment of the oxygen supply in the make-up water renews the bimetallic attack. The hydrogen has to be bled from central heating systems periodically.

The importance of maintaining low oxygen levels can be seen in the following case histories.

Case 13.3: Some sectors of the American motor industry suffered severe corrosion problems in engine blocks when the plastic overflow bottles in the cooling circuits cracked and allowed fresh oxygen to dissolve continuously in the cooling water.[8]

Case 13.4: Hastelloy B®, a nickel/molybdenum alloy designed for use in reducing environments, was specified for the manufacture of a pump in a process line transporting 35 per cent sulphuric acid. The pump failed in two weeks. The sulphuric acid was passed through a scrubbing jet which aerated the system; had this aeration not occurred, the material might have proved satisfactory. Hastelloy C®, a nickel/chromium/molybdenum oxidation-resistant alloy, worked well.[9] (See section 15.7.)

13.4 CONTROL OF THE AQUEOUS ENVIRONMENT IN STEAM GENERATORS AND COOLING SYSTEMS

The control of electrolyte quality is particularly important for water used in industrial heating and cooling systems, steam generating plant and condensers. The corrosion rate in these systems increases as the pH falls and as the oxygen content rises. Water may be used in 'once-through' processes, in which minimum control is exercised owing to cost, or in recirculating systems for which continuous monitoring is required with careful control of make-up water.

In low temperature, once-through systems using fresh water, a low cost inhibitor may be used, or the composition of the water may be adjusted to produce a thin protective scale on the surface of the metal. The scale is mostly composed of calcium or magnesium carbonates. It should be of negligible thickness and self-healing. It must not grow so that it impedes or blocks the flow of water, or significantly alters the heat transfer characteristics of the system. If the scale is damaged and does not self-heal, corrosion can occur on the bare metal surface. The production and maintenance of such a scale requires careful balance of water chemistry. The solubility of calcium carbonate in water is low, and the film is precipitated from the bicarbonate produced from dissolved carbon dioxide.

$$CO_2 + H_2O \rightarrow H_2CO_3 \qquad\qquad [13.4]$$
$$\text{carbonic}$$
$$\text{acid}$$

If calcium salts are present then:

$$\underset{\substack{\text{calcium}\\\text{carbonate}}}{CaCO_3} + H_2O + CO_2 \rightleftharpoons \underset{\substack{\text{calcium}\\\text{bicarbonate}}}{Ca(HCO_3)_2} \qquad [13.5]$$

® Registered trade marks of Cabot Corporation.

On heating, or if there is a slight deficiency of carbon dioxide in solution and the pH rises, the reaction is driven to the left and calcium carbonate precipitates.

The ability of the water to form a scale is given by the Langelier Index, which depends upon the pH, temperature, concentration of calcium salts and total dissolved solids. The calculation and application of the index is beyond the scope of this book, but a description of its use will be found in most water treatment books.[10] If there is just too little dissolved carbon dioxide to keep the bicarbonate in solution, calcium carbonate deposits on the cathodes owing to the slight rise in pH which accompanies the cathode reaction. If the film is complete and adherent, then the metal is separated from the water. Should there be too little carbon dioxide present, no scale is formed; if there is excess carbon dioxide, the scale already formed is redissolved in the acid solution leaving the metal unprotected.

In recirculating steam generating plant which uses steel as a major constructional material, the water is treated to maintain the pH above 11 in order to reduce the oxygen content and remove scale-forming salts. The main feed water consists of distilled condensate from steam which has already passed through the system. The impurities in the condensate will be carbon dioxide, oxygen and dissolved salts, mainly sodium based, which have been carried over by the steam. The additional water for 'making-up' the system will contain calcium and magnesium salts (producing the hardness value) and further quantities of dissolved oxygen and carbon dioxide. In marine installations and those using estuarine water as an initial feed source, large quantities of sodium chloride may also be present. Let us first consider the effect of these impurities on the corrosion of a steam generator system.

Carbon dioxide is very soluble in cold water, producing carbonic acid, eqn [13.4], with a pH of 5.5 to 6. When the water is boiled, the gas is ejected and passes through the system, redissolving in any condensate. Thus, condensate water in any part of the system will always have a pH well below that required to maintain a passive film on steel surfaces, Fig. 13.4.

Oxygen will increase the efficiency of the cathode reaction in the alkaline conditions which always exist in steel boilers. It can also cause oxygen or air bubble pitting if it is ejected from the water as the temperature rises and is allowed to collect in the system. When the boiler is working hard, the turbulence carries the gas out of the unit, but in quiescent periods or on shutdown, it rises to collect in air traps or on the top surfaces of the boiler. The development of an air bubble pit is shown in Fig. 13.5. The oxygen-rich area at the base of the bubble becomes a cathode in the alkaline water, while the surrounding metal corrodes. The corrosion product settles on the bubble wall. Once the bubble is completely encased in corrosion product (iron hydroxide) ingress of further oxygen to the bubble is prevented, although iron (II) ions can still diffuse through the corrosion product skin. Once the oxygen in the bubble is consumed, the area at its base can no longer support the oxygen reduction cathode reaction and becomes an

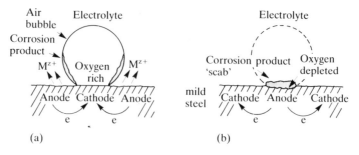

Fig. 13.5 Air bubble pitting.
(a) A differential-aeration cell is generated around the bubble, with the oxygen-rich area at the base forming the cathode, and the surrounding metal the anode. Iron (III) hydroxide corrosion products form on the bubble wall. When the wall is totally encased in corrosion product, the diffusion of oxygen into the bubble ceases, although metal ions can still diffuse through the wall.
(b) Once the oxygen content at the bubble base falls below that of the surrounding water, the differential aeration cell reverses. The small anode at the bubble base develops into a pit. The wall collapses to form a scab over the pit, below which differential-aeration continues to drive the cell.

anode, while the surrounding metal assumes the cathodic role: the cell has reversed. Generally, the skin on the bubble wall collapses to seal off the new anode and differential-aeration corrosion continues beneath the scab to produce a deep pit.

Other forms of scabbed pits develop on steel, again assisted by differential-aeration below the corrosion product.

The **dissolved salts** of magnesium and calcium precipitate from the water as it evaporates, producing layers of scale on the metal surfaces. As the scale thickens, heat transfer rates are reduced, leading to a loss of efficiency, increased risk of buckling or distortion and an increased risk of ash deposits forming on the hotter areas on the fire side of the boiler tubes.

The **treatment of feed and top-up water** for steam generators is a complex procedure which depends upon the level of impurities present and the operating conditions of the unit, especially the temperature and pressure of the steam.

Amines are added to control the harmful effects of carbon dioxide. Neutralising amines, such as morpholine, are added in very small quantities. They are volatile and weakly alkaline. The alkaline vapour condenses and dissolves in the condensate to neutralise the carbonic acid which forms as the carbon dioxide redissolves. Filming amines such as octadecylamine are volatile, but insoluble in water. The vapour is driven off by steam but condenses on the cooler metal surfaces to form an oily hydrophobic layer which separates the metal from the condensate.

Oxygen can be reduced to low levels by mechanical means. However, sodium sulphite, which oxidises to sulphate, or hydrazine can be used to scavenge oxygen chemically. In high-pressure systems with high steaming rates, hydrazine is preferred. None of the reaction products is harmful to the metal system, but hydrazine is toxic, especially in confined spaces.

$$N_2H_4 \quad + \quad O_2 \rightarrow \quad 2H_2O + \quad N_2 \qquad\qquad [13.6]$$
hydrazine

It decomposes if it does not react with oxygen:

$$3N_2H_4 \rightarrow \quad 4NH_3 \quad + \quad N_2 \qquad\qquad [13.7]$$
ammonia

The ammonia acts in a similar way to neutralising amines in combating the effect of carbon dioxide, but hydrazine and ammonia can be corrosive to copper and copper alloys in condensers. Therefore, the excess concentration of hydrazine is limited to very low levels, typically 0.1 ppm. The sulphite is used in low-temperature or closed down systems, but excess sulphite should be avoided as it is aggressive to steel surfaces. In modern plant, the maximum level of dissolved oxygen is kept below 0.03 ppm, while in high-pressure power stations oxygen in the feed water is kept below 0.007 ppm. Control of pH is achieved by adding either sodium hydroxide in conjunction with sodium hexametaphosphate or sodium orthophosphate. In hard waters the orthophosphate precipitates sludge, which necessitates periodic 'blowing down' of the boiler to remove the residue. In marine practice, Navy Boiler Compound is used to maintain the quality of the water in boilers. This compound consists of 38 per cent sodium carbonate, 48 per cent disodium hydrogen phosphate and 13 per cent corn starch. Sodium carbonate decomposes under the boiler pressure and temperature to give sodium hydroxide which provides the alkalinity to keep the pH above 11. Disodium hydrogen phosphate, in the presence of the carbonate, precipitates the calcium and magnesium ions. Starch acts as a catalyst and may produce flocculent precipitates of the neutralised salts. In seriously contaminated boilers, sodium carbonate (soda ash) is used alone.

Great care should be taken to avoid over treatment, which can lead to caustic cracking of steel components (section 10.1). You will recall from section 2.3 that the pH of neutral water is temperature dependent: it falls to 6 at 60 °C. This change affects the results produced by adding both pH and oxygen scavenging chemicals.

Stainless steel closed loops are used in nuclear power steam generating systems. The water in these loops is passed through ion exchange columns to remove ionic contamination, and hydrazine is used to scavenge oxygen. One of the main causes of failure in the stainless steel loops is stress-corrosion cracking. As early as 1967 it was well known that the chloride and oxygen content of the water passing around the loop controlled the time to failure of the metal[10] It has been shown[11] that sulphate ion contamination resulting from carry-over of the ion exchange resin also induces stress-corrosion cracking of sensitised stainless steel in the presence of oxygen. In these systems, the efficient operation of the ion exchange unit and the oxygen scavenging are of paramount importance.

13.5 SOIL AS AN ENVIRONMENT

The corrosion rate for metals buried in dry soil is negligible. However, as the moisture content rises, the corrosion rate for buried metal is controlled by the conductivity, pH, oxygen content, aggressive ion concentration and biological activity in the soil. The nature of the soil (sand, silt or clay) and the values of these parameters can vary considerably. Long pipelines or deep oil and gas well casings may pass through several different types of soil which generate anodes and cathodes on the metal surface, often separated by large distances.

The conductivity and pH depend upon the mineral content of the soil and any ions which are washed into it. Fertiliser spread on fields, or de-icing salts washed from the roads will affect the corrosivity of the surrounding soil.

Case 13.5: A buried stainless steel pipeline carrying steam condensate passed under a road. The pipeline gave excellent service, except for the areas directly beneath the two edges of the road. Here, de-icing salts applied to the road in winter ran into the soil and caused chloride stress-corrosion cracking of the hot pipe within two years.[12]

Differential-aeration cells can be set up between the top and bottom surfaces of pipes, or down buried piles and well-casings owing to the generally lower oxygen content at deeper soil depths. The same effect can happen where metal systems pass under roads which are relatively impervious to oxygen. However, the moisture content can also be reduced and the practical outcome may be difficult to predict, indeed, it may vary over a period of time depending upon the amount of rainfall.

Pipelines which have been buried for a long time become encased in corrosion products which may passivate the metal or slow down the rate of corrosion. Should replacement lengths of pipe be installed in the old line, the new metal is anodic to the corrosion product on the old. Severe pitting corrosion occurs at holidays (small pinholes in a paint coating) in the coating on the new pipe, which often fails in a much shorter period than the original ones. Indeed, in clay soils there were a number of instances where the old mild steel gas pipes corroded completely away leaving the gas flowing through holes in the impervious clay. The problem only became serious when natural gas, supplied at a higher pressure, was substituted for low pressure town gas.

Buried structures are normally protected, either with barrier coats, paint, bitumen or concrete, or by cathodic protection. These methods are discussed fully in Chapters 14 and 16. Short runs of pipe are often placed in inert ducting which should be sealed to prevent the ingress of water which cannot drain away. On longer runs in trenches, suitable backfills can ensure that there is good drainage and control of pH. Granite chips are used to improve drainage, and limestone to control pH.

Correct backfill is very important around anodes used in cathodic protection systems. The backfill provides a uniform environment for the anode and ensures that there is an even reaction over the whole of the active surface. It also prevents contact with soil which might passivate the anode and reduce its effectiveness. For zinc or magnesium anodes, a mixture of 75 per cent gypsum, 20 per cent bentonite (clay) and 5 per cent sodium sulphate is used. For small anodes the backfill and the anode are placed in perforated bags, while a loose filling is used for larger ones. The backfill attracts moisture and increases the conductivity of the area around the anode. It prevents local corrosion of the anode by aggressive elements in the soil, and passivation of the anode surface. Impressed current anodes are surrounded with carbonaceous material, usually a high carbon coke mixed with 5 per cent slaked lime. This has a low resistivity: less than $50 \, \Omega$ m. Owing to the efficient current transfer from the anode to the backfill, most of the reaction is on the outer edge of the backfill, increasing the apparent anode size and its life.[13]

13.6 REFERENCES

1. Porges J 1976 *Handbook of heating, ventilation and air conditioning* (7th edn). Butterworths, section V, p 14
2. Department of Industry and the Central Office of Information 1979 Corrosion of metals in wood, in *Guides to practice of corrosion control* (2)
3. Brasher D M 1969 Stability of the oxide film on metals in relation to the inhibition of corrosion. II. Dual role of the anion in the inhibition of the corrosion of mild steel. *British Corrosion Journal* **4**, May: 122–28
4. Shreir L L (ed.) 1979 *Corrosion* (Vol. 2). Newnes-Butterworth, p 18:19
5. Evans U R 1948 *Metallic corrosion, passivity and protection*. Edward Arnold, p 593
6. Shreir L L (ed.) 1979 *Corrosion* (Vol. 2). Newnes-Butterworth, p 21:54
7. Roller D, Scott W R 1961 *Corrosion Technology* **8**(3): 71–76
8. Siebert O W 1983 Classic blunders in corrosion protection revisited, *Materials Performance* Oct: 9–12
9. Siebert O W 1978 Classic blunders in corrosion protection, *Materials Performance* April: 33–37
10. Degremont Company Acfi, S Africa 1960. *Water treatment handbook* (2nd revised edn) S Africa: Elliott Ltd
11. Latanision R M, Staehle R W 1967 Stress corrosion cracking of iron, nickel and chromium alloys, in *Proceedings of conference on fundamental aspects of stress-corrosion cracking*, Ohio State University, pp 214–307
12. Ruther W E *et al.* 1984 Effect of sulphuric acid, oxygen and hydrogen in high temperature water on stress-corrosion cracking of sensitised AISI 304, *Corrosion* **40** (10): 518–27
13. Shreir L L (ed.) 1979 *Corrosion* (Vol. 2). Newnes-Butterworth, p 11:22 and 21:54

13.7 BIBLIOGRAPHY

Ailor W H Jr, Dean S W Jr, Haynie F H (eds) 1973 *Corrosion in natural environments.* ASTM-STP 558

Collie M J (ed.) 1983 *Corrosion inhibitors: developments since 1980.* Noyes Data Corporation, Chemical Technology Review No. 223

Dean S W Jr, Rhea E C (eds) 1980 *Atmospheric corrosion of metals.* ASTM-STP 767

Ranney M W 1976 *Corrosion inhibitors: manufacture and technology.* Noyes Data Corporation, Chemical Technology Review No. 60

14 CORROSION CONTROL BY BARRIER COATINGS

> There is no way to success in our art but to take off your coat, grind paint and work like a digger on the railroad, all day and every day.

(R W Emerson: *Conduct of Life, Power*)

Barrier coatings are applied to metal surfaces, either to separate the environment from the metal, or to control the micro-environment on the metal surface. Many different types of coating are used for these purposes including paints, organic films, varnishes, metal coats and enamels. By far the most common of these is paint.

Today, the technology of paint and its application is changing fast, spurred on by ever-rising costs of energy, raw materials and labour. Pollution is a problem too in a conservation-conscious society: it has been calculated that about 360,000 tonnes of volatile organic compounds are released into the atmosphere every year, as a result of the application of paint. This makes the use of other solvents, particularly water, more attractive. Dangerous lead additives have mostly been replaced with such materials as titanium dioxide, a material which is now one of the most common in everyday life.

The protection afforded by barrier coatings is remarkable in view of their thickness. A typical paint coating is between 25 and 100 microns thick; a ship's hull requires a heavy duty combination of coatings totalling about 250 microns, while a lacquer inside a can of food may be as little as 5 microns thick. When choosing a coating, consideration should be given to the service conditions which will be experienced by the particular part of the structure to which it will be applied, and the general environment in which it will operate. We have already seen (section 12.2) that temporary coatings may be applied to protect the metal, or even the service coat, during a hazardous stage of fabrication, storage or transit. Many cars are now coated with a water-soluble temporary paint to protect them from minor damage during delivery to the showrooms. When the car is safely in the dealer's hands, the paint is removed in a modified car-wash. A black gloss temporary paint has also been used to show up small dents in car body panels. The imperfections can then be knocked out and the temporary paint removed.

Many modern applications are found for coil-coated steel. This steel is factory produced by unrolling a massive steel coil and passing it through a paint sprayer. The steel is immediately baked to dry and cross-link the paint, after which it is recoiled and is ready for transport to a fabrication process. Refrigerator, freezer and washing machine panels can be pressed out of the sheet and are ready for immediate assembly. Alternatively, items such as guttering or railings can be formed on-site by installers.[1]

The discussion in this chapter will concentrate on the corrosion control mechanisms, the methods of application and the properties of paint, plastic and metallic coatings applied to steel. Some mention will be made of other coatings and metals.

14.1 THE COMPOSITION OF PAINT

The term 'paint' includes a number of different coating systems which are designed for a variety of tasks. Before a paint is used, it is necessary to determine the surface preparation, method of application and any constraints on its use.

A paint consists of:

(a) **A vehicle** – the liquid which gives the paint its fluidity and dries or evaporates to form a solid film.

(b) **A pigment** suspended in the vehicle. The pigment controls the corrosion reaction, or rate of diffusion of the reactants through the dry film.

(c) **Additives** which accelerate the drying process or enable the dry coating to withstand better the working environment.

The vehicle may dry by one of three processes:

(a) Evaporation of a solvent constituent of the vehicle.
(b) Chemical change, mainly oxidation of the liquid constituent of the vehicle, for example, linseed oil. The paint dries from the surface inwards and is applied in a number of thin coats to build a thick layer.
(c) Polymerisation, a chemical reaction between the vehicle and a curing agent which is mixed into the paint just before it is applied to the metal. The curing agent is kept in a separate container and the paints are referred to as **twin-pack systems**. The paint dries throughout the film which can be applied in thick layers. The curing process commences immediately the agent is added and there is a very limited time period after mixing, before the paint becomes unsuitable for application. This is called the pot life.

When the paint has dried the remaining solid portion of the vehicle forms the **binder**. It holds the pigment in place, keys the film to the surface and provides a barrier to restrict the passage of water, oxygen and aggressive ions to the metal surface.

Although pigments do give colour to the dry paint film, they fulfil two much more important roles. First, in a primer coating the pigment controls the corrosion process at the metal surface, either by inhibiting the reactions or providing sacrificial protection to the substrate metal. Second, in the top coats inert pigments increase the length of the diffusion path for oxygen and moisture penetrating the film, to delay the onset of the corrosion process and to slow the reaction rates. The mechanism of corrosion control by pigments is complex and only a brief summary of the processes can be given here.

When clean steel is exposed to dry air, a film of iron (III) oxide forms on the surface. This film is impervious to the diffusion of iron ions and protects the underlying metal. However, water (even the thin layers deposited from the atmosphere when the relative humidity exceeds 60 per cent) breaks down the integrity of the film and rusting occurs. The Fe(II) ions formed by the dissolution of the iron react with the hydroxyl ions produced on the cathode to form $Fe(OH)_2$ in the first instance, and finally rust, $FeO.OH$.

Some basic pigments, notably lead compounds but also those of zinc and cadmium, inhibit the corrosion reaction on steel by stabilising the iron (III) oxide. When ground with linseed oil the lead compounds form soaps which deposit metallic lead on the steel surface. At these points oxygen reduction proceeds more easily and generates a sufficiently high current density to maintain Fe(III) film production until the oxide coating on the steel surface thickens and it becomes impervious to iron ion diffusion.[2] Lead salts are more efficient inhibitors than zinc and cadmium, but they are very toxic and many authorities restrict or prohibit their use. Lead-based paints must never be used where they will come into contact with food or drinking water and operatives who handle the paint, or who clean or remove old lead paint, should be given frequent medical checks.

Another common class of pigments contain chromate salts which yield chromate ions in solution after water has penetrated through the binder. The chromate ions react with the air-formed iron(III) film to produce a complex chromium/iron oxide (a spinel) which forms an impervious barrier over the metal surface. Since chromate salts are thought to be carcinogenic, paints containing these pigments should be treated with care and advice on their handling obtained from the manufacturers.

Some soluble lead pigments increase the pH of any water which penetrates the binder and this in turn inhibits the corrosion reactions on the steel. Zinc phosphate, either alone or with red lead, also produces a tightly adherent barrier layer, although the exact mechanism of its formation has not yet been determined.

Anodic metallic pigments can protect the steel sacrificially. To be successful, there must be electrical contact between the particles of the pigment, and between the pigment and the steel surface. Zinc dust is the only commercially available pigment which is successful in marine environments and other aggressive locations, although there is some evidence that manganese may give adequate control at less aggressive rural or urban sites. The initial penetration of water through the binder causes the zinc to corrode sacrificially, producing zinc hydroxide. Carbon dioxide also diffuses through the binder to react with the hydroxide and form a carbonate. The zinc corrosion product fills the pores in the paint film and produces an impervious, compact and adherent layer. A high zinc content in the dry paint film is essential if the first part of the reaction is to succeed in blocking the pores; once the film is sealed the loss of electrical contact between the pigment particles is not so important. Tests have shown that paint with a zinc content of 95 per cent in the dried film gave two years' protection before rust appeared at a scratch on a steel plate immersed in sea water, but rust appeared after two days and twenty days when the coat contained 86 per cent and 91 per cent zinc respectively.[3]

Aluminium and micaceous iron oxide provide inert flake pigments which become orientated parallel to the metal surface as the paint is brushed out. The flakes increase the length of the moisture diffusion path through the binder and slow the rate of attack on the substrate metal. Glass beads fulfil the same function.

Small additions of other materials are made to modify the properties of the paint. **Organic metal salts** are added to accelerate the drying of the paint. **Anti-oxidants** prevent skins forming on the paint surface while it is still in store, but must not interfere with the drying of the applied paint film. **Surface-active agents** help to disperse the pigment uniformly through the paint, and to prevent segregation as the paint dries. **Thixotropic agents** reduce runs and drips in the wet coat.

14.2 THE CHARACTERISTICS OF PAINTS

A dried paint coating, approximately 0.1 mm thick, is expected to give a long life during which it will restrict the access of air, moisture and aggressive ions to the metal surface. While many paint films are impermeable to ions such as chloride, sulphate and carbonate, no paint film forms a complete barrier to either oxygen or water. These, in time, always penetrate to the metal surface and paint films cannot, therefore, inhibit the cathode reaction.

Paints which are exposed to the atmosphere rely upon pigments to control corrosion (see section 14.1). Such paints must be tolerant to wide variations and rapid changes in the ambient conditions. Not only will the temperature of the surface fluctuate throughout the day but thermal expansion of the substrate will stress the paint skin. Changes in the relative humidity will introduce wetting and drying cycles which can cause the paint to swell or crack owing to water uptake. Ultra-violet radiation will degrade the paint surface. Aggressive ions in polluted atmospheres may directly attack the paint or reduce the pH of rain water falling on the painted surface, leading to chemical changes in the pigment or binder which will decompose the coating. As the paint ages the continued oxidation of the binder and loss of gloss will increase the permeability of the paint and result in erosion of the coating and more rapid loss of pigments.

Paint systems which are used under water may, or may not, have a pigmented primer, an inorganic zinc, say, but the top coats rely upon very low water absorption and transmission coefficients to restrict the access of the electrolyte to the metal surface. The bond between the paint coats and to the metal substrate must be strong and continuous over the whole surface to prevent damage by osmosis (section 14.3). Thus a high quality surface preparation is required to ensure that the paint wets the whole of the metal surface as it is applied.

An ability to resist attack by alkali is a most important property for paints applied to immersed structures which are protected by impressed current cathodic protection (section 16.5). The cathode reaction on the metal surface releases hydroxyl ions which can soften many paints and lead to stripping and undercutting of the film. This in turn increases the area of metal requiring protection from the impressed current system. The large shields which surround impressed current anodes are to protect the area in the immediate vicinity of the anode from excessive alkali production.

The thickness of the dried paint film should be uniform over the whole of the surface to be treated, including the edges and corners of plates, bolts, exposed threads, rivets and joints. However, surface tension will always tend to reduce the thickness over sharp corners, as shown in Fig. 14.1. Where possible, edges should be rounded to maintain the film thickness. A failure at any point in the coating will concentrate corrosion in that area and lead to further deterioration as differential-aeration corrosion spreads beneath the paint film. If a stone chip in the paintwork of a car is neglected, not only will a rust spot form in the chipped area but the surrounding, apparently sound, paint work will easily peel off to reveal rust between the paint and the metal.

The vagaries of the climate and the conditions under which most painting is carried out require that a paint should also be very easy to apply using a variety of methods such as brushing, rolling, spraying and dipping. It should dry very quickly to facilitate the application of additional coats, to prevent rain, dust and debris from spoiling the surface and, on maintenance painting, to reduce down-time and costs. Paint should be easy to touch up or repair, be resistant to attack by fungi and bacteria, and should have a

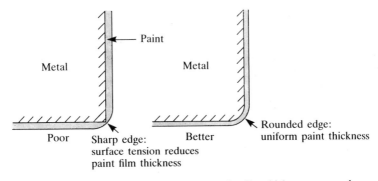

Fig. 14.1 The effect of surface tension on paint film thickness over a sharp corner.

good, long-lasting appearance! This is a good reason for the application of multiple coats: in general, the greater the thickness, the greater the protection. Small pinholes (holidays) in a paint coating can be very dangerous because of the principle that large cathodes cause high corrosion rates at small anodes. Occasionally, in systems containing a bimetallic couple, only the anode is painted because, it is thought, only the anode will corrode. However, the corrosion rates at holidays are even greater than usual because of the combination of the area effect and the bimetallic effect.

14.3 PAINT TYPES

Paint systems are divided into broad generic groupings named either from the use of the paint or the chemical nature of the binder. Within each group there are many types of paint which may be blended to achieve a particular set of coating properties or to suit the method of application. Some groupings are listed below and the durability of the coatings is summarised in Table 14.1.

Prefabrication primers

These are applied to clean, rust-free steel to protect the metal for some months during the fabrication and assembly stages of a structure. It is not necessary to remove sound prefabrication primer before the final service coating is applied. If it deteriorates or is damaged by the fabrication process, local touching up should be undertaken, the manufacture's procedures being carefully followed. Primer which has been carelessly applied can increase the rate of corrosion by trapping corrosion products beneath the paint film or by reducing the relative area of the anode. Typical primers consist of zinc dust or red iron oxide in an epoxide resin base. They dry within 2 to 3 minutes of application and protect the metal for up to 12 months.

Pretreatment primers

Paints which are used to condition the metal surface to ensure good adhesion and performance of the final paint system. They offer only limited corrosion protection to the metal and should be overcoated as soon as the primer is dry. To obtain good results the steel surface should be clean and free from rust; pretreatment primers must not be applied over mill scale, prefabrication primer or old paint. A twin pack paint for use on steel contains zinc tetrahydroxychromate in an alcohol resin solution, and an alcohol-based phosphoric acid. It dries in a very short time. It is recommended[4] that a pretreatment primer should always be used on zinc coatings, particularly when zinc chromate paint will be used as a primer for the final coating system. Aluminium and its alloys are treated with zinc chromate or zinc-oxide-based pretreatment coats, referred to as etch primers.

Table 14.1 Characteristics of Paint Systems

Type of binder	Mode of drying	Acid resistance	Alkali resistance	Water resistance	Solvent resistance	Exterior weathering resistance
Raw linseed oil Boiled linseed oil	Air oxidative polymerisation	Fair	Bad	Fair	Poor	Poor/ fair
Oleoresinous varnishes	Air/Stoving condensation/ oxidative polymerisation	Fair	Bad	Fair/ good	Poor	Fair/ good
Long oil length alkyd	Air oxidative polymerisation	Fair	Bad	Fair	Poor	Very good
Medium oil length alkyd	Air/Stoving oxidative/ condensation polymerisation	Fair	Poor	Fairly good	Fair	Very good
Short oil length alkyd	Stoving condensation polymerisation	Fair	Fair	Good	Fairly good	Very good
Urea formaldehyde alkyd blend	Stoving condensation polymerisation	Fairly good	Fairly good	Very good	Good	Fair
Melamine formaldehyde alkyd blend	Stoving condensation polymerisation	Fairly good	Fairly good	Very good	Good	Very good
Epoxide amino or phenolic resin blend	Stoving addition/ condensation polymerisation	Good	Good	Very good	Very good	Good
Polyester/ polyisocyanate blend	Air/stoving addition polymerisation	Fairly good	Good	Fairly good	Very good	Very good
Vinyl resin	Air solvent evaporation	Very good	Very good	Very good	Poor	Good
Chlorinated rubber	Air solvent evaporation	Good	Good	Very good	Poor	Good

Oil-based paints

Vegetable drying oils such as linseed or tung oil form the base for these paints. They take a long time to dry by oxidation and it is considered necessary to leave the paint for up to 48 hours between coats and up to 7 days before applying top coats to primers. Red lead in linseed oil is a typical example of an oil-based primer. Zinc phosphate may be added should a shorter drying time be required. The pigment forms a soap with the oil.

Oleo resinous paints (varnishes)

A drying oil and a natural or synthetic resin are used to form the vehicle in this broad class of paints. The resin improves the drying and binder properties of the coat which are an improvement over the simple oil-based types. Tung oil with 100 per cent phenolic resin paints can be used on structures immersed in water, including ships' hulls which are not protected by impressed current systems. The phenolic resin types resist abrasion but do not tolerate surface dampness during application. Conversely, coal tar resin types do tolerate some surface dampness but have poor abrasion resistance.

Alkyds

This is a widely used coating with an unlimited number of types. The basic paint is a polyester, prepared by reacting polyhydric alcohols with monobasic fatty acids and dibasic acids. A typical composition would include ethylene glycol, linseed oil and phthalic anhydride respectively. The oil length shown in Table 14.1 refers to the ratio of oil (fatty acid) to resin. Long oil paints dry by oxidative polymerisation of the oil and therefore have a high oil content, usually more than 65 per cent. Those which are dried by stoving do not require such high oil contents and are referred to as short length oils, say, less than 50 per cent oil.

Epoxide resins

This is a broad group of paints which dry by a polymerisation reaction between the epoxide resin and a curing agent. They can be divided into two classes, air-drying and stoving, each using a different type of curing agent. The air-drying varieties come in a twin pack system, the curing agent, an amine, being thoroughly mixed with the resin just before the paint is applied. The stoving types are sufficiently stable at room temperature for the curing agent, a phenolic compound, to be included with the resin in a single container; the polymerisation reaction is initiated only when the paint is heated in the stoving process. Recently, water-dispersed varieties have been introduced (see below).

A dry film of over 0.25 mm can be built with each application. The paint is tough and has excellent abrasion resistance. Although it adheres strongly to properly prepared substrates, inter-coat adhesion can be poor. It may be necessary to employ a tie coat when repairing old paint, or if there is a long delay between the application of coats. The properties of epoxides vary over a wide range. Different formulations are used in the atmosphere, immersed in water, acid and alkali; some are resistant to turbulent flow, while others provide hard wearing surfaces on walkways, heli-landing pads and in chutes. Great care should be exercised in selecting a type to meet a given set of service conditions.

Coal tar epoxides

The combination of coal tar and epoxide base yields a paint film which is highly impermeable to water while resisting attack by most chemicals. They are used extensively on submerged areas of marine structures, ships, oil platforms and piers, especially when an impressed current cathodic protection scheme is employed; the paints are resistant to attack by hydroxyl ions produced by the cathode reaction.

Polyurethanes

This is a group of paints which set by polymerisation, the reactants being an isocyanate curing agent and an alkyd polyester resin, possibly with pigments added. Air-drying systems come in two packs while the curing agent can be included with the resin in one pack for the stoving paints. The properties of the dry paint depend upon the ratio of alkyd to polyester in the resin. As a group, the paints are expensive and they do not tolerate high humidities or dampness during application or curing. A pretreatment primer is usually required to give good adhesion and this must dry thoroughly before the polyurethane is applied. However, the dry film has a high resistance to water, it is hard, abrasion resistant and retains a high gloss for long periods. Its performance often justifies the extra cost and care needed during application.

Vinyls

A very broad family of paints with many copolymers, they are suitable for a wide range of applications. A common type is polyvinyl chloride/polyvinyl acetate copolymer modified with maleic anhydride. The copolymer is dissolved in a solvent and the paint dries by evaporation of the solvent. However, the dry film is always soluble in the solvent and recoating is easy with new layers bonding to clean, old paint. The drying time is short: 2 to 5 minutes. In addition to the properties shown in Table 14.1, vinyls are resistant to oils and grease. They form very effective coatings for steel structures immersed in water. However, the paints are very difficult to apply by brushing; spraying is the normal method of application. Adhesion to bare steel is poor and a pretreatment primer is recommended. One suitable primer consists of polyvinyl butyral, zinc chromate and phosphoric acid. Surface moisture also impairs adhesion; it is not uncommon for whole areas of coating to peel off if the paint is applied to damp surfaces. Two other problems affecting their storage and use are the inflammability and toxicity of many of the solvents. However, if these paints are correctly applied and matched to the environmental conditions they give excellent service.

Chlorinated rubber

These paints are produced by dissolving chlorinated rubber in a particular group of solvents (aromatics). After application the evaporation of the solvent deposits the dried film in which there is very little polymerisation. The rubber alone forms very brittle layers and inert plasticisers are added to the paint to yield a tough, durable coating which retains a high resistance to water, acids and alkalis, with good weathering characteristics when exposed to the atmosphere. Adhesion to the substrate and between coats is very good; damaged areas are easily repaired. However, the paints are readily softened by many oils and greases.

Water-based paints

Water has a number of advantages as a base for paint, it is cheap, non-flammable, dries very rapidly and, in general, results in good adhesion to the substrate and between coats. Several paint systems are available, either dissolved in water, or as emulsions. Three of these are vinyls, acrylics and epoxides. The vinyls and acrylics are used on masonry for decorative and waterproof coatings. (Restriction of the penetration of water into concrete controls the corrosion of steel reinforcing bars, see section 14.7.) They can also be used as top coats over zinc-rich primers. Epoxide paints are used over inorganic zinc primers as maintenance coats for structural steel, tanks, and marine structures, as well as for initial paint treatments. Adequate ventilation must be provided while the paint is drying to maintain a low relative humidity over the surface. If the relative humidity is too high, water is trapped in the paint lattice as it dries, resulting in a poor performance from the coating.

The mass production car industry uses water-suspended paints for the intermediate coating on vehicle bodies. The process is known as electrophoretic coating. The body shell is first degreased and dipped in a phosphoric acid priming bath which deposits a tough, corrosion-resistant layer of iron phosphate, $Fe_3(PO_4)_2$, on the metal surface. It is essential that all the surfaces, internal and external, are treated, hence the need to leave large access holes in sills, box sections and pillars. The phosphate-treated body is immersed in a bath containing a water-dispersed paint formulated to give large, electrically charged paint particles. Figure 14.2 shows a car body leaving an electrophoresis tank. An array of electrodes are arranged around the body and a potential applied so that the particles of paint are attracted to and deposited on the metal, where they give up their charge. The neutral deposited layer acts as an insulator and by limiting the potential applied, the thickness of the coat can be controlled. The electrolyte resistance and the distance between the electrode and the metal surface also control the electro-deposited paint thickness. Complicated electrode arrays, and very large holes to increase the throwing power of the electrodes, may be required to ensure even thickness inside closed sections with limited access. (The large holes allow easier ingress of moisture and mud when the car is in service, increasing the risk of inside-out corrosion.) The electro-

Fig. 14.2 A Ford Sierra bodyshell emerges from an electrophoretic painting process where all surfaces have been coated with primer. Before immersion, shells are cleaned, degreased and then subjected to zinc phosphating. Bodies are then given a negative electrical charge, while the paint is positively charged, with the result that the paint is 'forced' on to all surfaces, particularly in hidden and other inaccessible areas. (Photograph kindly supplied by Ford Motor Company Ltd.)

deposited film is baked in a stoving process to form a coherent and adherent paint coating.

After stoving the external visible surfaces are primed and coated with final gloss coats while the underbody is treated with an underseal. Additional protection is given to the electro-deposited coats inside box sections and sills by wax injection. The wax forms a film approximately 0.05 mm thick which is drawn into crevices and gaps by capillary action. Poorly applied wax coatings that do not completely fill a crevice by capillary attraction but merely seal its mouth can cause increased corrosion damage. This occurs when an electrolyte enters the crevice by another route and is unable to drain away owing to the sealing of the normal drainage route.

On small volume luxury car production the expense of installing an electrophoresis coating line is not justified and many of these cars are hand-sprayed over dipped phosphate and primer treatments.

Inorganic zinc

These coatings are essentially a combination of zinc dust and complex silicates, the binder being either water-soluble or a solvent-based self-curing system. The dry film is tough, abrasion resistant, adheres firmly to the substrate, and appears to be unaffected by weathering, such as by sunlight, ultra-violet light, rain, or condensation, for example. Recoating is easy. Depending upon the formulation and environmental conditions, the paint is capable of giving 10 to 40 years' protection to steel from a one-coat application, protecting the steel sacrificially until the zinc corrosion products have plugged all the micro-pores and holidays in the coat. It has given excellent protection over extended periods when overcoated with a variety of paint types, such as epoxides, vinyls, chlorinated rubber, or polyurethane. Munger reported[4] that a coating on a 400 km-long pipeline, applied in 1942, was still giving good protection in 1984, despite being exposed to brush and grass fires, salt and marine environments, as it traversed bush land, salt marshes and marine areas.

Anti-fouling paints

These are applied to the immersed areas of marine structures, as the final coat in a paint system. They release toxins into the water to prevent living organisms from attaching themselves to the structure. A heavy infestation of barnacles will increase the drag on a ship or the legs of an offshore platform, limiting the speed of the vessel and increasing its fuel consumption, or raising the stress levels in the platform (see section 16.3).

Copper and tin are two toxins currently in use. Older mercury-based salts have been withdrawn because they produced too much pollution. The toxin is held as a pigment in the binder network and leached from it. Clearly, the leach rate from a freshly applied coating may be very high giving excess toxin, while an older coat is likely to have a surface layer of denuded binder which may impede the diffusion of the toxin. The dried paint should therefore be formulated to give a sufficiently high leach rate throughout the life of the coat, but not so high that it is exhausted before the next scheduled repainting.

Anti-fouling salts are likely to be cathodic to steel. The paints should never be applied directly to steel surfaces and care should be taken to see that they are compatible with the rest of the paint system. Copper-based anti-fouling paints must not be applied to aluminium or aluminium alloys; the risk of ion exchange resulting in pitting corrosion is too great. Tin-based paints are suitable for aluminium alloys.

A new generation of anti-fouling paints has been introduced, the erodible or self-polishing types. The paint matrix is an acrylic with a long molecular chain to which the toxin, tributyl tin, is loosely attached as a side chain. The paint is impermeable to moisture, but on the surface of the paint water hydrolyses the side chain to release the toxin. The matrix, weakened by the loss of the side chain, erodes away to reveal a fresh tin-containing surface. The leach rate remains reasonably constant throughout the life of the paint and the smooth surface reduces drag.

It must be emphasised that the selection of a paint system is a task for a specialist. We have discussed only the broad outlines of the characteristics of each group and there are many pitfalls for the unwary. A primer and top coat, even in the same generic group, may be incompatible; two apparently similar primers may require different pretreatments; a pigment may be incompatible with a substrate, or with some pollutant in the environment. Specialist help should always be sought before embarking on a costly painting exercise in which the actual paint may account for far less than 5 per cent of the charges (see section 14.5).

Case 14.1: A contract for painting two water tanks specified that the units should be blast-cleaned, pretreatment primed, vinyl primed, given two coats of a vinyl intermediate coat, and an aluminium pigmented top coat. The fabricator carried out the process as far as the vinyl primer in his factory and erected the tanks on site. The painting contractor was obliged to blast-clean the erection welds and areas damaged in transit. To save money he then applied to the bare steel a vinyl primer which did not require a pretreatment primer, and carried on with the intermediate coats and the top coat over the whole of the primed surfaces. The tank went into service. He started on the second tank but an inspector stopped him and demanded to know why he was not applying the pretreatment primer to the bare steel surface before the vinyl primer. He was made to comply with the specification. During removal of the scaffold from the second tank it was found that the coating had failed to adhere and was lifting from large areas of the surface before the tank was put into use. Inspection of the first tank revealed good adherence of the paint film and it subsequently gave a satisfactory life. Investigation showed that the contractor's vinyl primer, which he had used on both tanks, was incompatible with the pretreatment primer, hence the failure on the second tank. The fabricator's scheme which used a vinyl primer matched to the pretreatment primer was successful; the contractor's scheme which used a vinyl primer suitable for direct application to steel was also successful, but mixing the two paint schemes resulted in immediate failure owing to the incompatibility of the different coatings.[5]

14.4   PAINT FAILURE

The major causes of failure in a paint system which has been selected to match the environmental conditions are:

(a) poor or inadequate surface preparation;
(b) application of the paint coating under unsuitable atmospheric conditions or by inappropriate methods.

Some paints may be applied over light surface rust, indeed some coatings have a phosphoric- or tannic-acid-based component in the paint which converts rust to an inert, adherent layer, usually oxidising Fe(II) to Fe(III). The paint then keys to the inert surface. However, most paints, particularly those which will be exposed to aggressive environments, benefit from a thorough surface preparation. For outdoor work, wire brushing a surface is not a satisfactory preparation because it leaves contaminants and potentially active corrosion products on the metal.

Satisfactory surfaces for painting are obtained either by chemical treatments, such as pickling by immersion in acid solution, or by grit blasting to remove grease, dirt, corrosion products and scale. Pickling baths are normally inhibited to stop attack on the metal. Components treated by pickling must be thoroughly washed and dried after treatment to remove all the active chemicals from crevices and holes, and to ensure that the paint is able to key to the metal. Grit blasting can be combined with a water spray to wash away all the debris, dust and water-soluble contaminants which would reduce the adhesion of the paint. The surface profile produced by the grit is very important. If it is too rough, micro-peaks will be poorly covered by the paint and premature penetration of the film will occur, as illustrated in Fig. 14.3.

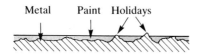

Fig. 14.3 Paint thickness is affected by roughness of the substrate surface. Grit used for surface blasting should be graded to produce a uniform roughness if holidays are not to occur.

For a satisfactory life, paint must be applied under the correct atmospheric conditions. If the relative humidity is too high a thin, invisible film of water will prevent proper keying to the metal surface and may interfere with the integrity of the dry paint film. Ambient temperature affects the drying or curing time. Solvent evaporation can be slow at very low temperatures and some of the twin-pack paints will not cure below a specified temperature.

Variations in temperature across a component, especially where the paint is heated or stoved in an oven to speed or complete the drying process, can cause solvent which has evaporated from one area to condense on adjacent, cooler surfaces which are below the dew point of the solvent. The paint dissolves in the excess precipitated solvent which runs off the surface leaving streaks in the finish and thin areas in the protective film, see Fig. 14.4.

Many failures of twin-pack paint systems have been attributed to inadequate mixing of the two components prior to the application of the paint, or the use of incorrect proportions of the two parts. Failures have also occurred when twin-pack paints, which rely upon a cross-polymerisation curing process to form the finished coat, have been applied below the minimum temperature for rapid and full curing. This can be as high as 15°C

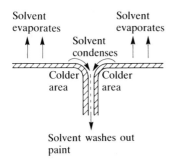

Fig. 14.4 Solvent wash-out. In a stoving oven the single thickness skin heats up more quickly than the double thickness in the joint. While the joint is below the dew point temperature for the solvent, condensation occurs in the joint and washes out the paint.

for some epoxy-based systems. Such paints also have a maximum pot-life after mixing and should be discarded at the end of that period.

When moisture diffuses through the paint film, salt solutions based on soluble corrosion products or compounds from the paint can form at any point where there is a lack of adhesion between the paint layers or between the paint and the substrate (see Fig. 14.5). The paint film acts as a semi-permeable membrane and as the concentration of the salt solution in the defect rises osmotic pressure forces water through the film to dilute the solution. This increases the water pressure in the defect and blisters the paint while spreading the area of delamination between the layers. Besides ruining the paint finish, the pH of the solution in the blister can become very

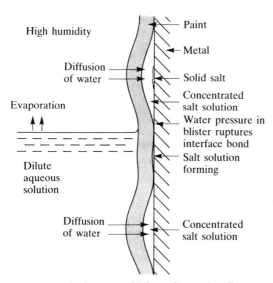

Fig. 14.5 Development of blisters in poorly adherent paint films owing to osmotic pressure. (After LaQue.[6])

255

high and cause caustic burns to anyone who carelessly pierces the blister to let the 'water' out prior to remedying the defective paint scheme.

A medium quality paint properly applied to a thoroughly prepared surface will give a better performance than the best quality paint applied to a poorly prepared surface.

14.5 COST OF A PAINT SYSTEM

Bearing in mind the discussion of the previous section, it is not surprising to find that the paint itself accounts for only a small proportion of the cost of coating a structure. The bulk of the charge is used to erect scaffolding to gain access to the structure, to provide shelter under which the correct temperatures and relative humidities can be maintained, to clean and dry the surfaces to be painted, and to provide adequate supervision and inspection of the prepared surfaces and the painting process. Any loss of production and the redeployment of staff temporarily displaced from their workplace must also be taken into account.

In the repainting of a large land-based plant, perhaps with large runs of open steel framing, the cost of the paint may be only 5 per cent of the total bill, while in marine locations it will be even less. Porter[7] found that contractors quoted a price of approximately £6 m^{-2} for painting girder work in 1977, excluding scaffolding, cleaning and transport. Duncan[8] estimated that the cost of repainting the tidal zone of a North Sea oil platform was £300 m^{-2} in 1984. This included such essential items as accommodation on a support ship, safety training on the platform, transport, and lost time owing to inclement weather and tide conditions. The paint cost was a trivial amount, but the overall cost for a programme covering three or four platforms could be over £5 m. in a single year. At that time he listed three factors which would significantly reduce costs:

(a) Development of coated blasting grit which deposits the primer coat as the blasting process continues and hence allows work to continue in damp conditions.

(b) Development of high-build, fast-curing top coats which could be applied at low ambient temperatures.

(c) Development of coatings which would build the full thickness in one application.

In these circumstances it would be beneficial to pay two or even three times as much for the basic paint if it would reliably tolerate wet steel and high relative humidities during application. While these factors resulted from a consideration of conditions on North Sea oil platforms, they would also be relevant to any painting task on a metal structure in the open air.

14.6 PLASTIC COATINGS

Thermoplastic and elastomer coatings are applied to many relatively cheap metal substrates to combine the mechanical properties of the metal with the corrosion resistance of plastics. Coatings have been developed which will operate in a wide variety of environments, such as acids, alkalis, abrasive slurries, continuous immersion in sea water, or in the splash zone of ships and marine structures. Some will withstand temperatures of up to 250°C for continuous operation. They are found as linings for tanks used to hold and transport acids, on strainers in power station intakes, on caissons, helidecks for offshore platforms, as water box and tube plate liners for condensers, as liners for pipes, chutes and launders. In some cases catastrophic and rapid failure of the substrate would occur if the coating should be breached, for example in lined steel tanks carrying acids. Clearly, the integrity of the coating and the bond to the metal are of paramount importance in using these materials. It is advisable to consult the manufacturer's literature, or seek direct advice before the application of a coating.

Many of the materials can be built to a thickness of 10 mm or more in a single application, although the coating properties often limit the thickness which can be used. However, in all cases the article to be treated must be thoroughly cleaned, free from corrosion products and correctly treated to give the optimum surface finish for the coating material and process to be used. An adhesive/primer coat may be necessary to ensure good bonding to the substrate. Some plastics can release volatile components in service under the effect of heat or ultra-violet radiation, which can attack adjacent metal surfaces. For example, PVC releases hydrogen chloride gas, while nylon releases acetic acid.

Plastics are coated on to metals in a number of ways:

(a) **Dipping**. The heated component is dipped into a fluidised bed of finely powdered coating material. The powder adheres to the heated surface. Subsequently the component is heated to higher temperatures to fuse the powder into a smooth coat. PVC, nylon and polythene are applied in this way. In some processes a vacuum is used to avoid the entrapment of air in the coating. Masking can be applied to leave desired areas of the surface free from the coating.

(b) **Spraying**. This is a labour-intensive coating method, but one which can be applied on site or to large structures which cannot be dipped. It is very easy to apply the coating to selected sections of the structure. Spraying techniques include airless spraying, electrostatic spraying, in which a high voltage is used to charge the powder while the component is earthed so that the charged powder adheres to the surface, and flame spraying, in which the powder is heated in the gun. Post-spraying heat treatment may be required to produce a sound bond and a smooth coating. Spray coatings are gradually built up to the desired thickness, time being allowed between coats for the material to cure. The pattern of spraying for each coat should be varied to ensure that the entire surface is covered and any holidays (small holes in the coating) are sealed by subsequent coats. Spray coatings are easily built up to 10 or 15 mm, although it may be necessary to use an inter-coat adhesive layer if long periods, say more than 12 hours, elapse between applications.

(c) **Rolling, trowelling and brushing**. Many twin-pack systems can be applied by these manual techniques.

Competition to supply plastic coating materials is as intense as that in the paint industry and there is a wide range of materials available, most sold under trade names. Some of the basic materials are described below.

Nylon
The low water absorption types are preferred. They are easily coloured, do not chip and have good resistance to oils and solvents. Nylon can be used at temperatures up to 120 °C. This enables it to be sterilised, which makes it useful in the food processing industry. Coating thickness is normally up to 1 mm and it is applied by dipping or spraying. While a good bond is formed on both steel and aluminium, an adhesive primer may be necessary on copper alloys.

Polyethylene (polythene)
The coating is applied either by dipping or spraying, but only in thin films. It is prone to stress-corrosion cracking in some environments, for example, certain detergents, alcohols and silicones. It is used for coating domestic ware, wire shelves and display units, while high density polythene has been applied to ducting, chemical tanks, industrial shelving and pipework.

Polyvinyl chloride (PVC)
This is a most versatile coating with properties which can be varied by changing the plasticiser content to meet differing service conditions. An adhesive/primer coating is necessary to give good bonding to the substrate. The material is applied by spraying and fluidised bed techniques, including the liquid plastisol process in which a heated component is dipped into cold liquid PVC. On the hot surface the polymer and plasticiser cross-link to give a gelatinous deposit. This is cured in a higher temperature treatment to yield a tough coating. Coatings up to 12 mm thick are laid down in one treatment, but the powder method is more suitable when a thinner coating is required. For continuous use, the ambient temperature should not exceed 60–70 °C.

Polytetrafluoroethylene (PTFE)
This is an expensive but highly corrosion-resistant material. It is well known for its high temperature stability (up to 250°C) and its resistance to attack by solvents, acids and alkalis. It does not absorb water. However, corrosion control on a substrate cannot be guaranteed owing to the difficulty of building a sufficiently thick layer to eliminate micro-porosity in the film.

Polyurethane

This twin-pack material can be applied by airless spraying, trowelling, pouring or centrifugal casting. Different formulations will operate in the temperature range $-70\,°C$ to $90\,°C$, although the maximum operating temperature will be restricted when the structure is exposed to very high relative humidities or immersed in water. Elastomer coatings have given satisfactory service in sea water, lubricants, detergents, and mild concentrations of acids and alkalis.

14.7 CONCRETE COATINGS

The construction industry uses large quantities of steel reinforcing bars and load carrying girders to strengthen concrete structures. The highly alkaline environment which exists in the concrete inhibits the corrosion of the steel by maintaining a passive film on the metal surface. However, if water, oxygen and carbon dioxide are able to penetrate into the concrete, the carbon dioxide reacts with components of the concrete to precipitate carbonate in place of the hydroxides which confer the alkalinity. As the pH falls, the passive film breaks down and the water and oxygen cause the formation of rust which occupies a larger volume than the metal from which it formed. Thus, the formation of rust generates tensile hoop stresses in the concrete surrounding the steel and the concrete often cracks. (The tensile strength of concrete is much lower than the compressive strength.) Once the concrete has cracked, the free access of water, or the freezing of water in the cracks in cold weather, accelerates the deterioration of the concrete. The carbon dioxide reaction with concrete is referred to as carbonation, and the cracking as spalling. Chloride ions from a marine environment or from de-icing salts also contribute to the deterioration. Figure 14.6 shows an example of spalling.

In the 1950s calcium chloride was added to concrete mixes to accelerate the setting of the material. The calcium chloride not only accelerates the setting of concrete, but also the rate of corrosion of the steel, and leads to premature cracking and spalling of the concrete matrix. Such additions are now prohibited.

Case 14.2: A multi-storey office block built in the 1950s using a reinforced concrete frame with brick infill panels began to spall 10 years after completion. Calcium chloride had been added to the mix to speed setting. Although a surface sealant was applied to the concrete to restrict the ingress of moisture, the deterioration continued. In 1983 the damaged areas of concrete were removed, the steel reinforcing either replaced or grit blasted back to clean, sound metal, and then coated with epoxy. The outer layer of epoxy was dusted with sand to provide a key for the repair mortar which was applied, to build up the surface to match the surrounding levels.[9]

Fig. 14.6 Concrete spalling to reveal corroded steel reinforcing bars. This effect on a concrete pavement on a bridge deck in the US was caused by the use of de-icing salts during winter months. (Photo courtesy of Taywood Engineering Ltd.)

An impervious, dense surface layer of concrete will maintain the alkaline conditions at the metal/matrix interface. Clearly, the exposure conditions will govern the thickness of such a layer, but for structures immersed in sea water a minimum of 50 mm is recommended. On oil drilling platforms, the recommended minimum coating thicknesses are 75 mm for submerged areas and 100 mm for those in the splash zone.[10]

Galvanising has been suggested as a suitable coating to protect steel reinforcing bars. However, there is still considerable controversy over such a course. Certainly, zinc corrodes in damp, alkaline concrete, but an impervious surface film will probably form on the zinc if carbon dioxide is present.

The film will restrict the corrosion of the underlying steel. The zinc corrosion product from galvanising is unlikely to generate sufficient stress to spall the concrete, although the greater bulk of the products from solid plate, rods or tubes of zinc may do so. Cathodic protection is often applied to reinforcing bars in concrete to control corrosion, especially that arising from chloride ions (see section 16.5), although inhibitors can also be added to concrete to reduce the corrosion of the reinforcing steel.

14.8 METALLIC COATINGS

Many of the commonplace objects around us are finished with metallic coatings to preserve and give lustre to the basic substrate metal which provides the strength, rigidity and formability to produce the object. Thus, for example, dustbins are galvanised, cans are coated with tin, and the bright parts of cars are chromium plated (although many of these have now been replaced with plastics).

The metallic coating interposes a continuous barrier between the metal surface and the surrounding environment. The ideal properties of a metal coating are summarised as follows:

(a) It should resist attack from the environment to a greater degree than the substrate metal.

(b) It should not promote corrosion of the substrate at any flaws or deliberately introduced breaks in the coating.

(c) Its physical properties, such as elasticity or hardness, should be adequate to meet the operational requirements of the structure.

(d) Its method of application must be compatible with the fabrication processes used to produce the completed product.

(e) It should be uniform in thickness and free from porosity and holidays. (This requirement is almost impossible to attain.)

A pre-requisite for all metal coating processes is thorough preparation of the substrate surface to:

(a) remove all surface contamination, grease, oil, dirt and process debris;
(b) remove surface corrosion products;
(c) control the physical characteristics of the surface.

The simplest method of removing surface contamination is to dip the component in a tank of solvent such as acetone, trichloroethylene, carbon tetrachloride or benzene at room temperature. A more efficient process is to use hot solvent or solvent vapour. Many of the solvents are toxic or carcinogenic and most are highly flammable: the treatment plant must be correctly designed and maintained. Alkaline solutions are also used in immersion baths, followed by thorough washing of the component to remove the solution. The effectiveness of both solvent and alkaline baths can be improved by ultrasonic agitation during the cleaning process.

Acid or alkaline solutions are used to remove corrosion products, the choice depending upon the metal involved. However, during these cleaning processes hydrogen may be generated on steel surfaces and may diffuse into the metal causing embrittlement of the matrix. High strength steels are particularly susceptible to hydrogen embrittlement (see section 10.3). Grit blasting is also used to clean steel surfaces before a coating is applied.

While most of the coating processes can be applied to cast, wrought, polished or grit-blasted surfaces which have been thoroughly cleaned and degreased, spray coatings adhere only to roughened surfaces. A uniform roughness is required and this is obtained by grit blasting with a grit size appropriate to the substrate metal. On steel surfaces the spray coat is put on immediately after blasting to prevent re-oxidation of the metal.

14.9 METALLIC COATING METHODS

The following methods are commonly used to apply metallic coatings described in the previous section.

Electroplating

In this method, the component, together with rods or plates of the metal to be plated, is immersed in an electrolyte containing salts of the plating metal. When a potential is applied to the cell so that the component becomes the cathode and the rod or plate the anode, ions of the plating metal deposit from the solution on to the component surface while more ions dissolve from the anode. Using correctly formulated solutions and anodes, alloys as well as pure metals can be plated. Good practice produces coatings of controlled thickness, a fine grain size and relative freedom from porosity. **Throwing power** is a term used to describe the ability of a plating solution to produce an even coating thickness as the distance between the anode and the component surface varies when plating complicated shapes. Chromium has poor throwing power and requires complicated anode arrays to give an even plating thickness, especially on curved or convoluted surfaces, or in blind holes. Hydrogen embrittlement can occur during the plating process and post-plating heat treatment may be specified to diffuse the hydrogen and prevent cracking of the substrate.

Hot dipping

In this method, the structure is dipped into a bath of molten coating metal. A good metallurgical bond is formed with the substrate, owing to interfacial alloying. There is less control over the coating thickness in the dipping process; the coat tends to be uneven, thicker on lower surfaces and thinner on top. However, all surfaces exposed to the molten metal are coated. The process is limited to low melting point metals such as tin, zinc and

Fig. 14.7 Galvanising by hot dipping in a bath of molten zinc. The component is the sports car chassis depicted in Fig. 12.7. (Reproduced by permission of Lotus Cars Ltd.)

aluminium. Figure 14.7 shows the sports car chassis of Fig.12.7 being galvanised by hot dipping in a zinc bath.

Spray coats

Wires of the coating metal are fed into a torch where they are melted and blown out under pressure as fine droplets. The droplets, travelling at 100 to 150 m s^{-1}, are flattened on striking the substrate surface and adhere to it. For a given thickness, the coat will be more porous than dipped or electroplated coats, but thick layers are easily built up by repeated spraying. The process can be applied on site after final erection of a structure. However, it is labour intensive and requires grit blasting just before spraying. The coating metal is usually anodic to the substrate; aluminium or zinc is applied to steel by this method.

Clad coatings

Corrosion-resistant metal skins can be laminated with other metals which, while they have the particular engineering properties needed in a structure, lack the necessary corrosion resistance in the working environment. Skins have been applied by rolling, by explosive welding, and by building up a welded coat on the substrate, a process known as **buttering**. Normally, diffusion across the interface between the metals produces an alloy-bonded layer which has very good adhesion. The main difficulty in obtaining a sound bond is the lack of diffusion in the presence of surface contaminants, mostly oxides. The pressure in the rolling mill must be sufficiently great to break

263

up the oxide. Explosive welding scours the two metal surfaces at the moment thay are forced together. Additional protection is required at cut edges, holes, or other breaks where the substrate is exposed to the environment. Of course, the skin must be able to cope with the fabrication processes used to make the final product.

Aluminium roll-bonded to duralumin is marketed as Alclad; cupro-nickel has been butter-welded on to steel pipes used in sea-water services, and lead has been explosively bonded to steel.

Diffusion coatings

A number of processes exist for diffusing a coating metal, or even non-metals, into the surface layer of a substrate to form an alloy layer on a component. A very good bond is produced, but the process is limited to relatively small objects. Components to be treated are first degreased and cleaned. Then they are heated, either in contact with a powdered coating metal in an inert atmosphere (the solid route), or in a gaseous stream of a volatile compound of the coating metal (the gaseous route).

In the solid route, the components and the metal powder are mixed with sand and sealed in a drum. Next they are heated to just below the melting point of the coating metal and rotated for several hours. This technique is used to coat steel with zinc, a process known as **sheradizing**, or with aluminium, in which case it is called **calorizing**. The gaseous process uses halides to coat steel with chromium, called **chromizing**, or for silicon coatings. Both types of coating improve corrosion resistance; chromium also improves wear resistance, while silicon increases resistance to acid attack. Chromium can also be applied to certain alloy steels, while both aluminium and chromium are coated on to nickel and cobalt alloys (the superalloys).

14.10 THE BEHAVIOUR OF METALLIC COATINGS

The behaviour of metallic coatings is influenced by many factors. The nature of the electrolyte, oxygen concentrations, polarisation characteristics, relative areas of anode and cathode, and surface deposits on the coatings will all affect the performance of any corrosion cells which develop. It is not unknown for cell reversal to occur (see section 5.1) or for insoluble corrosion products to block defects in coatings which might otherwise be expected to offer little corrosion protection.

If an anodic (sacrificial) coating is used the conductivity and continuity of the electrolyte will control the size of the surface defect which can be tolerated before corrosion occurs on the substrate. Thus, for a zinc coating on steel, a defect as small as 3 mm wide will produce corrosion in distilled or soft water, whereas in sea water the steel will be protected at a defect several decimetres across (but the rate of loss of zinc will be much higher). The following discussion relates to the performance of specific metals.

Zinc

Although zinc is more anodic than iron in the galvanic series, the zinc corrosion products, such as oxide, hydroxide and basic carbonates, form a protective film on the metal surface which reduces the corrosion rate of the zinc to levels well below those for iron or steel. Thus a zinc coating has a long life, but provides sacrificial protection at any breaks in the coating. Zinc is applied by all the coating methods described in section 14.9. The corrosion resistance of a zinc coat is independent of the method of application, although those containing 5.8 per cent iron are said to be less prone to pitting attack than pure zinc. Post-coating heat treatment can be used to encourage diffusion across the interface between the metals to increase the iron content of the coating. Small additions of magnesium, aluminium and titanium improve the performance of galvanised coatings in flowing sea water.

The useful life of a coating largely depends upon its thickness and the environment to which it is exposed. For most of its life, zinc simply forms a corrosion-resistant barrier between the steel substrate and the environment. However, owing to the sacrificial protection of the zinc, large areas of the coating can be lost before the steel is attacked, assuming the electrolyte has good electrical conductivity and extends across the bare substrate to the remaining coating. Under these conditions serious rusting will be delayed while as little as 10 per cent of the original coating remains. In general, a coating of thickness 0.03 mm exposed to the atmosphere will last 11 to 12 years in rural areas or 8 years in marine locations, while sulphur oxide pollution in industrial areas will reduce the life to 4 years. Immersed in sea water, each 0.03 mm of coating thickness lasts for approximately 1 year, but pollution, especially hydrogen sulphide arising from sewage contamination in estuaries, will increase the rate of deterioration of the coating. Figure 14.8 shows the life to first maintenance for specified minimum thickness of zinc coatings in different environments.

Case. 14.3: Telford's suspension bridge over the Menai Strait in North Wales is exposed to an aggressive marine environment with high wind velocities and a wet climate. During 1938–40 all the steel work of the bridge was renewed, including the deck structure and the suspension chains. The deckwork was scratch-brushed and painted with two lead-based priming coats and one finishing coat. The tensile steel suspension chain links were grit blasted and a flame-sprayed zinc coat applied to a thickness of 125 to 150 μm. They were then primed with one coat of red lead paint. After erection a further red lead primer coat and a finishing coat were applied to the chains. Zinc spray was used because the individual chain links (approximately 6 m long) were too large to galvanise by dipping at that time. Furthermore, spraying, a cold process, would not affect the temper of the heat-treated steel links.

Galvanised coating, specified minimum weight (g m^{-2})

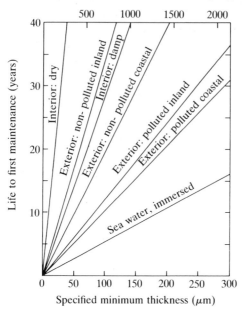

Fig. 14.8 The effect of environment and thickness on the life of a zinc coating.

In 1947 a detailed examination showed that the deck structure had corroded badly, especially on the under surfaces. It has required a continuous repainting scheme since that date. On the other hand, the chains showed no sign of rust. The top coat of paint was thin in places, but the primer and underlying zinc-sprayed coat were intact. In 1950 the complete bridge was repainted. The chains required only a fresh top coat to match the new colour scheme. A further examination in 1980, after 40 years' service, showed that there was still no sign of rusting on the chains, although areas of the deck needed replacement because of corrosion damage.[11]

A problem with zinc coatings which must never be overlooked is cell reversal. This is promoted by HCO_3^- and NO_3^- ions, especially in hot electrolytes, while Cl^- and SO_4^{2-} ions maintain the normal sacrificial role of the zinc.[12]

Cadmium

An examination of standard electrode potentials (Table 4.1) reveals that cadmium should be cathodic to iron, promoting its dissolution in a corrosion cell. However, it is found that cadmium behaves anodically to iron and steel and protects them sacrificially. This is reflected in the practical galvanic series, Fig. 5.5. Again, for most of its life, a cadmium coating acts as a physical barrier between the substrate and the environment. The sacrificial role becomes operative only at breaks in the coating, or towards the end of its life when the coating becomes porous. Unlike zinc, the nature of the film which develops on the surface of cadmium is strongly dependent upon the atmospheric conditions. In rural or industrial locations soluble sulphates form part of the film; rain water washes these away and the corrosion rate of the cadmium remains high. Insoluble carbonates and chlorides form in marine areas and the corrosion rate is much lower. For a given metal thickness cadmium offers better protection to steel than zinc in marine environments, but the situation is reversed at rural or industrial sites.

Cadmium is normally applied in thin coats by electroplating. It has a bright finish, but is more costly than zinc and is mainly used to coat small articles such as fasteners. Cadmium is preferred to zinc as a coating by the electronics industry because it is easier to solder. It is also useful as a coating on high strength steels owing to the lower risk of hydrogen embrittlement from the plating bath[13] (see section 10.3). Unfortunately, cadmium salts are highly toxic and its use can form a significant pollution hazard in some instances. Cadmium and its corrosion products must be kept well away from food and food processing areas.

Aluminium

Coatings of aluminium are normally produced by spraying or hot dipping, although small articles are diffusion coated. Silicon is added to hot dip baths to retard the formation of brittle intermetallics at the interface between the coating and the substrate. Such intermetallics reduce the formability of the coated sheet and impose restrictions on the subsequent fabrication process. Sprayed coatings contain much higher levels of oxide than dipped coatings, and they are more porous. However, the corrosion product rapidly fills the micro-holes in the coating and forms an adherent, compact and impervious coating. Some corrosion of the substrate may occur during the sealing process, in which case unsightly rust stains may appear on the outer surface of the coat. This can be avoided by sealing the coat with a lacquer during the early stages of its exposure, although the rusting is not harmful. Lead-based paints must **NEVER** be used on aluminium, but zinc chromate pigmented paints are suitable for use on aluminium coatings.

It has been shown that when aluminium coatings on steel are exposed to industrial atmospheres they give a longer life than zinc.[14] In one test, a zinc coating of thickness 0.08 mm, for example, lasted about 7 years, while a similar aluminium coating gave a life of 12 years.

Both dipped and diffusion coatings of aluminium are used to protect steel exposed to temperatures up to 1100 °C in air. The coatings also offer good protection from attack by sulphur compounds and are used in the chemical industry, on some gas turbine blades and for car exhausts.

Nickel and chromium

Both of these metals are cathodic to steel and act as a barrier coating to

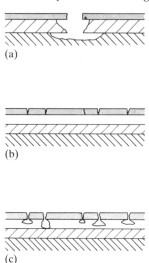

Fig. 14.9 Corrosion in normal and micro-cracked chromium plating:
(a) Normal chromium plate: steel is anodic to both coating layers.
(b) The usual arrangement of layers in micro-cracked chromium plating.
(c) Corrosion in micro-cracked chromium plating: the corrosion is confined for long periods to the bright nickel layer which is anodic to both the chromium and the nickel layer.

protect the substrate surface. Once the coating is breached, the corrosion rate of the steel is faster than it would have been without the coating. There are many different types of coating arrangement involving nickel and chromium. Figure 14.9 shows two of these which will be discussed below. Some coatings have chemical additions to modify the different layers in the system. Thus, the addition of sulphur compounds to nickel produces a layer with a bright finish which is anodic to pure nickel. If pure nickel is plated on to steel and overcoated with the bright nickel, corrosion is initially restricted to the bright coat and penetration to the steel is delayed.

During prolonged exposure outdoors, nickel coatings tend to lose their surface brightness. To maintain a bright and therefore attractive surface, thin coatings of chromium are plated over the nickel. This is the traditional chromium plating used on articles intended for outdoor use.

Chromium coatings are highly stressed and it is difficult to deposit a coating that does not contain micro-cracking or porosity. When water enters the cracks, micro-cells are generated. The chromium is the cathode in the micro-cell, it has a large surface area compared to the anode (the nickel plate at the base of the crack) and therefore the corrosion rate of the nickel is very high. The propensity of the chromium to crack is turned to advantage in the production of micro-cracked chromium plating. The steel is first flash-coated with copper to produce a smooth surface and facilitate buffing. This is followed by a pure nickel coating over which a bright nickel coat containing sulphur compounds is deposited. The process is completed with a thin overlay of chromium (Fig. 14.9(b)). The normal cracking of the chromium is enhanced by the sulphur compounds in the nickel coat to give a cracking density of the order of 10^8 m^{-2}. The bright nickel is anodic to both the pure nickel and the chromium, therefore corrosion proceeds sacrificially in this layer (Fig. 14.9(c)). However, the larger anode/cathode ratio produced by the increased cracking density reduces the tendency for pitting to occur. Gradually, flakes of the overlying chromium drop off, dulling the surface, but the appearance of unsightly rusting is postponed. Micro-cracked chromium coatings are used extensively for the bright trim on cars.

Nickel coatings are used to replace metal lost by wear, while chromium coatings can be used to produce hard surfaces, but these are outside the scope of this book.

Tin

Large quantities of tin plate are used for cans in the food industry. The coating is cathodic to steel on the outside of the container and promotes corrosion if damaged. However, cell reversal occurs with many organic acids such as citric acid found in fruit juices, and sacrificial protection occurs on the inside of the can occurs.

14.11 REFERENCES

1. Nicholson J 1986 Paint is only skin deep, *New Scientist* **110** (1510): 40–43
2. Mayne J E O 197˘ The mechanism of the protective action of paints, in Shreir L L (ed.), *Corrosion* (Vol. 2). Newnes-Butterworths p 15:29
3. Uhlig H H 1971 *Corrosion and corrosion control.* John Wiley, p 248
4. Munger C G 1984 Zinc silicate primers – forty years' experience, in *Proceedings of UK corrosion 1984.* Institution of Corrosion Science and Technology
5. Seibert O W 1978 Classic blunders in corrosion protection, *Materials Performance* April: 33–37
6. LaQue F L 1975 *Marine corrosion.* John Wiley, p 293
7. Porter F C 1978 *Comparative costs of protecting steel.* London: Zinc development Association
8. Duncan J 1984 Cost implication of offshore platform structural protection by coatings, in Lewis J R, Mercer A D (eds) *Corrosion and marine growth on offshore structures.* Ellis Horwood
9. Taywood Engineering Ltd 1984 *Fact sheet No. 4.* Southall, Middlesex: Taywood House
10. Sharp J V 1979 The use of steel and concrete in the construction of North Sea oil production platforms, *Journal of Materials* **14**: 1773–99
11. Editorial 1983 *Corrosion Prevention and Control* **30**(1): 2

12. Chamberlain J 1985 Cell reversal of the zinc/iron system in sodium bicarbonate and sodium nitrate solutions, *Surface Technology* **25**: 229–57
13. Hole H G 1979 Cadmium coatings, in Shreir L L (ed.), *Corrosion* (Vol. 2) Newnes-Butterworths, 14: 31–5
14. Uhlig H H 1971 *Corrosion and corrosion control.* John Wiley, p 240

14.12 BIBLIOGRAPHY

Gaber D R 1972 *Principles of metal surface treatment and protection.* Pergamon Press

Slater J E 1983 *Corrosion of metals in association with concrete.* ASTM-STP 818

Tonini D E, Gaidis J M (eds) 1980 *Corrosion of reinforcing steel in concrete.* ASTM-STP 713

15 THE CORROSION PROPERTIES OF SOME METALLIC MATERIALS

Gold is for the mistress – silver for the maid –
Copper for the craftsman, cunning at his trade.
'Good!' said the Baron, sitting in his hall,
But Iron – Cold Iron – is master of them all.

(Old tale)

It is clearly impossible to list materials to match every conceivable combination of environmental conditions. Aspects of the selection of the correct materials for particular applications have been discussed in Chapter 11. Many factors are likely to restrict the range of materials from which a choice may be made. Outside the oil and chemical industries, most large structures will be made from mild or low-alloy steel, aluminium, or concrete with steel reinforcing, owing to the cheapness, availability and strength of these materials. The choice between them will mainly be decided on the pattern of stress within the structure, the fabrication and joining techniques to be used, and the availability of labour with the skills to undertake the construction. The inherent corrosion resistance of the material will generally play a minor role in the selection process, the engineer looking to some barrier coating or other technique to control the loss of metal. Sophisticated corrosion-resistant alloys will only be used where there are specific hazards, as may be found in the oil and chemical industries, or where reliability is of paramount importance.

We have already seen that the expected life of a structure and the environments in which it will be fabricated, stored and used, must be specified early in the design process. Certainly, these factors should be known before any decision over the choice of materials is made.

Case 15.1: The use of closed loop, dry cooling towers at an electricity generating station at Rugeley in the UK was expected to eliminate the need for 5700 m³ of make-up water a day. The water coolers in the towers, which at the time were the largest natural draught shells in the world, were made from louvred aluminium plate. The only other towers operating on this principle were in Hungary, where they had given several years' satisfactory service. Within 18 months of installation, the Rugeley coolers had corroded and perforated. The problem was attributed to the difference in atmospheric environment between the Hungarian site and that at Rugeley. The latter was much more moist and contaminated with chloride ions. The aluminium coolers were the most expensive part of the plant and had been expected to last for the life of the plant. The problem was solved by modifying the design and applying an epoxy resin coat to the replacement units.[1]

Many metal manufacturers supply data sheets on their products which include information on the behaviour of the metals in various environments. Often this information is derived from accelerated testing procedures. These data sheets should be treated with some caution, especially concerning the environmental and test conditions under which the information was obtained. We have already seen that changes in the environment can alter the corrosion characteristics of the metal and cause pitting, selective phase or localised attack, and environment-sensitive cracking. All of these shorten the useful life of the metal. Accelerated test data must be correlated to that from field trials in the expected environment. Field trials should include elements to represent the real structure, such as the relative anode to cathode area ratio, joint configurations and heat treatments, particularly those changes which will occur on welding.

Incorrect testing techniques can produce false results which can be financially disastrous.

Case 15.2: In a process plant, chlorides were stripped from a strongly acid mixture with some residual chlorine and a very high salt content. The cold liquor was heated to 80 °C in a glass-lined steel vessel to boil off the chlorides. The glass-lined vessels had a life of one to two years. In a search for a longer life material, nickel, Inconel, Hastelloy, stainless steel and titanium were tested, but only titanium proved to be resistant to the process conditions. A complete titanium unit was installed but failed after two days' operation with a corrosion rate of 38 mmpy. In the laboratory tests, all the alloys had been placed in the same reaction chamber, and the iron(III) ions produced by some of the materials had proved an excellent corrosion inhibitor for titanium under these conditions. The iron(III) ions were absent in the real plant. The use of individual test cells for the laboratory tests would have prevented the false result.[2]

On the other hand, potentially useful materials may be missed.

Case 15.3: A large chemical scrubber was installed to operate in 40 per cent sulphuric acid at 82 °C. After some failures, a PVC-lined fibre-reinforced plastic was selected for the constructional material. It gave a life of one year. A stainless steel stack band had fallen into the scrubber and was observed, immersed in the acid, during an internal inspection following a PVC failure. Normally, it would be expected that stainless steel in hot 40 per cent sulphuric acid would corrode badly, but in this case it was bright and shiny and appeared to be in pristine condition. Tests showed that the sulphuric acid actually contained sufficient nitric acid to passivate the stainless steel under the conditions in the scrubber. A replacement scrubber made from type 304 stainless steel gave more than 12 years' maintenance-free service.[2]

This chapter is concerned with the general corrosion characteristics of the metals described in each section. It must be emphasised once again that if there is any doubt about the performance of a metal in a given environment, the advice of the manufacturers should be obtained, or a test programme run to evaluate the metal's behaviour.

15.1   PLAIN CARBON AND LOW ALLOY STEELS

Plain carbon steels corrode in most atmospheric environments when the relative humidity exceeds 60 per cent. Once a moisture film has formed, the rate of corrosion depends upon many environmental factors, the most important being oxygen supply, pH and the presence of aggressive ions, particularly oxides of sulphur, and chlorides.

The composition of the steel, its surface condition and the angle of exposure also affect the corrosion rate. Increases in carbon, manganese and silicon content tend to reduce the corrosion rate, although the presence of manganese sulphide in free machining steels produces local galvanic cells (section 7.2) and lower corrosion resistance. Copper additions have a beneficial effect which is discussed later.

Rolled or pickled steels corrode faster than machined or polished samples. Mill scale, a layer of iron oxides which is produced during heat treatment, hot rolling and forging processes, forms a very effective cathode on steel surfaces if it is not removed before the metal is put into service. The layer consists of FeO adjacent to the metal with Fe_3O_4 between it and an outer skin of Fe_2O_3. Although mill scale may act as a barrier coat to reduce atmospheric corrosion in the short term, it cracks and spalls under shock and vibratory loading, and differential thermal expansion, to leave bare areas of metal which corrode rapidly. Paint films applied over mill scale are useless as they spall with the scale.

Vertical plates are attacked more slowly than those at 45°, where more than half the metal loss is on the underside of the plate. Initially, the corrosion rate on bare metal is high, but it falls to fairly constant rates after the first year's exposure because of the formation of surface oxide films.

Table 15.1 lists some typical corrosion rates for different localities and types of atmosphere.

The corrosion rate for steel immersed in water is influenced by the interaction of many factors which have already been discussed in Chapters 12, 13 and 14. The geometry and fabrication techniques may introduce poor design features and control the flow rates, while the quality of the water also plays a major role. Increasing the flow rate, especially where turbulent flow results, increases the corrosion rate.

Table 15.1 Corrosion rates for mild steel plates (100 × 50 × 3.2 mm) freely exposed for one year in a vertical position[3]

Atmosphere type	Location	Corrosion rate (mmpy)
Rural	Godalming	0.048
	Teddington	0.070
	Khartoum	0.003
	Delhi	0.008
Marine	Brixham	0.053
	Calshot	0.079
	Singapore	0.015
	Lagos (beach)	0.615
Industrial	Woolwich	0.095
	Derby	0.170

In fresh water a corrosion rate of 0.05 mmpy is typical, although it may fall to 0.01 mmpy when calcareous deposits are formed. In sea water the average rate is likely to be in the region 0.1 to 0.15 mmpy, but pitting atack may penetrate much deeper. The presence of mill scale or exposure at half-tide locations where a repeated wet/dry cycle occurs, increases these figures considerably. The fastest corrosion rate for mild steel in marine environments occurs in the splash zone where there is an abundant oxygen supply. Here the rate of metal loss is likely to be four or five times faster than that for a fully immersed plate at the same location. A typical corrosion allowance on painted steel structures exposed in the North Sea is an extra 12 mm on plate thickness for a life of 15 to 20 years.[4] The factors affecting the corrosion of carbon steel in sea water are given in Table 15.2.[5]

Generalisations about corrosion rates for mild steels in soils are even more difficult. Soil quality is very variable, not only from one district to another, but with the seasons of the year. It is more profitable to consider the conditions required to maintain a sound barrier coat or cathodic protection system (Chs 14 and 16) and if a corrosion rate in a particular situation raises doubts about the safe operation of a system trials should be made in that area.

Low-alloy steels contain up to 2 or 3 per cent of various alloying elements, notably Cr, Ni, Cu, Mn, V and Mo. These alloying additions improve the mechanical properties, but have marginal effects on the corrosion rate of immersed or buried components where mild steel, low-alloy steel and wrought irons corrode at approximately the same rate. Slightly larger additions of chromium can produce marked improvements in corrosion behaviour, while small amounts of copper, up to 0.3 per cent, are said to reduce the pitting corrosion of steels in boilers, giving rise to lighter general corrosion instead.

The commercial pressure to develop new oil and gas fields has led to a

Table 15.2 The effect of changes in the environment upon the corrosion of carbon steel in sea water[5]

Factor in sea water	Effect on iron and steel
Chloride ion	Highly corrosive to ferrous metals. Carbon steel and common ferrous metals cannot be passivated. (Sea salt is above 55% chloride.)
Electrical conductivity	High conductivity makes it possible for anodes and cathodes to operate over long distances, thus corrosion possibilities are increased and the total attack may be much greater than that for the same structure in fresh water.
Oxygen	Steel corrosion is cathodically controlled for the most part. Oxygen, by depolarising the cathode, facilitates the attack; thus a high oxygen content increases corrosivity.
Velocity	Corrosion rate is increased, especially in turbulent flow. Moving sea water may (1) destroy the rust barrier and (2) provide more oxygen. Impingement attack tends to promote rapid penetration. Cavitation damage exposes fresh steel surface to further corrosion.
Temperature	Increasing ambient temperature tends to accelerate attack. Heated sea water may deposit protective scale, or lose its oxygen; either or both actions tend to reduce attack.
Biofouling	Hard shell animal fouling tends to reduce attack by restricting access of oxygen. Isolated hard shell animals promote localised pitting by differential-aeration under the shell. Bacteria can take part in corrosion reaction in some cases.
Stress	Cyclic stress sometimes accelerates failure of a corroding steel member. Tensile stresses near yield also promote failure in special situations.
Pollution	Sulphides, which normally are present in polluted sea water, greatly accelerate attack on steel. However, the low oxygen content of polluted waters could favour reduced corrosion.
Silt and suspended sediment	Erosion of the steel surface by suspended matter in the flowing sea water greatly increases the tendency to corrode.
Film formation	A coating of rust and mineral scale (calcium and magnesium salts) will interfere with the diffusion of oxygen to the cathode surface thus slowing attack.

significant increase in the number of sour (i.e. containing hydrogen sulphide) gas and oil wells. The low-alloy steels used in these drilling rigs and production wells have suffered higher rates of general corrosion, together with sulphide-promoted stress-corrosion cracking problems. It has been shown[6] that, of the elements listed above, only copper in amounts up to 0.3 per cent reduces the corrosion rate in hydrogen sulphide, while vanadium and molybdenum appear to be beneficial in resisting stress-corrosion cracking.

Where components are exposed to the atmosphere, up to 0.2 per cent of copper is very beneficial, yielding a dense, compact corrosion product which inhibits further attack. However, the conditions of exposure are important. The improvement is lost unless the steel is exposed in the open: in sheltered conditions where the steel is protected from rain the corrosion rate is little different from mild steel. Small additions of chromium also improve the corrosion resistance of steels.

Despite all the limitations imposed by the corrosion behaviour of mild and low-alloy steels, they have well-known engineering properties and with a sound, properly maintained protection system in a well-designed structure they will give a long life.

15.2 STAINLESS STEELS

Stainless steels are iron alloys which depend upon a very thin, transparent passive surface film of chromium oxide to resist corrosion. A supply of oxygen and a minimum level of 11 per cent chromium dissolved in the matrix material are necessary to maintain the surface film, which is self-healing in air at room temperature. If the film is not repaired after damage, the corrosion of stainless steel will be very rapid in most environments.

The main corrosion problems associated with stainless steels are environment-sensitive cracking (Ch. 10) and crevice corrosion (Ch. 7). Both processes are accelerated by chloride ions which are very aggressive to the chromium oxide film. Too high a carbon content (>0.03 per cent) can be detrimental to the corrosion behaviour of stainless steels owing to the precipitation of chromium carbide which leaves some areas of the matrix with chromium levels below the critical minimum to maintain the oxide film. The depleted areas become the anodes in a cell with the remaining oxide-coated metal forming the cathode (section 6.2). Although the carbon content of the steel, *per se*, may be low, additional carbon may dissolve in the steel during any heating process which carbonises oil, paints or greases on the metal surface, or even from heating where gas flames are used. All surface coats which may provide a source of carbon should be removed before stainless steels are welded, heated for forging or pressing, or for general heat treatment. Do not forget to remove colour code paints!

In the 1970s, stainless steel making was revolutionised by the introduction of argon oxygen decarburisation processing, AOD. This allowed carbon to

Table 15.3 Some stainless steels and their approximate compositions

Material type (AISI no.)	Approximate % composition (Fe remainder)					
	Cr	Ni	Mo	C	N	Other
Austenitic:						
201	16–18	3.5–5.5	—	0.15	0.25	5.5–7.5 Mn
202	17–19	4–6	—	0.15	0.25	7.5–10 Mn
205	16–18	1–1.75	1–1.7	0.25	0.4	14–15.5 Mn
304	18–20	8–10.5	—	0.08	0.10	—
304L	18–20	8–12	—	0.03	0.10	—
304N	18–20	8–10.5	—	0.08	0.1–0.16	—
304LN	18–20	8–12	—	0.03	0.1–0.16	—
316	16–18	10–14	2–3	0.08	0.1	—
316L	16–18	10–14	2–3	0.03	0.1	—
316N	16–18	10–14	2–3	0.08	0.1–0.16	—
316LN	16–18	10–14	2–3	0.03	0.1–0.16	—
317	18–20	11–15	3–4	0.08	—	—
317L	18–20	11–15	3–4	0.03	—	—
321	17–19	9–12	—	0.08	—	<0.4 Ti
254 SMO	19.5–20.5	17.5–18.5	6–6.5	0.02	0.18–0.22	0.5–1 Cu
20 MO–6	22–26	33–37	5–6.7	0.03	—	2–4 Cu
Ferritic:						
403	11–13	—	—	0.15	—	1 Mn
409	10.5–11.75	—	—	0.08	—	—
410	11–14	1–2	—	0.15	—	1 Mn
430	16–18	—	—	0.12	—	—
434	16–18	—	0.75–1.25	0.12	—	—
440A	16–18	—	0.75	0.7	—	1 Mn
444	17.5–19.5	—	1.75–2.5	0.025	0.035	Ti + Nb
446	23–27	—	—	0.20	0.25	—
AL 29–4–2	28–30	2–2.5	3.5–4.2	0.01	0.02	—
Duplex:						
3 RE 60	18–19	4.25–5.25	2.5–3	0.03	—	—
SAF 2205	21–23	4.5–6.5	2.5–3.5	0.03	0.08–0.2	—
329	23–28	2.5–5	1–2	0.08	—	—
Ferralium	24–27	4.5–6.5	2–4	0.04	0.1–0.25	1.5–2.5 Cu
DP–3	24–26	5.5–7.5	2.5–3.5	0.03	0.1–0.2	0.2–0.8 Cu 0.2–0.5 W

be reduced to very low levels, while the amount of other alloying elements could be controlled within very narrow limits. It became practical to specify nitrogen as a specific alloying element at levels of about 0.1 per cent, while keeping the carbon content at 0.03 per cent. The amounts of elements such as sulphur, bismuth, tin and lead which affect the fabrication and rolling properties could also be controlled. These improvements in production techniques allowed much tighter control on the specifications for existing steels and the introduction of a whole new range of higher alloyed steels with improved corrosion resistance. Many stainless steels are sold under trade names, but typical compositions are given in Table 15.3. The discussion which follows summarises separately the general features of each group.

There are many stainless steels and only the broadest discussion can be given here. Three categories which will be described are: **austenitic stainless steels**, in which sufficient nickel and nitrogen are added to stabilise the face-centred cubic austenite phase at room temperature; **ferritic stainless steels**, in which the matrix is body-centred cubic ferrite; **duplex steels**, which contain some nickel (about 5 per cent), and have a mixture of ferritic and austenitic phases.

Austenitic stainless steel

Owing to the high solubility of carbon in austenite, the very low carbon contents of the L grade austenitic steels permit welding without the danger of sensitising the metal to weld decay. (See Section 6.2)

The higher alloyed stainless steels are more resistant to crevice and pitting corrosion, especially in sea water. In the early 1970s, it was shown that type 316 could be used in fast-flowing sea water when crevices were absent from the structure; it could not be used in stagnant water, or where differential-aeration cells might occur. Intermediate grades of stainless steel with 4 to 5 per cent molybdenum gave improved performance, but were still subject to corrosion in crevices or beneath biofouling. More recently, 6 to 7 per cent molybdenum steels have proved to be resistant to crevice corrosion, even under adverse conditions. Highly alloyed steels containing molybdenum are resistant to stress-corrosion cracking.

Nitrogen can be substituted for nickel in these steels as an austenite stabiliser. It reduces the possibility of precipitating second phases (i.e. sigma phase) which are harmful to the general corrosion behaviour. It also improves the pitting resistance of molybdenum-containing austenitic steels, but adversely affects their stress-corrosion cracking performance. Austenitic steels which include nitrogen as an alloying element usually have higher molybdenum contents to offset the deleterious effects of the nitrogen on stress-corrosion cracking behaviour.

Molybdenum-containing austenitic stainless steels have been widely used in flue gas scrubbers attached to North American coal-fired power stations. The scrubbers utilise limestone slurries to neutralise the sulphurous and sulphuric acids in the flue gases, which may also contain chlorides and fluorides. In tests to simulate flue gas conditions, type 316 stainless steel developed pits 0.26 mm deep, type 317 pits 0.017 mm deep, while 254 SMO had no visible corrosion or pitting after the test. The introduction of stricter controls on effluent pollution from the paper manufacturing industry has also led to the use of highly alloyed stainless steels for the construction of waste processing systems. Although the use of highly alloyed austenitic steels greatly increases the capital cost in such systems, the reliable, long-term performance of these materials has justified the initial investment.

Ferritic stainless steels

In contrast to the relatively high solubility of carbon in austenite, the very low solubilities of carbon and nitrogen in ferrite cause the ferritic stainless steels to be easily sensitised to intergranular corrosion by the precipitation of carbides and nitrides. These stainless steels suffer both weld decay and reduced toughness unless the levels of carbon and nitrogen are kept very low. The new AOD and its extension to vacuum oxygen decarburising, VOD, allow these elements to be controlled while producing a wide range of chromium and molybdenum levels. Additions of titanium and niobium also help to keep chromium depletion by grain boundary precipitation to a minimum in the ferritic stainless steels.

Ferritic stainless steels are finding a wide variety of uses in modern industry. Type 409 has proved successful in trials for catalytic converters to be fitted to motor vehicle exhausts, while small additions of nickel, molybdenum and aluminium to 409 make it a financially viable alternative to mild steel for the manufacture of car exhausts. If the carbon and nitrogen contents are kept to suitable low levels so that the material can be welded and fabricated without sensitisation to weld decay, it can be used in tasks where abrasion would quickly damage normal barrier coatings, for example, in chutes, waggons, ducting and shields. Although the initial cost is higher than for mild steel, the total life cost may be lower. When fresh water is used in condensers and heat exchangers in the food industry, and chemical and refining plants, the higher levels of chromium, with some molybdenum, make type 444 a satisfactory material, combining good corrosion resistance with resistance to stress-corrosion cracking.

The very highly alloyed ferritic steels are resistant to severely corrosive environments. AL 29-4-2, for example, is able to withstand boiling 10 per cent sulphuric acid. These materials have been used extensively in caustic evaporators and in heat exchangers in the chemical and petrochemical industries. In some cases, they offer satisfactory replacements for more expensive nickel, nickel alloys, and titanium.

Since 1973, both ferritic and austenitic stainless steels have found increasing use in sea-water-cooled condensers at electricity generating stations. Initially, they replaced copper alloy tubes which had suffered sulphide attack owing to pollution of the water, but latterly they have been put in as initial installations.[7] The ratio of chromium to molybdenum is very important. In austenitic steels, a chromium level of 20–22 per cent requires 6 per cent of molybdenum, while ferritic steels with 25–28 per cent chromium require 3 per cent molybdenum. These stainless steels offer a high corrosion resistance, relative freedom from crevice corrosion in low velocity sea water, and in power station cooling systems they are operating at temperatures below that at which stress-corrosion cracking becomes a problem. They are not limited by upper critical water velocities at the flow rates it is practicable to use in such condensers.

The duplex stainless steels

These materials broadly combine the toughness and weldability of the austenitic types with the strength, corrosion resistance and stress-corrosion cracking resistance of the ferritic types. The solubility of the various alloying elements differs for the two phases, and this can produce variations in properties and corrosion resistance following the rapid temperature changes which occur during welding. However, they are readily produced in heavy sections, possess excellent toughness and corrosion resistance. A small increase in chromium and molybdenum content to ensure adequate concentrations of each element in both phases is a slight cost to pay for their excellent properties. Duplex stainless steels have proved very resistant to the general corrosion and stress-corrosion conditions which are found in sour gas wells.[8]

15.3 CAST IRONS

The corrosion resistance of grey cast iron immersed in water is relatively good when compared with that of mild steel. A slight improvement in corrosion resistance and an increase in strength is obtained by adding up to 3 per cent nickel. Further improvements in the impact resistance of the iron can be obtained by modifying the casting procedure to produce graphite spheroids instead of the normal flakes.

The graphite flakes in cast irons are inter-connected and noble to the surrounding matrix. When an iron corrodes, the graphite flakes are often left standing proud of the surface, gradually forming a noble, carbon-rich layer on the metal. This noble layer, acting as a cathode, promotes the corrosion of many metals when they are connected to iron, although an inspection of the galvanic series would lead one to expect that the iron would sacrificially protect the second metal. This reversal of the role of the iron is called **graphitisation**. It is less of a problem in spheroidal graphite irons where the spheroids are not inter-connected.

Additions of nickel in the range 13–36 per cent yield an austenitic matrix in the iron to produce the NiResist series of cast irons. These irons have greatly increased corrosion resistance when compared with grey irons. For example, in flowing sea water, the corrosion rate for NiResist is only one-tenth that of a grey or low-alloy iron. Austenitic irons are used for valve and pump bodies, and for condenser water boxes in refinery, marine and electricity generating systems. They are almost immune to graphitisation.

In concentrated sulphuric acid (greater than 65 per cent) a layer of insoluble iron(II) sulphate forms on the surface of grey iron to protect it from further attack. However, anything which destabilises the film, including increased flow rates, fluctuations in temperature or the presence of abrasive particles, can lead to pitting attack on the surface. Dilute sulphuric acid, and other acids readily attack unalloyed irons.

The corrosion rate of cast iron in dilute caustic solutions, below 30 per cent concentration, is negligible. Above this level, the addition of 1–3 per cent nickel is said to be beneficial in controlling the rate of attack. Nevertheless, cast iron is used for caustic fusion pots where the thick walls compensate for the initial high corrosion rate which falls to acceptable levels after a few days.

The austenitic irons have better corrosion resistance than the unalloyed or low-alloy irons in dilute sulphuric acid and many other acids, as well as in caustic solutions, although they show poor resistance to solutions of nitric acid.

15.4 ALUMINIUM AND ITS ALLOYS

Aluminium is a very active metal. Exposed to a source of oxygen, it reacts to form a thin, transparent oxide film over the whole of the exposed metal surface. This film controls the rate of corrosion and protects the substrate metal, allowing the production of long-life components in aluminium and its alloys. If the film is damaged and cannot be repaired, corrosion of the substrate is very rapid, as expt 3.14 shows.

Case 15.4: A series of 18 mercury-in-glass manometers were used to check pressures in an experimental condenser system. Just before the end of the trial, one manometer was broken and mercury spilled on to the aluminium alloy platform which surrounded the rig. To save an expensive test programme, the trial continued. Fifteen minutes later, when it was finished, the technicians were surprised to find that large areas on the platform had grown long filaments of white powdery oxide where their shoes had ground mercury into the aluminium surface. The facility was closed for 48 hours for cleaning-up operations.

Aluminium alloys can be divided into two groups, casting alloys and wrought alloys. The wrought ones can be subdivided into those which can be hardened and strengthened only by mechanical working, and those in which the properties can be varied between given limits by heat treatment alone. They are referred to respectively as non- heat-treatable (N) and heat-treatable (H). Typical compositions of aluminium alloys are shown in Table 15.4.

The corrosion performance of wrought aluminium alloys can be improved by cladding the material with thin sheets of pure aluminium or an aluminium alloy, often an aluminium/1 per cent zinc alloy. The cladding is pressure welded to both sides of the material by rolling. The core material provides the desired mechanical properties for the components being manufactured, while the cladding is chosen to be inherently more corrosion resistant than the core, but anodic to it, so that sacrificial protection is given at any break

Table 15.4 Some aluminium alloys and their approximate compositions

| Specification | | Approximate % composition | | | | | |
UK	USA	Cu	Mg	Si	Mn	Cr	Other
Wrought alloys:							
	2011	5.5	—	0.4max	—	—	0.5 Pb,0.5 Bi
H15	2014	4.4	0.4	0.8	0.8	0.1max	—
L97	2024	4.5	1.5	—	0.6	—	—
	2218	4.0	1.5	0.9 max	—	—	2.0 Ni
N3	3003	—	—	0.6 max	1.3	—	—
	4032	0.9	1.0	12.5	—	—	0.9 Ni
	5050	—	1.2	0.4 max	—	—	—
N4	5052	—	2.5	0.4max	—	0.25	—
N8	5083	—	4.5	—	0.7	—	—
	5086	—	4.0	—	0.5	—	—
N5	5154	—	3.5	—	—	0.25	—
H20	6061	0.25	1.0	0.6	—	0.25	—
	6063	—	0.7	0.4	—	—	—
	6101	—	0.5	0.5	—	—	—
	6151	—	0.6	1.0	—	0.25	—
DTD5074	7075	1.5	2.5	—	—	0.3	5.5 Zn
	7079	0.6	3.3	—	0.2	0.2	4.3 Zn
	7178	2.0	2.7	—	—	0.3	6.8 Zn
Cast alloys:							
	13	—	—	12.0	—	—	—
	43	—	—	5.0	—	—	—
	108	4.0	—	3.0	—	—	—
	A108	4.5	—	5.5	—	—	—
	D132	3.5	0.8	9.0	—	—	0.8 Ni
	319	3.5	—	6.3	—	—	—
	356	—	0.3	7.0	—	—	—
	380	3.5	—	8.5	—	—	—

in the cladding. Aluminium alloy tube clad on the bore has been used as water piping for several irrigation systems on farmland. If the coating is penetrated, the corrosion is confined to the cladding layer. This protection is particularly important at cut edges or drilled holes where the core is exposed. Among the factors which control the life and performance of the cladding are the following:

(a) Its thickness, which is directly proportional to its life.

(b) The conductivity of the electrolyte. This dictates the size of defect which can exist in the coat before corrosion of the substrate occurs; the better the conductivity, the larger the defect.

(c) The oxygen supply, which maintains the corrosion resistance of the coating.

Anodising may also be used to enhance the corrosion resistance of aluminium and its alloys. An electrolytic process is used to thicken the surface film and produce a hard, compact, strong and tightly adherent layer. The process can be applied to complicated shapes. Various finishes are available and the film may be dyed to give attractive colours to the finished object. (Details of the anodising of titanium to yield an attractively coloured finish are given in section 15.6.) The final process in anodising is to seal the surface by boiling in water or exposing the metal to steam.

Both anodising and cladding are frequently applied to the copper-rich 2000 series of alloys, which have poorer corrosion resistance than the other alloys. Without some form of additional protection, the 2000 series should not be exposed outdoors, immersed in water or buried in soil.

Care is needed in the heat treatment of aluminium alloys. Intermetallic phases are easily precipitated if the temperature or duration of treatment is outside those ranges specified for the alloy. Some phases, for example $CuAl_2$, are cathodic to the matrix and produce corrosion damage in the areas around the precipitate particles, while others, such as Mg_2Al_3, are anodic to the matrix. Continuous grain boundary films of Mg_2Al_3 are particularly dangerous because the large cathode size compared to the anode causes corrosion to penetrate deeply into the cross-section of the material for a relatively slight loss of weight.

Stress-corrosion cracking (Ch. 10) affects some alloys in specific heat treatment conditions. Those in the 2000 and 7000 series appear to be particularly at risk, especially in thick cross-sections. The heat treatment and environmental conditions which produce stress-corrosion cracking are complex and a specialist text should be consulted should problems of this nature occur (see the references and bibliography of Ch. 10). One method of reducing SCC in high-strength aluminium alloys is to use cladding of pure Al.

Most aluminium alloys suffer a higher rate of attack during the first year or so of exposure to the open atmosphere, but once they have developed a weathered surface, the rate falls, except in the case of the 2000 series. Rain washing can be beneficial to alloys exposed to the atmosphere, especially in marine locations where aggressive ions and deposits are washed away. Alloy 3003 is often used for building and constructional work outdoors. It may be clad when sulphur oxides and condensation are likely to be present, to prevent the formation of loose, voluminous corrosion products. Alclad is also beneficial in controlling differential-aeration corrosion in gutters, channels and roofing panels where leaves and other debris may collect. The 5000 series are also resistant to atmospheric corrosion, especially in marine or splash-zone locations.

When immersed in water, including sea water, most aluminium alloys exhibit good corrosion resistance. The major exceptions are the 2000 series, and some of the 7000 alloys which must be Alclad or protected with a barrier coating. Polluted water can promote pitting corrosion on many aluminium alloys. Copper and its ions, especially in the presence of chloride and carbonate, are particularly troublesome in this respect, the copper plating on to the surface and causing severe pitting attack (see case 12.5). The magnesium-containing alloys of the 5000 series offer a good combination of strength and corrosion resistance when immersed in water. Higher strength, at the expense of some corrosion resistance, may be obtained by using alloy 6061. When water temperatures are above 50 °C, the increased rate of oxide formation tends to prevent the initiation of pits, but in cold, static water, most alloys benefit from cladding.

Bimetallic couples involving aluminium should be treated with great caution and insulated where possible. Owing to the ease of ion exchange processes, copper and its alloys should not be used in the same system as aluminium, even where there is no direct contact between the two metals. The severe damage which results from ion exchange also rules out the use of copper and mercury anti-fouling paints on aluminium hulls.

Although aluminium and stainless steels are widely separated in the galvanic series, the oxide films present on both metal surfaces can act as a natural insulator and reduce corrosion to acceptable levels when the metals are coupled together. However, trials should always be undertaken before such couples are put into service.

Aluminium alloys are not suitable for handling inorganic acids, except concentrated nitric and sulphuric acids, although they do have a satisfactory performance with many organic acids. Alkalis attack aluminium, but dilute alkali solutions can be inhibited with silicates to reduce the rate of attack to acceptable levels.

A new class of alloys using lithium as the major addition are under development. Although additions of up to 12 per cent Li are being considered, current interest seems to be centred on 3 per cent Li with smaller amounts of zinc and magnesium. It has been predicted[9] that these alloys will supplant both 2024 and 7075 (see Table 15.4) in the construction of aircraft wing skins and fuselages over the next 10 years. The alloys offer a weight saving of up to 10 per cent and the corrosion resistance appears to be satisfactory. However, there is some loss of ductility when compared to 2024 or 7075.

The cast alloys are very similar in performance to the wrought ones. Again, the presence of copper in an alloy reduces corrosion resistance, especially in marine or other aggressive environments. However, the selection of a casting alloy will depend upon many factors, including the casting process (sand, die, or permanent mould), the fluidity required to produce complicated shapes, weldability when sections have to be joined, as well as the characteristics required in the final product – good wear-resistance, heat transfer, thermal expansion, and mechanical properties. In general, the alloys containing silicon and magnesium, or combinations of both elements, have good castability and excellent corrosion resistance.

15.5 COPPER AND ITS ALLOYS

Copper and some of its alloys are the oldest tool-making metals known to man. They have given excellent service in a wide variety of environments and the artefacts to be seen in many museums testify to the long-lasting qualities of these materials.

Pure copper is a very soft, malleable metal. It is alloyed with small quantities of metals such as Be, Te, Ag, Cd, As and Cr to modify the properties for particular applications, while retaining many of the characteristics of the pure metal. Much larger alloying additions of Zn, Sn and Ni are made to improve the mechanical properties of the metal, and to retain its excellent corrosion resistance under more arduous service conditions. Nickel permits increased flow rates in water systems; zinc gives increased resistance to sulphide attack. Typical compositions of some copper alloys are given in Table 15.5.

Table 15.5 Some copper alloys and their approximate compositions

Alloy	Approximate % Composition						
	Cu	Zn	Sn	Ni	Al	Fe	Other
Wrought alloys:							
Arsenical copper	99.6	—	—	—	—	—	0.4 As
Aluminium bronze	89	—	—	1	6–8	2	—
90/10 cupronickel	88.5	—	—	10	—	1.5	—
Aluminium brass (inhibited)	75.9	22	—	—	2	—	0.02–0.1 As
70/30 cupronickel	69.5	—	—	30	—	0.5	—
Admiralty brass (inhibited)	69	29	1	—	—	—	0.04 As, <1 Pb
Manganese bronze	65.5	23.3	—	—	4.5	3	3.7 Mn
Muntz metal	60	39	—	—	—	—	<1 Pb
Naval brass (inhibited)	58	38	1	—	—	—	0.04 As, <3 Pb
Cast alloys:							
Phosphor bronze	89.5	—	10	—	—	—	0.5 P
Aluminium bronze	88	—	—	—	9	3	—
Gunmetal	85.5	2	10	1	—	—	1.5 Pb
	83	5	5	2	—	—	5 Pb
Nickel aluminium bronze	82.5	—	—	4.5	9	4	—
Manganese bronze	57.9	40	—	1	1	—	0.1 Mn

The applications of copper and its alloys are numerous. They are used for structures open to the atmosphere, for example in architecture and sculpture. They are used immersed in fresh water and sea water heat exchangers and condensers, as well as in industrial, chemical and power-generating plant, and buried in the earth for water distribution systems.

In the open air, copper forms a green patina which, in its most stable form, consists of basic copper sulphate, $CuSO_4.3Cu(OH)_2$, although in marine environments it may contain chloride, or carbonate in industrial areas. This decorative long-lasting coating makes copper an ideal material for low-maintenance roof coverings, and for gutters and channels. A small amount of copper dissolves in water which runs over the metal surface, and this can precipitate on to other less noble metals downstream in the water cycle, leading to galvanic corrosion. Cast iron gutters and pipes used in conjunction with copper roofs benefit from a bituminous or other impervious coating to reduce the possibility of galvanic corrosion. General corrosion or stress-corrosion cracking may become a problem in industrial areas if ammonium compounds are present in the atmosphere (see case 13.2).

Copper and its alloys can safely be buried in most soils, although high corrosion rates have been experienced in those containing cinders or acid peat. If it is expected that corrosion will be a life-limiting factor in the use of the material, it can be protected with bituminous, plastic or paint coatings. Dezincification can be a problem in brasses with high zinc levels and it is best to avoid the use of these alloys unless they are specifically required to counter the difficulties which may result from high sulphide levels in the soil.

Copper and its alloys are used extensively in water distribution systems, and in treatment units, condensers and heat exchangers where fresh or salt water is used for cooling. Many components in valves, pumps and taps, as well as pipes and pipe fittings, are made from copper alloys. A distinction is often made between corrosion in fresh water and in sea water. However, the same types of corrosion problem are found in both environments, and such a clear-cut distinction cannot be made. Rather, the change from pure to salt water with varying degrees of pollution should be regarded as a gradually increasing aggressiveness in the environment, aggravated by increases in flow rates, and changes in the temperature and oxygen content of the water.

The main problems which beset copper alloys in water systems are differential-aeration corrosion, erosion corrosion, stress-corrosion cracking, and demetallification. Differential-aeration corrosion is mainly a design problem, although pitting may occur with very slow flow rates which starve the metal surface of oxygen. Erosion corrosion is a function of flow rate. Pressure changes in a liquid on passing through valves and pumps give rise to cavitation damage, while entrained air or abrasive particles disrupt protective surface films to produce shallow, horseshoe-shaped pits. The deterioration can be very rapid (see Ch. 8). Ammonia and its salts, together with mercury-based compounds, are the prime cause of stress-corrosion cracking; a more detailed description is given in Chapter 10.

Dealloying affects many of the alloys, the commonest being dezincification of brasses containing more than 15 per cent zinc, although dealuminification and denickelification have been reported for aluminium bronzes and cupronickels. Dealuminification is most prevalent in aluminium bronzes containing the γ-2 phase in the microstructure, and is most serious when the γ-2 forms a continuous grain boundary network. Rapid cooling from above 600 °C, additions of 1 to 2 per cent iron, or more than 4.5 per cent nickel, control dealuminification, but microstructural changes which occur during welding can still lead to corrosion problems in the heat-affected zone around the weld. Re-heat treatment to remove unsatisfactory post-weld microstructures can pose serious problems to the engineer; heating, handling and quenching large units without introducing distortion while achieving uniform properties throughout the material may be impossible.

Fittings made from brass may suffer dezincification (section 6.3), especially when the β phase is present. The loss of zinc is accelerated by high temperature, increased chloride content, low flow rates, and in situations where differential-aeration occurs. Additions of 1 per cent tin and about 0.04 per cent arsenic, phosphorus or antimony inhibit dezincification. However, phosphorus can lead to intergranular corrosion and most manufacturers concentrate on arsenic as an inhibitor in brasses. Inhibited α-brasses are largely immune to dezincification in most waters, but the effect of tin and arsenic additions to α/β brasses is not so reliable or predictable in controlling the problem. There have been many cases of dezincification in the duplex brasses in both fresh and sea water. In some instances in potable water distribution systems, duplex fittings which have given many years' service suddenly begin to lose zinc when only a slight change in water chemistry occurs.

A protective film of carbonate may be deposited on the metal surface from water containing carbon dioxide and oxygen (see section 13.4.) However, residual carbon deposited during manufacture must be avoided or cleaned from the internal surfaces of pipes. (Remember case 7.3.)

As the flow rate increases, copper and brass tubes become more prone to impingement attack. Aluminium brass and cupronickel offer a greater resistance to higher flow rates, but both have maximum limits which must not be exceeded or the surface film on the metal will be destroyed. While the maximum velocity for inhibited Admiralty brass and aluminium brass is lower than that for cupronickels, they both give better service should sulphide be present in the water, either as a pollutant in rivers or estuaries, or in chemical and oil production plant. However, work is in hand to develop coatings for copper-based alloy condenser tubes in land-based and marine systems. These coatings will offer protection against sulphide-polluted water while the natural film is established. Experimental programmes have shown that the coatings will also be effective on both cupronickel and aluminium brass tubes.[10]

Tin bronzes and phosphor bronzes have good resistance to flowing sea water. The alloys containing 8 to 12 per cent tin are less susceptible than brasses to stress-corrosion cracking and have excellent resistance to impingement attack and to attack in acid waters.

The aluminium oxide film on both aluminium brass and aluminium bronze which confers the additional corrosion protection, reduces the dissolution rate of copper ions from the alloy and makes them less effective in resisting bio-fouling.

The resistance of some copper alloys to erosion corrosion is improved when small quantities of iron are present in the alloy or the water. The iron apparently produces a tougher surface film. This has led to the use of iron sacrificial pieces, in preference to the normal zinc sacrificial anodes, in the water boxes of condensers and heat exchangers which use copper-based tubes and tube plates. The iron ions are absorbed into the film from the water and confer the additional resistance to impingement attack. Zinc ions do not have this beneficial effect.

A very clear distinction can be made between acids which can be safely handled in copper-based equipment and those which cause catastrophic attack. Non-oxidising acids such as acetic, phosphoric, dilute sulphuric and hydrochloric can be safely handled providing the concentrations of oxidising agents such as entrained or dissolved air, chromates and iron(III) ions are kept very low. Oxidising acids, nitric or concentrated sulphuric, and those containing oxidising agents must not be handled in copper-based systems. Small additions of oxidising agents are particularly dangerous in hydrochloric acid, causing a dramatic increase in the rate of metal loss. Before using copper alloys in acid systems, tests should always be made with the particular liquid to be processed, reproducing the actual plant conditions as closely as possible.

In general, copper and its alloys are resistant to attack by alkalis except ammonium hydroxide and those containing ammonium or cyanide ions. Ammonium ions promote stress-corrosion cracking and both ions form complex species such as $Cu[NH3]_4^{2+}$ and $Cu[CN]_4^{2-}$ which do not allow the double layer to develop to polarise the corrosion cell, hence the corrosion rate remains high.

Iron(III) and tin(IV) salts are aggressive to copper alloys. Copper itself suffers general corrosion, to thin the cross-section, in sulphur compounds.

Case 15.5:   When British Rail converted from steam to electric traction, copper overhead cables were installed over all the running rails and sidings which would still be required under the new system. During the transition period steam engines always stopped at a particular spot in a station. In a matter of months, the overhead cable at that one spot was reduced to half its original section by the sulphur-based gases in the steam engine exhaust.[11]

15.6 TITANIUM

Titanium has excellent resistance to atmospheric corrosion in both marine and industrial environments, and to corrosion and erosion corrosion in fresh water and sea water at normal ambient temperatures. Indeed, it could be said that titanium and titanium alloys offer the best corrosion resistance in many everyday environments. As a result, titanium is used extensively in military applications and by the aerospace, aircraft, nuclear, chemical, oil production and processing, electroplating, food processing and photographic industries. Titanium is one of the most abundant metallic elements in the Earth's crust, but a large amount of energy must be expended to separate it from its ore; this makes it expensive, as those female readers who have wanted to buy titanium jewellery will know.

Surprisingly, it was the very good reputation which titanium enjoyed as a corrosion resistant material that led to unsatisfactory performance, and even failure in some instances, when the alloys were first introduced, because of inadequate pre-service testing. It was incorrectly assumed that the metal would be unaffected by the environment. In fact, titanium is a highly active metal, but its excellent corrosion resistance over a wide temperature range in a variety of environments is conferred by the formation and maintenance of a thin, adherent oxide film. The presence of water, even in trace amounts, in the environment in which the film is formed, influences the ability of the film to protect the substrate metal. Those films developed in the absence of water in strongly oxidising atmospheres are often non-protective and may even result in pyrophoric reactions. Dry chlorine attacks titanium, but the metal is quite resistant to chlorine containing more than 0.01 per cent water.[12]

The oxide film which is almost always present on titanium and its alloys in air has the most exciting property of colour: thus, oxide films of different thicknesses can be made which have a wide range of attractive colours. This property has recently been used to great advantage in the titanium jewellery, mentioned earlier, and also certain works of art made from titanium.

Experiment 15.1

If you have access to some titanium (any convenient size that will fit into a medium size beaker), abrade it using wet and dry papers so that a good surface finish is obtained. Then, using a d.c. power supply capable of supplying up to 100 V, but **NO MORE**, and not more than 1 A, connect the polished specimen to the positive terminal of the power supply, and another unpolished piece to the negative terminal as cathode. Wearing rubber gloves, set the power supply to 40 V, immerse both electrodes in a solution of 10 per cent ammonium sulphate and switch on the power for a few seconds. Gas (oxygen) will be evolved and a beautiful colour will develop on the immersed part of the anode. By experimenting with different times and potentials, fascinating results can be achieved. (A used specimen can be abraded and re-used.)

Try soldering an electrical wire to the metal part of an artist's ordinary soft paint brush. Connect the other end of the wire to the negative terminal of the power supply; then connect the freshly abraded specimen of sheet titanium to the positive terminal. Wear rubber gloves; dip the brush in the electrolyte and switch on the power, setting the supply to 40 V. 'Painting' with the brush on the metal will produce startlingly beautiful effects. Different colours will be obtained depending on the length of time that the brush is in contact with the metal and the potential used. With practice, all sorts of *objets d'art* can be made. Always switch off the power before removing the rubber gloves.

Titanium can exist in two allotropic forms: α having a close-packed hexagonal structure, and β with a body-centred cubic lattice, the latter phase being stable above 882 °C in the pure metal. Alloying additions, which tend to stabilise one or other of the two phases at room temperature, are made principally to modify the mechanical and physical properties of the alloy, but the structural change also modifies the corrosion behaviour. Aluminium and tin are α stabilisers, while vanadium, molybdenum, chromium and copper are β stabilisers. Normally, the α/β and β alloys have superior mechanical properties but slightly inferior general corrosion resistance compared to the α alloys.

Turning our discussion to specific environments, the metal is very resistant to attack in oxidising solutions, the major exception being dry, fuming nitric acid. Although there may be an enhanced corrosion rate in the first few hours of exposure in normal nitric acid solutions, this quickly disappears and the metal can be safely used to handle nitric acid in all strengths and at temperatures up to 250 °C. Additions of silica compounds and silicone greases appear to inhibit the initial enhanced corrosion rate in nitric acid and mixtures of nitric acid and hydrochloric acids, even at temperatures as high as 200 °C. Other oxidising acids, such as chromic and hypochlorous acids, do not affect the metal significantly.

Reducing acids do attack the metal. Hydrofluoric acid and aqueous solutions of acidic fluorides produce rapid general corrosion of titanium and its alloys. Unless precautions are taken, hydrochloric, phosphoric and certain concentrations of sulphuric acid attack titanium.[13]

Four techniques are in general use for controlling corrosion in reducing acids other than hydrofluoric:

(a) Oxidising inhibitors can be used. Nitric acid or oxidising ions such as Fe^{3+}, Cu^{2+} or Ti^{3+} are added to hydrochloric or sulphuric acid.[14] Small quantities of inhibiting ions are often present as impurities in electrolytes actually being used in production flow lines, but not in laboratory reagents, which are used in test or monitoring programmes. Therefore, all testing to determine the reliability of titanium as a constructional material should be carried out in the process liquors where possible and not in artificial solutions.

(b) Aeration of the solution has been used to maintain the passive film on titanium, particularly in organic acids. Aerated, hot formic acid, for example, does not attack titanium, but deaerated solutions attack the metal quite readily.

(c) Alloying additions can be made. Some alloys are particularly resistant to attack in reducing acids, particularly the alloy containing 0.2 to 0.5 per cent Pd and the one containing 30 per cent Mo. The high molybdenum alloys are said to resist hot, concentrated hydrochloric acid, sulphuric acid or phosphoric acid.[15] An explanation for the corrosion resistance of the alloy with palladium can be found in section 5.2.

(d) Anodic impressed current protection (section 16.6) can be used to stabilise the oxide film on the metal surface. It is claimed that the corrosion rate in 40 per cent sulphuric acid at 60 °C is reduced by a factor of 33,000 after application of an anodic impressed current system.[16]

Although it is attacked by hot concentrated alkali solutions, titanium is resistant to attack by dilute alkali. It is also very resistant to crevice corrosion, although this has occurred in solutions containing chlorides, iodides and bromides at temperatures above 60 °C. The problem is most likely to be found in condensers and an upper limit of 130 °C has been suggested[17] for condensers operating in strong aqueous solutions of these ions.

In specific environments, titanium and its alloys may suffer stress-corrosion cracking. The damage occurs when the oxide film is disrupted and the crack growth rate exceeds the rate at which the film is repaired. Thus, the active metal at the base of the crack is continuously exposed to the local environment. At faster strain rates, normal ductile failure ensues. In most cases a pre-existing crack is necessary to initiate the SCC failure mechanism, but in certain environments, notably methanolic chloride solutions, pitting attack by the electrolyte is the initiating mechanism. Pre-existing cracks are often developed by fatigue and there can be a long delay while a fatigue crack initiates; its subsequent and much faster propagation is then assisted by the environment. Stress-corrosion cracking is rather more common in titanium alloys than in commercially pure titanium.[18]

15.7 NICKEL AND ITS ALLOYS

While nickel and its alloys resist corrosion in many environments, including aggressive ones in the oil and chemical industries, better than many of the materials we have considered so far, they are very expensive. Only when

Table 15.6   Some commercial nickel-containing alloys and their approximate compositions

| Alloy | Approximate % composition | | | | | | | | | | | |
	Ni	Cu	Cr	Fe	Mo	Ti	Mn	Co	Al	W	Nb + Ta	C_{max}
201 (pure nickel)	100	—	—	—	—	—	—	—	—	—	—	0.15
Inconel 600 *	77	—	15	8	—	—	—	—	—	—	—	0.15
Hastelloy B2 **	68	—	—	2	28	—	1	1	—	—	—	0.015
Hastelloy C4 **	68	—	16	3	16	0.7	—	2	—	—	—	0.015
Monel 400 *	66	30	—	2.5	—	—	1	—	—	—	—	0.3
Monel K500 *	66	30	—	—	—	0.5	—	—	3	—	—	0.15
Inconel 625 *	63	—	21.5	2.5	9	0.2	—	—	0.2	—	3.65	0.10
Hastelloy C276 **	57	—	16	5.5	16	—	—	2.5	—	4	—	0.02
Hastelloy G **	44	2	22	20	6.5	—	2	2.5	—	1	—	0.05
Incoloy 825 *	40	2	21	33	3	1	—	—	—	—	—	0.05
Incoloy 800 *	33	—	20	46	—	0.6	—	—	0.6	—	—	0.1
Carpenter 20 Cb ***	30	3.3	20	44	2.5	—	—	—	—	—	—	—

* A registered trade mark of the Inco Group.
** A registered trade mark of the Cabot Corporation.
*** A registered trade mark of the Carpenter Technology Corporation.

a cheaper alternative has failed to produce an economic life will a nickel alloy be selected.

Most of the alloys are sold under trade names which tend to run in a series differentiated only by a number or type letter. A limited selection of commercial nickel alloys and their typical compositions is given in Table 15.6. The most important alloying additions are copper, molybdenum, chromium and iron, each combination with nickel having a significant corrosion resistance in a particular environment. Great care should be taken in the selection of the correct alloy in a series; the response to an environment may be quite different for two alloys in a series – remember case 13.4. Information on specific alloys is given in the bibliography, or should be sought from the trade literature and the manufacturers.

Commercially pure nickel is resistant to attack in air at normal temperatures. However, if sulphur dioxide is present and the relative humidity exceeds 70 per cent, a very common set of circumstances in industrial or urban areas, fogging of the metal surface occurs. Fogging is a tarnishing process in which the nickel acts as a catalyst in the conversion of sulphur dioxide to sulphuric acid, and finally to a surface film of basic nickel sulphate. Fogging spoils the aesthetic appearance of the metals. To maintain a bright finish, nickel is frequently coated with a thin layer of chromium (see section 14.10).

Nickel is resistant to attack by hot or cold alkali solutions and fused (molten) sodium hydroxide. It is slowly attacked by reducing acids, but dissolved oxygen increases the rate of attack. Oxidising acids and ions such as HNO_3, Cu^{2+}, and Fe^{3+} rapidly attack the metal; in sea water pitting may result.

When sulphur is present in the atmosphere, the metal fails by grain boundary attack at temperatures above 320 °C, but in sulphur-free atmospheres it gives satisfactory service to 850 °C and is often used at even higher temperatures. The sulphur embrittles the grain boundaries at lower temperatures, with the greatest rate of penetration occurring in the range 550–650 °C. Above 645 °C a low melting point eutectic forms in the grain boundary and results in catastrophic failure. The maximum solubility of sulphur in nickel is 0.005 per cent and any larger amount will form the eutectic. Trace amounts of sulphur acquired from oil and grease residues, marking paints and even finger prints have caused failure in nickel alloy components at 650 °C. Additions of chromium and iron produce alloys which are resistant to sulphur attack at temperatures up to 1200 °C and these alloys are used in furnaces and chemical processing plant.

The wet corrosion resistance of the metal is also improved by alloying additions, notably copper, chromium, molybdenum and iron. Monel is a nickel alloy containing 30 per cent copper. At room temperature the metal fogs in sulphur-polluted industrial atmospheres, and it is attacked by oxidising acids and ions which destroy the passive surface film. However, it is more resistant than the commercially pure metal to reducing acids and alkalis, although hot concentrated caustic and aerated ammonium chloride will attack the alloy. In stagnant sea water the alloy is less prone to pitting or crevice attack than commercially pure nickel. Monel gives excellent corrosion and erosion resistance in fast-flowing (i.e. > 2 m $^{-1}$), well-aerated sea water. It is often used to make valve trims and pump components for flowing sea-water systems in land-based power stations and chemical plants, as well as ships and offshore platforms. Deaerated hydrofluoric acid does not attack Monel, even in boiling solutions, but aerated acid will rapidly attack the metal. Uhlig[19] gives the rate of attack in 35 per cent HF at 120 °C as 0.025 mmpy when saturated with nitrogen and 3.75 mmpy when saturated with air.

Several alloys contain chromium. A minimum of 14 per cent chromium must be added to reduce the risk of attack by nitric acid and oxidising ions. Thus both Inconel and Hastelloy C are resistant to such acids. They also give an excellent performance in fast-flowing sea water. In stagnant sea water Inconel tends to pit, but the addition of molybdenum to Hastelloy C in alloy C276 confers outstanding resistance to pitting and crevice attack in sea water (see section 7.3 and Fig. 7.7(b)). On the other hand, Hastelloy B is not resistant to attack by oxidising acids or ions. However, it withstands attack from hydrochloric acid over a wide composition and temperature range; the corrosion rate is 0.26 mmpy in boiling 10 per cent HCl and 0.05 mmpy in 35 per cent HCl at 65 °C.[20] Hastelloy B is also resistant to deaerated sulphuric and phosphoric acids (see case 13.4).

15.8 MATERIALS FOR PARTICULAR APPLICATIONS IN WATER SYSTEMS

Pipes and valves

Pipe systems and valves have to contend with water velocities which may vary from prolonged stagnation in dead legs, to 3 or 5 m s^{-1}, or higher, in pumped flow lines. Within a valve, or on bends, the local velocity may be much higher than the design velocity of the system. The turbulence which results from this increased flow rate can disrupt the surface films on some metals and lead to premature impingement failure. Such metals have maximum flow rates which must not be exceeded. Other materials, notably stainless steels, require a supply of oxygen to maintain the surface film which protects the metal from corrosion. For these materials there are minimum flow rates which ensure an adequate oxygen supply.

Case 15.6: It took about an hour to transfer 98 per cent sulphuric acid through a pipe line. In order to reduce the transfer time to a more acceptable 15 minutes, a stronger pump was installed. This also increased the flow rate from 0.7 m s^{-1} to over the 1.6 m s^{-1} critical velocity for the pipe material in sulphuric acid. The pipe failed in less than a week after the modification.[21]

All bends, transitions to smaller bore pipe and the internal surfaces of welds should be as smooth as possible.

Valve bodies are normally made anodic to the trim to give sacrificial protection to the internal components. Lists of typical pipe and valve materials are given in Tables 15.7 and 15.8.

Condensers and heat exchangers

These units are often the critical components for processing plant in the electricity generating, oil and chemical industries. Although a valve change can be a great inconvenience, the down-time involved in retubing or changing a condenser can impose severe financial penalties on the user.

Plastic or metal pipe inserts are frequently placed in the ends of condenser tubes where impingement damage is most severe. These can be cheaply replaced during routine maintenance close-down periods.

Although sacrificial zinc or iron pieces may be incorporated in the water box to protect the whole unit, the box is normally made from a metal which is anodic to the remainder of the system, especially the tubes. Cast iron and steel are often used for water boxes, although graphitisation of cast iron (see section 15.3) may ultimately make the metal cathodic to the tubes, hence the sacrificial anodes. Internal nylon coatings are applied to ferrous water boxes to reduce metal loss. However, this reduces the sacrificial protection available to the tubes and generates high corrosion rates at any small break in the coating. NiResist cast iron, aluminium brass, aluminium bronze and cupronickels have also been used for water boxes, as well as

Table 15.7 Material characteristics for water pipe systems[5]

Material	Characteristics	Limiting* velocity ($m\ s^{-1}$)
Cast iron	Cheap; gives good service at low flow rates; suitable for water distribution systems; thickness of pipe often controls life	Depends upon the oxide which forms on surface; with adherent impermeable film, 6 or above
Galvanised steel	Cheap; must protect welded joints; life depends on coating thickness; use low flow rates; poor resistance to turbulence; contact to copper fittings or copper ions in water can cause premature failure	
Copper	Poor resistance to turbulence; toxic to biofouling organisms	1.0
Inhibited Admiralty brass	Resists dezincification	2.0
Aluminium brass	Better resistance to turbulence than copper; better resistance to sulphide pollution than cupronickel; less resistant to biofouling	4.0
90/10 cupronickel + 1.5% iron	Very good overall performance; resists stress-corrosion cracking	3.6
70/30 cupronickel + 0.5% iron	Best corrosion resistance of any copper alloy; resistance to sulphide pollution inferior but improved by Fe and Mn additions	4.6
Highly alloyed stainless steel	High capital cost (see section 15.2)	no limit
Titanium	High capital cost; requires more support owing to lower Young's modulus	no limit

* In smaller diameter pipes (< 75 mm) the critical velocity may be only two-thirds of that quoted.

Table 15.8 Materials for valves in water systems[5]

Body	Trim	Pipe material	Notes
Cast iron	Brass	Cast iron/steel	Old systems; requires constant maintenance
SG iron	Brass	Cast iron/steel	Reduces cracking in body
Leaded tin bronze	Alloy 400 Ni/Cu	Copper base	Good performance; some limits on velocity
Nickel aluminium bronze	Alloy 400 Ni/Cu	Copper base	Tolerates higher flow rates

some non-metallic materials such as glass-reinforced plastics and thermo-plastics. The tube plate is normally made from the same metal as the tubes to avoid galvanic corrosion; in some areas it may be made anodic, but never cathodic, to the tubes. Common metals employed for tube plates include Admiralty brass, aluminium brass, aluminium bronze and cupronickels. When titanium tubes are used for maximum reliability, the tube plate may also be made from this metal (see Table 15.9). A comparison of the performance of different copper-based alloys used to make condenser tubes operating in a marine environment was undertaken by Todhunter[22] for the Harborstream plant in 1967. The results are shown in Fig. 15.1.

Pump

Pump materials are subject to a wide variety of corrosion problems. It is common practice to make the body of the pump anodic to the critical

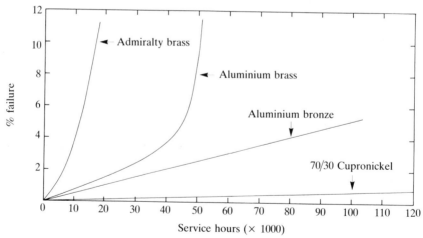

Fig. 15.1 Comparison of the performance of copper-based alloys used for condenser tubes operating in sea water at the Harbourstream plant (after LaQue[23]).

Table 15.9 Condenser tube materials[5]

Material	Critical velocity ($m\ s^{-1}$)	Notes
Aluminium	Min 0.3	Resistant to attack from CO_2 and O_2 in steam condensate; limited to maximum temp 150 °C; very low levels of copper ions initiate pitting (< 0.1 ppm)
Copper	Max 1.0	Fair performance
Admiralty brass	Max 1.8	Good performance; inhibited against dezincification
Aluminium brass	Max 2.5	Good performance; not so resistant to biofouling; better than 70/30 Cu/Ni if sulphide present
90/10 cupronickel + 1.5% Fe	Max 3.0	Better than aluminium brass in waters polluted with ammonium ions
70/30 cupronickel + 0.5% Fe	Max 3.6	Best flow rate of copper alloys; prone to failure when sulphide present
316 stainless steel	Min 1.5; no max limit	Very good in polluted environments (H_2S) but suffers crevice corrosion at low flow rates
High-alloy stainless steel	No limits	Replacing 70/30 cupronickel for sea and estuarine water-cooled systems; very good in sulphide-polluted water
Inconel 625	No limits	Excellent performance; tolerates sulphide pollution; expensive
Titanium	No limits	Expensive; very reliable; tolerates a variety of pollutants; requires additional support owing to lower Young's modulus

components, the impeller, shaft and seals, to ensure that they are sacrificially protected. Both the impeller and shaft undergo crevice-type attack, while the impeller suffers impingement and cavitation damage, especially on the blade tips. Again, the liquid velocity in some parts of the pump may be considerably higher than the output velocity of the pump. See Table 15.10.

Propellers
For the sake of completeness, a table of ships' propeller materials is

Table 15.10   Pump materials[5]

Component	Characteristics	Low Cost	Preferred	Maximum Life
Body	Should be anodic to other components. Resistant to high velocity flow and cavitation	Cast iron G or M bronze Cement or epoxy lined cast iron $2\frac{1}{2}\%$ Ni cast iron	Ni resist iron Nickel aluminium bronze 316 Stainless steel	Ni resist iron 316 Stainless steel
Impellers	Resistance to high flow rates and cavitation damage	$2\frac{1}{2}\%$ Ni cast iron G or M bronze	Nickel aluminium bronze 316 Stainless steel Alloy 400 Ni/Cu	316 Stainless steel Stainless steel 29 Cr 11 Ni + Mo
Shafts	Resistant to pitting and crevice corrosion with good torsional strength	410 Stainless steel Mild steel	316 Stainless steel Alloy 400 Ni/Cu	Alloy 400 Ni/Cu Alloy 20
Wear rings	Must be cathodic to other components owing to small size	G or M bronze	316 Stainless steel	Alloy 506 Ni Cu Si Alloy 20
Bolts	Cathodic to pump body	Silicon bronze Carbon steel	Alloy 400 Ni/Cu	Alloy 400 Ni/Cu

1. Beware of graphitisation in cast irons.
2. Materials should be matched to flow rates, temperatures and pollutants.
3. Stainless steels require thorough flushing to avoid crevice corrosion on shut down.

included, Table 15.11. Figure 15.2 is an old photograph showing the erosion corrosion of a propeller made from manganese bronze. Today, materials such as nickel aluminium bronze provide better resistance.

15.9   THE SELECTION OF MATERIALS

When a designer begins to choose the material from which a component or structure will be fabricated, he must consider a large number of factors other than corrosion resistance. Figure 15.3 shows in schematic format some of the many decisions which must be made. The cost and availability of materials are as important as their properties: a gold-plated car would have

Table 15.11 Typical materials for marine propellers[5]

Material	Comment
Cast iron	Low initial cost but requires frequent replacement; used on slow vessels; additions of nickel improve corrosion resistance
Manganese bronze	Medium cost; suffers dezincification, cavitation and crevice corrosion; suitable for low-powered ships
Nickel manganese bronze	Higher cost but better performance than manganese bronze
Nickel aluminium bronze	Copper/aluminium alloys with additions of nickel and manganese; very durable; resists cavitation and erosion damage; aluminium content limited to 8.4 to 9% to gain maximum benefit without forming corrosion susceptible γ–2 phase; welding or hot repairs require caution; used for high performance propellers.
Stainless steel	Used for harbour vessels; may suffer differential-aeration corrosion if buried in mud

Fig. 15.2 Erosion corrosion of a manganese bronze propeller.

The corrosion properties of some metallic materials

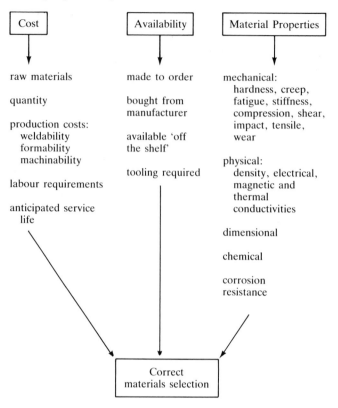

Fig. 15.3 The contribution of corrosion resistance to the process of materials selection.

excellent corrosion resistance, but would be rather expensive and hard to procure. Engineering design must always be concerned with compromise so that the best match of design criteria can be converted into an economic, readily available and durable product.

Nevertheless, corrosion resistance is an essential part of materials selection, but is often given scant consideration relative to other mechanical properties which might, at first, seem more important. Poor corrosion resistance will always lead to premature failure no matter how well the component has been designed in other respects. The previous sections have provided little more than the most general coverage of five major groups of alloys. However, one of the aims of this chapter will have been achieved if you have begun to appreciate the enormous number of materials which are available and the complexity of their behaviour in different environments. No solution has yet been found to the problems posed by the need to select the correct material for a particular corrosive environment; if it had, the corrosion problems described in this book would not exist.

At present, the best available method is by means of a massive compilation of material properties, such as can be obtained in the *Fulmer*

Materials Optimizer.[24] Comprised of four large volumes, this work is an essential part of a designer's library, but still covers only a fraction of the criteria listed in Fig. 15.3. The complete design process remains arduous and littered with pitfalls for the inexperienced designer. However, at a more basic level, the approach adopted by this book is that elementary errors are the cause of the majority of corrosion problems, and that these can be eliminated by the application of relatively simple ideas, combined with a general knowledge of material properties.

The best hope for the future is in the development of an entirely computer-aided materials selection process. The vast amount of data storage and high speed computing required for such a task is only now becoming available, though the first steps along the road towards achieving optimum selection have already been taken. In many areas of engineering, computers are being used as data bases for materials selection in restricted applications and environments. The setting up of national facilities for on-line information retrieval has also been successful and will undoubtedly be considerably extended in the future.

15.10 REFERENCES

1. Christopher P J, Forster V T 1969–79 Rugeley dry cooling tower system, *Proceedings of the Institute of Mechanical Engineers* **184**: 197–211
2. Seibert O 1978 Classic blunders in corrosion protection, *Materials Performance* April: 33–37
3. Shreir L L (ed.) 1976 *Corrosion* (Vol. 1). Newnes-Butterworths, p 3:9
4. Lewis J R, Mercer A D 1984 *Corrosion and marine growth on offshore structures.* Ellis Horwood, p 104
5. Waterman N A, Pye A M 1980 *Fulmer materials optimizer.* (Vol. 4) Fulmer Research Institute. Reproduced with permission.
6. Yoshino Y 1982 Low alloy steels in hydrogen sulphide environments, *Corrosion* **38**, March: 3
7. Fairhurst W 1984 Some recent developments in stainless steels and their applications to sea water cooled condensers, *Materials and Design* **5**, Nov: 235–43
8. Davison R M, Laurin T R, Redmond J D, Watanabe H, Semchyshen M 1982 A review of worldwide developments in stainless steels, in *Conference on speciality steels and hard materials.* South Africa: Pretoria
9. Webster D 1984 Aluminium lithium alloys. *Metal Progress* Apr: 33–37
10. BNF Press release, Apr 1985. BNF Technology Centre, Grove Laboratories, Oxford
11. Private communication, 1974. Mr Hare, Area scientist, British Rail Research Department, Rail House, Crewe
12. Shreir L L (ed.) 1979 *Corrosion* (Vol. 1). Newnes-Butterworths, p 5:39
13. Ibid., p 5:42
14. Bomberger H B 1968 The general corrosion resistance of titanium, in *Titanium and titanium alloys source book.* ASTM, p 161
15. Ibid., p 164
16. Ibid., p 165

17. Shreir L L (ed.) 1979 *Corrosion* (Vol. 1). Newnes-Butterworths p 5:43
18. Miska K H 1974 Titanium and its alloys, in *Titanium and titanium alloys source book*. ASTM, p 11
19. Uhlig H H 1971 *Corrosion and corrosion control*. John Wiley, p 356
20. Ibid., p 360
21. Seibert O W 1983 Classic blunders in corrosion protection revisited, *Materials Performance* Oct: 9–12
22. Todhunter H A 1967 Condenser tubes in sea water service, *Power* **8** (3): 57–59
23. LaQue F L 1975 *Marine Corrosion*. John Wiley, p 269
24. The Fulmer Materials Optimizer (2nd edn.) 1981. Fulmer Research Institute Ltd., Stoke Poges, Slough, UK.

15.11 BIBLIOGRAPHY

ASTM 1969 *Materials performance and the deep sea*. ASTM-STP 445

Fontana M G, Greene N D 1978 *Corrosion engineering*. McGraw-Hill, Ch. 5

Friend W Z 1980 *Corrosion of nickel and nickel-base alloys*. John Wiley

Haynes G S, Baboian R (eds) 1985 *Laboratory corrosion tests and standards*. ASTM-STP 866

Leidheiser H Jr 1971 *The corrosion of copper, tin and their alloys*. John Wiley

Rogers T H 1968 *Marine corrosion*. Newnes, Chs 6, 7 and 8

Sedriks A J 1979 *Corrosion of stainless steels*. John Wiley

Shreir L L (ed.) 1979 *Corrosion*. Newnes-Butterworths, Chs 3 and 4

16 CATHODIC AND ANODIC PROTECTION

It is better to wear out than to rust out.

(Bishop Richard Cumberland: 1631–1718)

To someone with only a passing interest in corrosion, the concept of using an electrical technique to control corrosion would probably seem as sensible as building one of the world's largest steel structures and immersing it unpainted in the sea. The individual would be shocked to learn that both suggestions are not imaginary but reality. Indeed, it is only the fact that the first is possible that makes the second practicable.

The principles and applications described in most of this chapter relate to **cathodic protection**, a name which is almost self-explanatory. This important and widely used method of corrosion control involves two techniques: the **sacrificial anode** method, in which the important principle of bimetallic corrosion is utilised; and the **impressed current** method which is an electrically controlled process.

In recent years, a similar electrical technique called **anodic protection** has been developed for some metal/electrolyte combinations. This will be described in section 16.6.

16.1 THE THEORETICAL BASIS

By far the most important application of electrical techniques to control corrosion is for steel. Ships' steel hulls, offshore drilling platforms and oil and gas undersea pipelines are all protected against attack in one of the most aggressive natural environments by methods which will be discussed in this chapter. Additionally, the steel reinforcement within concrete structures, and the containers which hold potentially corrosive chemicals, have also been protected by such techniques.

Except in section 16.6, which is devoted to anodic protection, this chapter will deal only with the protection of carbon steel. Let us begin by considering the E/pH diagram for iron in water,[1] Fig. 16.1, which shows the regions of thermodynamic stability under different environmental conditions.

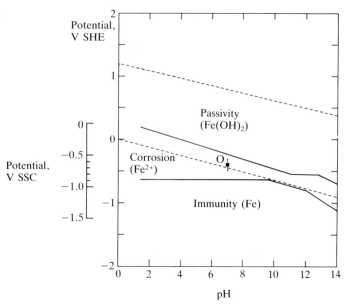

Fig. 16.1 E/pH diagram for iron in water.

The free corrosion potential, E_{corr}, of iron in aerated water is in the range −600 to −700 mV SSC at a pH of 7. It is represented by point O in the figure. In sea water, the only significant change to the figure is a lack of passivation below pH 5. This is largely irrelevant in natural sea water which has a pH of 8.2–8.5 approximately. Point O is thus shifted to the appropriate pH value, but the range of values of E_{corr} remains about the same. Whether it is in a chloride-containing environment or not, point O is situated within a zone of corrosion and the iron will rust rapidly if a supply of oxygen is maintained.

Let us consider four different ways of affecting the thermodynamic equilibrium which exists at O.

Option (a) Decrease the pH: the solution is made more acidic. Inspection of Fig. 16.1 shows that the specimen remains in the corrosion zone at all values of pH < 7, i.e. the zone in which the soluble iron ions are the most stable species. The rate of corrosion increases as pH falls.

Option (b) Increase the pH: even a small increase moves the iron into a region of passivity where the most stable species is an insoluble iron hydroxide (or hydrated oxide). If it is allowed to coat the iron we should expect a reduction in the corrosion rate because the film will separate the iron from its corrosive environment. However, it cannot be assumed that an insoluble film will behave in a protective manner. If the film is not uniform, non-conducting and impervious, or if it is damaged by flow of electrolyte across the surface or mechanical means, then corrosion will continue. Indeed, the rate may well be accelerated because the exposed anode area is smaller. Pitting corrosion is quite likely under such conditions.

Option (c) Apply a more negative potential. The metal condition moves into a zone of immunity. Remember that the distinction between a corrosion zone and an immunity zone is merely one of definition, i.e. a concentration of iron ions at equilibrium of 10^{-6} M determines the boundary between the two zones. From Fig. 16.1, this occurs at potentials less than -800 mV SSC at pH 7. Although the metal is in the immunity region according to our definition, a corrosion reaction may still occur. Consult Fig. 4.9(a) and remind yourself how the anodic and cathodic reactions vary with potential. The more negative the potential, the smaller the anodic reaction becomes but the greater the cathodic reaction becomes: the metal is more cathodic. This is the principle of cathodic protection of metals.

Option (d) The potential is made more positive: the metal is once again brought into the passive region in which there is a chance that the rate of corrosion may be reduced by the formation of a barrier between metal and electrolyte. Although it was said that iron is difficult to passivate, this principle has been used effectively for certain steel/electrolyte combinations, as well as for other metals and electrolytes. The principle is known as anodic protection (see section 16.6).

In the example above we have found that options (b), (c) and (d) suggest three theoretical ways of protecting against corrosion. With regard to option (b), the corrosion engineer very often has no control over the electrolyte. The offshore industry engineer cannot change the pH of the sea, neither can the process engineer alter the composition of the product he is trying to manufacture but which corrodes his containers and pipework. Sometimes, inhibitors can be added to the electrolyte, and examples in which option (b) can be used were given in Chapter 13. Very often, however, engineers must consider other methods whereby their expensive metal structures can be safeguarded. Options (c) and (d) form the basis of corrosion control by cathodic and anodic protection.

Figure 16.2 is based upon the work of LaQue[2] and shows, in schematic form, the variation of cathodic potential against current density for carbon steel in sea water. It also shows the effect of potential and current changes upon the rate of corrosion, measured in terms of weight losses. LaQue found a good correlation of a corrosion rate/currrent density curve with a potential/current density curve. Let us now consider how this information can be used to advantage.

Figure 16.2 shows us that less of the metal corrodes as potential is made more negative. This is hardly surprising, for a metal which is the cathode in a wet corrosion cell does not, in general, corrode. It is possible for small local anodes to exist on the metal surface, for reasons which have been expressed in other chapters, and it is not until the anode reaction has been completely suppressed that all corrosion will cease. A designer of a cathodic protection system begins by stating a maximum acceptable corrosion rate,

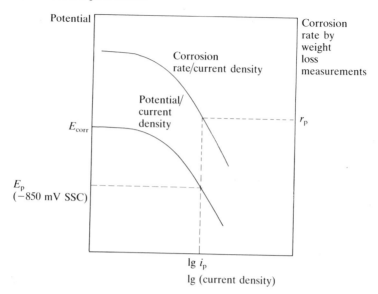

Fig. 16.2 Schematic diagram showing the variation of cathodic potential with current density for steel in sea water, and the correlation with corrosion rate measured by weight loss data. (Adapted from data obtained by LaQue[2])

r_p, and uses a graph such as Fig. 16.2 to obtain a value of current density, i_p, which will lead to the desired corrosion rate. This, in turn, yields a protection potential, E_p. The actual protection potential used in cathodic protection systems is dependent upon the application[3]; more discussion of this will occur in following sections. From Fig. 16.1, we have seen that it is probably more negative than -800 mV SSC, and many designers choose -850 mV SSC. There have been many instances where the choice of potential required for adequate protection has been arrived at by trial and error, and to some degree, uncertainty remains. As a general rule, the protection range is -800 to -900 mV SSC.

The following example shows how the expected corrosion rate is calculated for different applied cathodic potentials.

The corrosion rate at a given cathodic polarisation is expressed by the equation:

$$i = i_o \exp(\alpha \eta \, zF/RT) \qquad\qquad [16.1]$$

where we assume that $\alpha = 0.5$ (see section 4.7), $z = 2$ (for iron), $F = 96{,}494$ C mol^{-1}, $R = 8.3142$ J mol^{-1} K^{-1} and $T = 283$ K (a sea-water temperature of 10 °C). For a polarisation of -200 mV, we get:

$$i = i_o \exp\left(\frac{0.5 \times (-0.2) \times 2 \times 96{,}494}{8.3143 \times 283}\right)$$
$$= i_o \exp(-8.20)$$
$$= 0.00027 \, i_o$$

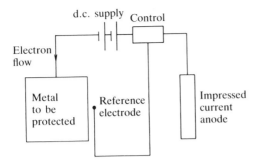

Fig. 16.3 The principle of impressed current cathodic protection using a potentiostat.

Thus, if we assume the free corrosion potential for steel in sea water to be −650 mV SSC, then the corrosion rate at −850 mV SSC will be reduced to 0.03 per cent of the rate for unprotected steel. At −750 mV SSC, the rate is reduced to only 2 per cent of the rate at the free corrosion potential.

In order to carry out such protection, an electrical system similar to that shown diagrammatically in Fig. 16.3, is set up. The components shown have already been described in detail in section 4.8 which concerned the three-electrode cell. The control system is essentially a potentiostat, but with an automatic compensation device; the role of the control unit will become more apparent during the course of the discussion which follows.

The control unit is set to impress a current, I_p, so that, for a given wetted surface area, A, the current density is $I_p/A = i_p$, and the chosen protection potential, −850 mV SSC, say, is measured at the metal surface. Assuming there are no further changes in the system, the metal surface is protected against corrosion.

Question: *Potentials more negative than −850 mV SSC show even less metal loss. Why not use −1000 mV SSC, or lower?*

There are two reasons why it is not a good idea to use more negative potentials. The evolution of hydrogen, which occurs at more negative potentials, is often found to cause damage in other ways, notably hydrogen embrittlement (see Ch. 10). In addition, the large currents associated with more negative potentials produce high local concentrations of hydroxyl ion which often damage barrier coatings such as paint, if present. The potential of −850 mV is a convenient value which gives efficient protection, while reducing the likelihood of these other forms of damage.

The system of cathodic protection described above is also beneficial because, with a suitable control unit, it can be made self-regulating, as the following example shows.

If the surface were protected by a barrier coating or paint film, the currents observed would be much lower, but the current density for protection of the bare metal areas would still be the same. Let us now consider what happens if the protective coating is damaged, causing an increase of effective surface area. If the control unit continued to impress current, I_p, then the current density would fall. Figure 16.2 shows that this would result in a change of potential back towards the free corrosion potential, together with a corresponding increase in corrosion rate. In an impressed current cathodic protection system, the instrument has an automatic compensation circuit which senses the onset of a potential change and counters it by maintaining the potential at the set level, E_p. In order to achieve this, the control unit must maintain current density, i_p, and must therefore impress a greater current.

The opposite effect is equally important. If there is a reduction in the effective surface area, there would be a tendency for the current density to increase, and the potential to become more negative. As was mentioned above, this is also undesirable. By continuously sensing the potential of the metal surface, the control unit is able to maintain the constant potential required for protection and impress a smaller current. Instruments which maintain constant applied potentials are known as **potentiostats** It is important to appreciate that the magnitude of the current varies considerably with time and the circuitry must be able to cope with such changes.

Remember also (with the aid of Fig. 16.2) that in order to maintain the corrosion rate below the maximum acceptable value, r_p, the same current density, i_p, is required to protect steel, whether painted or bare metal. The problem is that the area is not known accurately. In Chapter 14 we saw how paint films are imperfect and usually contain defects or holidays which allow restricted access of an electrolyte. Bare metal when immersed in the sea becomes covered in marine growth and, as we shall see below, is coated with films called calcareous deposits when cathodically protected. These effects mean that current density cannot be determined precisely. The great advantage of applying a given potential using a system such as that described in Fig. 16.3 is that **current** is varied until the reference electrode senses that the structure is at the required potential. In this way, the necessary current density for protection has been achieved without the input of an accurate area value and irrespective of the presence or absence of films on the surface. When current densities are quoted in the discussion which follows, the limitations introduced by not knowing the precise area of metal exposed to the electrolyte should be remembered.

16.2 CATHODIC PROTECTION: THE SACRIFICIAL ANODE METHOD

Cathodic protection can be achieved in two ways:

(a) the sacrificial anode method;
(b) the impressed current method.

This section describes the sacrificial anode method and involves a knowledge of the contents of Chapter 5, as well as elements of the theory of Chapter 4.

Perhaps the simplest way of describing how sacrificial anode cathodic protection works is to recall the basic wet corrosion cell, Fig. 4.2, and the general rule which was established in section 4.2: in any cell it is the anode which corrodes and the cathode which does not. An engineer, faced with the problems of preventing a metal structure from corroding, apparently has a ready solution: if he makes his structure the cathode in a corrosion cell, the structure will not corrode. Sometimes, cathodes may be damaged if too great a current density occurs. Usually, this results in damage to paint coatings but hydrogen embrittlement of the metal is also a possibility if the potential is sufficiently negative (see section 10.3).

Question: *If the metal structure is the cathode, what is the anode?*

The engineer uses his knowledge of the galvanic series (Fig. 5.5) to select a material which, when coupled to the metal he wishes to protect, becomes an anode. Because engineers are most frequently concerned with the protection of iron and steel, a glance at the galvanic series shows that any of the metals at more active potentials than iron will, in theory, be suitable. In practice, it does not make sense to try to protect iron by coupling it to sodium, for example: sodium reacts explosively with water. Sodium will protect iron, but for such a short time as to be useless. It is therefore necessary to find another anodic metal which corrodes more slowly than sodium. Metals such as zinc, magnesium and aluminium are suitable and much used. Hence, we conclude that if the engineer bolts a piece of zinc to his metal structure, the zinc will corrode preferentially and leave the iron unaffected.

Question: *Is it really that simple?*

Almost. It is obvious that a 1 kg lump of zinc on an offshore drilling platform will not have a significant effect in reducing corrosion on the platform. We need to take into account the relative areas of anode and cathode. If the zinc and iron were of similar area, then there would be a significant effect. However, there comes a time when the zinc has corroded away completely and the iron corrodes once more. Thus it is necessary to consider the rate at which the zinc corrodes in order to predict when a replacement will be necessary. By the same token, the output current density supplied by the corrosion of the zinc at any instant must be sufficient to suppress the corrosion of the iron. Sacrificial anodes are evaluated partly by means of their **capacity**, a term which gives the number of ampere-hours which can be supplied by each kilogram of material. Example E18.6 in section 18.1 illustrates the determination of this parameter. Other useful parameters can also be determined, such as the current output per unit of exposed surface area, or so-called **wastage** rates which express the rate of loss of metal, either by volume or by mass.

Another factor affecting the use of sacrificial anodes is the **throwing power** effect. The term 'throwing power' is used as a general description of the effect on the protection level of the distance of the anode from the metal it is protecting. If the distance is great, part of the potential is used up in overcoming the resistivity of the electrolyte. The methods used to determine the positioning of anodes on a steel structure cannot be described here. Most designers rely upon experience gained in the field and it is often more a matter of art than science. However, calculations involving potential field gradients are possible when the structures are simple and these give some scientific basis for the location of anodes where the aim is to maintain a uniform potential field over the whole structure.

Figure 4.15(a) is an Evans diagram which shows the effect of coupling metals such as zinc and copper together. The potential of the copper cathode becomes more negative as current is drawn, while the potential of the zinc becomes more positive. The potentials of both metals converge to the corrosion potential, E_{corr}, at which there is a negligible resistance in the cell and the limiting current is passed. To obtain this particular E_{corr} value, the areas of the two electrodes represented in Fig. 4.15(a) must be the same. In practice, variations in the ratio of the electrode areas change the value of E_{corr}. This was discussed in Chapter 5 and illustrated in Fig. 5.8. If there is a large cathode/anode ratio, protection is unlikely to be satisfactory because the cathode has not been sufficiently polarised away from its free corrosion potential. Remember that the anodic reaction still occurs at potentials more negative than the free corrosion potential and for this reason it is usual to define a protection potential – the least negative potential necessary to achieve a satisfactory level of protection. For steel in aerated sea water, this is considered to be about -800 mV SSC.[3] The correct amount and type of anode material when distributed over the surface of a structure will result in its cathodic polarisation to a potential more negative than this. Anodes connected to structures for the purpose of effecting corrosion protection in this way are termed **sacrificial anodes**. Because they rely upon the galvanic effect, it is essential that they make a good electrical contact with the structure being protected. Anodes are usually welded to special lugs which are integrated into the structure at pre-determined points.

The traditional sacrificial anode material for steel in sea water is zinc. Sir Humphrey Davy reported in 1824[4] the successful use of zinc anodes to protect the copper sheathing on warships. Although sacrificial anodes were used prior to the First World War to prevent condenser tube corrosion in warships, it was not until the 1950s that the technique was used on a significant scale. Since then, a zinc alloy known as C-Sentry ®, containing 0.1–0.5 per cent aluminium and 0.025–0.15 per cent cadmium, has been used extensively throughout the marine industry.

® Trade mark of Impalloy Ltd.

The protective action of zinc would be excellent if it dissolved at a reasonably constant rate. Unfortunately, this is not usually the case. Normal commercial purity zinc corrodes in sea water with the formation of an impermeable skin which severely limits its current output. Of the principal impurities: iron, copper and lead, the most detrimental to anode performance is iron. Its solubility in zinc is so low (<0.0014 per cent) that any excess is present as discrete particles. These, in turn, yield local galvanic cells which produce a coating of insoluble, non-conductive zinc hydroxide/zinc carbonate, which eventually renders the anode ineffective. Addition of aluminium is beneficial because a less noble aluminium/iron intermetallic is formed which reduces the effect of the local corrosion cells. The addition of cadmium acts in a similar way to reduce the adverse effect of lead impurity.

Aluminium normally undergoes pitting corrosion in sea water because of the cathodic oxide layer which always exists on the metal in air. The unpredictable nature of this form of corrosion makes the pure metal most unreliable for use as a sacrificial anode, so alloying additions are made to prevent the formation of a continuous, adherent, protective oxide film, and permit continued and regular galvanic activity. As a result, aluminium alloys containing zinc and mercury or zinc and indium have been developed. These have much greater electrical power/weight ratios than zinc alloys and the use of aluminium anodes has begun to eclipse the use of zinc in some applications, notably the offshore industry.

The highly negative free corrosion potential of magnesium means that it dissolves rather too vigorously in sea water. Its use is thus restricted to the protection of pipelines in soil, or structures in estuarine waters where the resistivity is high enough to limit the effectiveness of zinc or aluminium alloys. The protection of storage tanks containing fresh or brackish water is another suitable application for magnesium anodes. It should be noted that magnesium represents a significant fire hazard which precludes its use in certain applications.

Although the theoretical free corrosion potential of magnesium is -2.12 V SCE, it is found in practice to be about -1.7 V SCE. This is reflected in the metal's low (50–60 per cent) efficiency: the calculated current yield should be about 2200 Ah kg^{-1}, but in reality is rarely more than 1200 Ah kg^{-1}. This compares unfavourably with zinc or aluminium alloys which have efficiencies greater than 90 per cent. The two commonly available magnesium alloys, one containing 6 per cent Al, 3 per cent Zn, 0.2 per cent Mn, and the other, a high purity alloy with 1 per cent Mn, are the result of attempts to increase the efficiency by alloying additions, with limited success. The reasons for the inefficiency are too complex to be described here, but are thought to be concerned with changes in anion and cation concentrations close to the metal surface, as well as the evolution of hydrogen at local cathodes in the metal.

A complete review of sacrificial anodes and their properties has been published by Schreiber.[5] Table 16.1 lists three typical sacrificial anode materials and their properties.

Table 16.1 Sacrificial anode materials and their properties

Property	Zinc alloy* (C-sentry)[†]	Aluminium alloy (Galvalum I)[‡]	Magnesium alloy (Galvomag)[‡]
Composition (%)	Al: 0.4–0.6 Cd: 0.075–0.125 Cu: < 0.005 Fe: <0.0014 Pb: <0.15 Si: <0.125 Zn: rem	Al: rem Cu: <0.006 Fe: <0.1 Hg: 0.02–0.05 Si: 0.11–0.21 Zn: 0.3–0.5 Others: each <0.02	Al: <0.01 Cu: 0.02 Fe: <0.03 Mg: rem Mn: 0.5–1.3 Ni: 0.001 Pb: <0.01 Sn: <0.01 Zn: <0.01
Density	7060 kg m^{-3}	2695 kg m^{-3}	1765 kg m^{-3}
Capacity	780 Ah kg^{-1}	2640 Ah kg^{-1}	1232 Ah kg^{-1}
Wastage (weight)	10.7 kg Ay^{-1}	3.2 kg Ay^{-1}	4.1 kg Ay^{-1}
Wastage (volume)	1518 ml Ay^{-1}	1180 ml Ay^{-1}	2296 ml Ay^{-1}
Output	6.5 A m^{-2}	6.5 A m^{-2}	10.8 A m^{-2}
E_{corr} (SSC)	− 1050 mV	− 1050 mV	− 1700 mV

* US Dept of Defense spec for zinc sacrificial anode material requires stricter control of impurity levels than in this alloy
[†] Trade mark of Impalloy
[‡] Trade mark of Dow Chemical Company

16.3 PRACTICAL APPLICATIONS OF SACRIFICIAL ANODE CATHODIC PROTECTION

The attachment of sacrificial anodes to structures in both marine and soil environments has been common practice for decades and continues to be a very important means of corrosion protection. Sacrificial anodes are relatively inexpensive, easy to install, and, in contrast to impressed current techniques, can be used where there is no power supply. The method has the added advantage that there is no expensive electrical equipment to buy and current cannot be supplied in the wrong direction, as in the case of HMS *Blackwood*, section 16.5. Sacrificial anodes are most suitable in small-scale applications, though they are used extensively, with equal effect, on large structures. They do, however, need frequent replacement and, if large amounts are necessary, extra stress may be placed upon the structure. More will be said of this below.

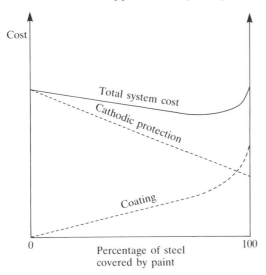

Fig. 16.4 Comparative cost of cathodic protection and applied coatings.

It has been shown[6] that a combination of cathodic protection and protective coating is the most economical means of protecting a steel structure. No paint coating is perfect; there will always be a number of defects in the coating which leave areas of bare metal exposed to the environment. When extra care is taken to reduce the number of such defects, the cost of painting the structure rises. If attempts are made to eliminate the last few defects, the cost rises dramatically. Conversely, the cost of cathodic protection falls as the surface is coated because less anode material is needed. Logically, there exists a point at which the cathodic protection, together with a good but not perfect coating, is the most economic option. This is shown on Fig. 16.4.

The use of sacrificial anodes to protect ships' hulls has become less favoured than using impressed current techniques, but it is still to be found on smaller vessels where the latter method is uneconomic. Zinc is the commonest anode material, and is used preferably in conjunction with a paint coating (section 14.3). Anodes are welded or bolted to fixtures on the hull, often in arrays located in the stern area where protection is most likely to be needed. Here, the extreme turbulence caused by the propellers damages protective coatings and results in impingement corrosion. Cavitation corrosion is also common in this vicinity. Furthermore, fittings such as propellers are often made from non-ferrous (copper-based) alloys and provide such excellent cathodes that protection, in addition to paint, is essential at the stern of a ship. Anodes may be found further forward on bilge keels, if they are present, and are also extensively used inside the numerous sea-water intakes for the machinery systems.

313

Question: *How does a designer determine the the number of sacrificial anodes he must use to protect a ship's hull?*

Various standards have been laid down for the current density which is necessary to protect steel surfaces. A surface which has been given a fresh coal tar epoxy paint coating is deemed to be well protected if the current density is between 20 and 30 mA m^{-2} to accommodate holidays in the coating, in contrast to a bare steel surface which is reported to require in excess of 100 mA m^{-2}.[3,7] There is much debate about optimum current densities and the figures quoted are subject to considerable variation, depending upon the severity of the environment. The reasons for the uncertainty stem from the problems in knowing the area of metal exposed to the electrolyte. (This was explained at the end of section 16.1.) In current practice, a designer, knowing the surface finish to be applied to a structure, must make assumptions about the current density required to protect it and the area of steel which will be exposed to the sea. (The number of defects in a paint film will increase as the structure ages, and even bare steel surfaces may become coated with scales or other barriers which will alter the area needing protection.) Having made these assumptions and knowing the output of the anode material he plans to use, he can calculate the necessary weight of material. Such methods are discussed in examples in Chapter 18. On simple geometrical structures, the distribution of anodes is uniform, but on other structures, the placement is more difficult and often based on previous experience, rather than scientifically determined data.

Regular dry-dockings are essential so that consumed anodes may be replaced. In the Royal Navy, dockings are typically every 18 months to two years, and it is usual to find that, after emptying the dock, most of the anodes have been consumed. Complete replacement of all sacrificial anodes is thus a routine maintenance procedure. It is worthwhile to point out that during such operations anodes are sometimes painted by over-zealous or ill-informed painters during the application of the various hull coatings, as on the stern of the commercial ship shown in Fig. 16.5. This, of course, renders anodes useless.

The need for frequent docking means that sacrificial systems are unpopular with shipping operators who maximise economy by keeping their ships at sea for as long as possible. For this reason, impressed current systems which use non-consumable anodes tend to be favoured.

In the North Sea, the tapping of the vast reservoir of natural resources has been possible only by giant feats of engineering achievement. Some of the world's largest steel structures have now been operating in extremely aggressive environments for 15 to 20 years and the offshore technology which began in the relatively tranquil Gulf of Mexico has 'come of age' in one of the world's most hostile marine environments. Figure 16.6[8] gives an indication of the extent of the development in 1984, by which time as many as 200 platforms had been commissioned in depths ranging from 25 to over 150 m. Of these, about 90 per cent were of bare steel, with large numbers of sacrificial anodes welded all over the submerged parts of the platforms.

Fig. 16.5 Painted sacrificial anodes on a ship's stern.

In several cases, as much as 30 per cent of the submerged mass was zinc anode material!

The need to find alternative fossil fuel supplies during the period of instability in the Middle East in the mid-1970s led to the urgent development of the North Sea resources. New technology had to be developed to enable this exploitation to take place, but the corrosion protection of the vastly expensive structures had to be based upon cathodic protection systems about which the available scientific data were sparse. Designs were often based more upon inspired guesswork than the application of science. This remains true in 1985, especially in the field of impressed current techniques, about which operators remain wary, and is a probable reason for the preference for copious quantities of inexpensive 'zincs' in the belief that overprotection is safer than the risk of underprotection. The complexities of the environment have made very difficult the determination of reliable quantitative data. Section 16.5 highlights some typical problems which have been experienced with impressed current systems.

One of the earliest platforms in the North Sea was the West Sole B platform which was set in 25 m of water in 1966. It was protected by a combination of coal tar epoxy coating and 20 tonnes of zinc sacrificial anodes; it is thus not entirely typical of the majority of the North Sea structures which have no protective coating, but is one rig for which long-term test data are available. The distribution of anodes over the structure is, however, akin to that used on uncoated structures, and is shown in Fig. 16.7.

In a detailed survey covering the first 16 years of its life, Fairhurst[9] reported that the mean current density over the first five-year period of 11 mA m^{-2} rose to 17 mA m^{-2} over the following nine years. The anodes were of rectangular block and triangular geometry, and all of identical

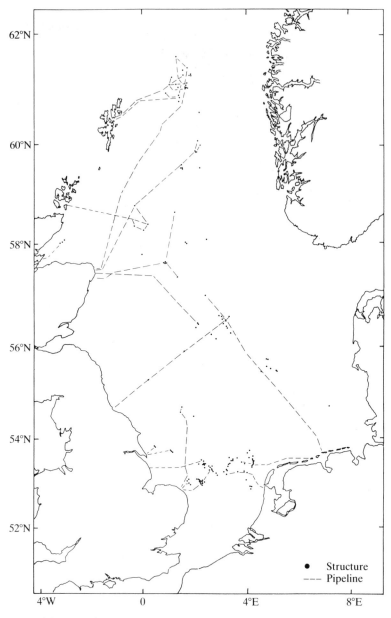

Fig. 16.6 The exploitation of fossil fuels in the North Sea, showing major structures and pipelines, 1985. (Drawn from data published elsewhere[8] and reproduced with permission of the Hydrographer of the Navy.)

composition. They showed a marked and unexplained difference in consumption rates, the triangular ones being almost exhausted at the time of the survey. It was also noted that far less wastage than expected had occurred in the region immediately below the waterline. One benefit which

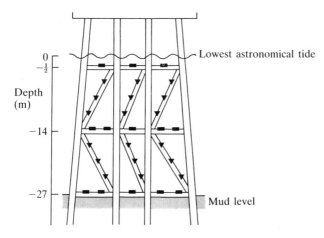

Fig. 16.7 Distribution of sacrificial anodes over the West Sole B platform (after Fairhurst[9] and reproduced by kind permission of BP Petroleum Development Ltd)

was ascribed to the protective coating was that, despite the severe wastage of the triangular anodes, the structure was in good condition and no marked depolarisation had occurred. The overall impression of the survey was that the cathodic protection system had worked well, though in a way which was different from that which had been expected at the design stage. This high-lights the difficulties in making the basic design assumptions for a sacrificial system noted earlier.

On West Sole B the unknown parameters have operated in favour of the designer. As rigs age over the next 10 to 20 years there must surely be instances to the contrary. Many systems in current use were designed and installed after the basic design of the platform had been formulated. (The use of the word 'retrofit' is common.) The extra surface area of the anodes, coupled with that caused by the growth of marine organisms, results in an increase in the impact energy of the waves. Stresses on the platforms may sometimes be greater than the designers anticipated.

A favourable natural phenomenon has also assisted the designer of marine structures. Calcium, magnesium and other metal ions are present in significant quantities in sea water. The negative potentials of cathodic steel surfaces generate hydroxyl ions (eqn [4.21]) which cause precipitation of insoluble calcium and magnesium salts, known as **calcareous deposits**. A strongly adherent film is formed which reduces the current required for protection and protects extensive surfaces against corrosion at any local anodes which may be present. A study of the effects of the calcareous films on environment-sensitive cracking has produced favourable conclusions.[10] It is believed that as films form, they block cracks and prevent the opening and closing which can lead to propagation.[11]

Recently, greater use has been made of aluminium alloy anodes for their better performance to weight ratio. In 1982 Wyatt[6] reported that the cost of an aluminium anode sacrificial system for a bare steel deep water oil

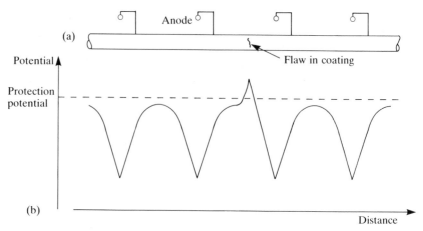

(a) Anode

Potential

Protection potential

Flaw in coating

(b)

Distance

Fig. 16.8 The protection of buried steel pipelines using sacrificial anode cathodic protection.
(a) Schematic plan view of buried pipeline.
(b) Distribution of potential along pipeline with a coating defect.

platform was £3.3 m. and was no cheaper than the cost of a system which combined the benefits of sacrificial anodes and coated steel. However, the benefits of an applied coating, compared to the well-documented benefits of calcareous deposits, are not yet known with certainty, and the proven performance of the latter seem to be favoured, at least in the North Sea.

Pipelines, too, need protection. Steel pipelines buried in soil have been successfully protected by sacrificial anodes for many years. Anodes are buried at fixed intervals along the length of the pipe, and at a constant distance from it, as shown in Fig. 16.8. The method relies upon the existence of a conducting path through the soil to the pipe. Electrical connection is made from the anode to the pipe, the anode dissolves and a current flux is created which polarises the pipe to potentials shown diagrammatically in Fig. 16.8. The importance of the correct anode spacing is obvious from the figure, because if the distance is too great, the polarisation at points furthest from the anodes will not be sufficient to give protection. The potential profile is extremely soil-dependent and reliable performance requires care, both at the design stage, and in subsequent monitoring. A defect in the coating can also cause a localised potential which is outside the protection range. Such a situation may lead to perforation of the pipe and render the cathodic protection useless. The reaction which occurs at the steel surface is dependent upon the type of soil: in acid soils, the preferred cathode reaction is the reduction of hydrogen ions to hydrogen gas, eqns [4.19/20], while in well-aerated, non-acid soil, the cathode reaction is one in which oxygen is reduced to hydroxyl ions, eqn [4.21].

In practice, the monitoring of pipeline potential is fraught with difficulties. For many years, serious errors were consistently made because of lack of consideration of *IR* drops between the protected surface and the reference electrode. Buried pipelines, in particular, suffer from this problem because of their inaccessibility. Methods in which a **stabbing electrode** takes readings as close to the metal surface as possible have been most successful for subsea pipelines (see below), though this, too, is difficult if the pipelines are buried or cased in concrete. In the past, overdesign of underground pipelines probably assisted in the generally acceptable performance levels, but current trends towards higher operating pressures, higher stress levels and thinner wall thickness have increased the importance of better monitoring of pipeline corrosion resistance. Numerous other new methods are presently being developed for underground pipelines and a complete review is published elsewhere.[12]

Since the 1970s the installation of 4000 km of subsea steel pipelines in the North Sea (see Fig. 16.6) has been achieved with the almost universal use of zinc alloy sacrificial anodes combined with an insulating coating. Until 1985, aluminium alloy had been used only rarely. Subsea pipelines usually employ a different anode arrangement from their land-based counterparts. Anodes, mostly in the form of 'bracelets' weighing 300 to 400 kg, are fixed around the circumference of the pipes at intervals of about 150 m. A pipeline may stretch along the seabed, either exposed or buried, for hundreds of kilometres and link the platforms with the mainland.

The fear of pollution arising from a pipe leak caused by corrosion resulted in both the UK and Norway requiring oil companies to take safeguard action against such an event. This has led to the formation of teams whose task is to traverse and survey the length of the line, sometimes as much as 250 km and often in appalling weather conditions. Backhouse and Holt[13] have described the requirements of a survey as follows:

(a) The most important parameter to measure is the local cathodic protection potential of the pipe steel along its entire length. Steel which is at a potential of −850 mV SSC is adequately protected and it is of paramount importance that accurate potential measurements be made. This is, however, not as simple as it seems. Local variations are found at anodes, flanges, joints and coating defects. They can only be measured accurately at points close to the pipe. Superimposed on the local variations are long distance changes which result from a number of sources. A platform has considerable influence on potential when connected to a pipeline. At platform approaches, when no isolation joints are fitted, the average pipe potential may change by 100–200 mV in a few kilometres. Variations will also be caused by a poorly coated pipe connected to a well-coated pipe or a pipeline with a poor sacrificial system connected to one with a good system. Many methods used to determine pipeline potentials involve errors which are sufficient to undermine greatly the value of the survey.

(b) Measurements of the output currents and potentials of bracelet anodes are of secondary importance but can be used to predict the life of anodes.

(c) Determination of the field gradients along a pipeline can allow the detection of coating defects. While, at first sight, it might appear that a

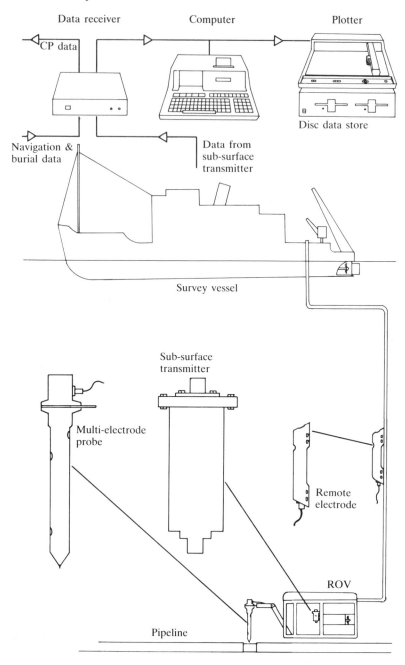

Fig. 16.9 The measurement of potential profiles for subsea pipelines.
(a) The equipment used in the survey.
(b) The result of a typical potential survey on a subsea pipeline. (Reproduced by kind permission of Subspection Ltd.)

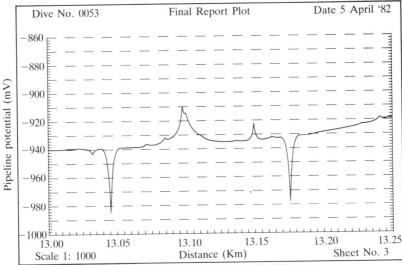

Final C.P. Potential Plot after Data Analysis.

(b)

coating defect is serious, it should be remembered that the sacrificial anodes have been used for that very purpose. A defect will cause the adjacent anodes to be consumed at a faster rate than would otherwise be the case, but the steel should be protected.

Inspection of the performance of the corrosion protection of a pipeline is achieved using the equipment illustrated in Fig. 16.9(a). A multi-electrode probe is attached to a remotely operated vehicle (ROV) which, in turn, is attached to a survey ship by an umbilical. An electrode which is remote from the probe is attached to the umbilical so that field gradient measurements can be made. The probe travels under water along the length of the pipe, periodically stabbing the coating on the pipe to take a local measurement of the potential field gradient. An analogue-to-digital converter connected to a transmitter on the ROV then sends the data to an on-board computer which is coupled to the ship's navigational computer. Thus it is possible to produce an accurate chart of the potential of each of the positions chosen for stabbing, Fig. 16.9(b).

16.4 IMPRESSED CURRENT CATHODIC PROTECTION (ICCP)

Although the principles underlying impressed current cathodic protection described in section 16.1 are very similar to those of the sacrificial anode method, there are some important differences:

321

(a) In the sacrificial anode method, the anode material and the structure must be in electrical contact. In the impressed current method, the anode, when mounted on the structure it protects, is surrounded by an insulating shield which protects the surrounding metal from excessive current densities in the immediate vicinity of the anode. Anodes for the few ICCP systems on oil rigs may, however, be sited on the seabed, about 100 m away from the structure. This is thought to give better potential distribution, though there are those who would argue that this is not a better system than having anodes mounted on the structure.

(b) A major advantage of the impressed current method is that it can use anodes which are virtually non-consumable. In electrolytes of pH 7 or less, the anode reaction is the oxidation of water (section 4.8) rather than metal dissolution:

$$2H_2O \rightarrow O_2 + 4H^+ + 4e^- \tag{4.54}$$

In alkaline solutions, it is the oxidation of hydroxyl ions:

$$4OH^- \rightarrow O_2 + 2H_2O + 4e^- \tag{16.2}$$

The use of permanent anodes obviates the necessity to renew large amounts of anode material at regular intervals and is cheaper to run.

In sea water, it is usually found that oxidation of chloride ions to chlorine gas is the favoured anode reaction:

$$2Cl^- \rightarrow Cl_2 + 2e^- \tag{16.3}$$

although in the diluted waters of estuaries, the oxidation of water, eqn [4.54], may predominate.

(c) The use of electronics enables the system to be self-regulating, as described in section 16.1. This is a significant advantage over the sacrificial method where there is no control of the current supply, once the anodes have been installed.

The power supply of the control unit is a transformer rectifier which converts a locally available alternating current supply into direct current of the required voltage. Such sources are usually custom built for each application: there may be separate control units for each anode or group of anodes, depending on the size of the system to be employed.

Early anode materials consisted of large pieces of scrap iron or steel which were slowly consumed by the normal process of anode dissolution. Today the use of consumable anodes is usually restricted to sites in buried mud or seabed sand where the dispersal of the gaseous anode reaction products from the non-consumable anodes is difficult. Most modern ICCP applications use anode materials such as lead/silver alloy, platinised titanium and platinised niobium. Table 16.2[14] lists some materials and their properties. Although they are capable of supplying large current densities, the anodes are usually distributed at regular intervals over the whole of a struc-

Table 16.2 Impressed current anode materials and their properties (from Brand[14])

Material	Consumption $(kg\ Ay^{-1})$	Recommended uses
Platinum and platinised metals	8×10^{-6}	Marine environments and high purity liquids
High silicon iron	0.25–1.0	Potable waters and soil or carbonaceous backfill
Steel	6.8–9.1	Marine environments and carbonaceous backfill
Iron	Approx 9.5	Marine environments and carbonaceous backfill
Cast iron	4.5–6.8	Marine environments and carbonaceous backfill
Lead–platinum	0.09	Marine environments
Lead–silver	0.09	Marine environments
Graphite	0.1–1.0	Marine environments, potable waters and carbonaceous backfill

ture, rather than in small numbers which must protect large areas. There are two main reasons for this:

(a) The large current density in the immediate vicinity of an anode is damaging to many types of paint film. The use of more anodes reduces the current density of each anode and lessens the damage to the protective coating.

(b) In the complex geometrical arrangements of offshore platforms, it is difficult to predict the distribution of potentials. Consequently, it is safer to employ more anodes to protect smaller areas. If any doubt exists about the ability of an anode to protect a particular part of the structure, sacrificial anodes may be used in conjunction with the impressed current system.

When used in ICCP systems, reference electrodes are either of the zinc, silver/silver chloride (SSC), or copper/copper sulphate (CSE) types, the last of these tending to be favoured in applications involving reinforced concrete. Reference, or control electrodes, are vital components which determine the current to be provided by the power supply. Malfunction is a fault which must always be guarded against: physical damage to the anode system or reference electrodes by any one of the hundreds of daily operations around an offshore platform is always a possibility. So also is the damage to an electrode caused by numerous pollutants of estuarine waters.

If systems are found to be inadequate it is not necessarily the result of bad design. The complexity of the structure and the lack of proven research in this area make the designer's task even more difficult than it should be.

An ICCP designer should always work alongside the structural engineer to decide the areas which are most critical to structural integrity and ensure they are adequately monitored. Even so, most users regularly inspect both structures and systems to ensure correct functioning of the equipment and that no untoward corrosion is occurring.

16.5 APPLICATIONS OF IMPRESSED CURRENT CATHODIC PROTECTION

The first recorded application of ICCP is uncertain, although the protection of buried pipelines was probably the first. One of the earliest applications which was actually demonstrated to be immediately successful occurred in the power industry.

Case 16.1: In the late 1920s, pitting of condenser tubes was experienced at a UK power station. It was decided to apply impressed current cathodic protection using cast iron anodes. These were fed from a d.c. generator which supplied power into the water box via a cable through a wooden stuffing gland. The problem was completely solved.[15]

Similar techniques have been used to protect ships' hulls since the early 1950s, though not always with such a successful outcome.

Case 16.2: One experimental installation in the Royal Navy vessel, HMS *Blackwood*, was destined to become an ignominious disaster. Two control systems were used, one to protect each side of the ship. Unfortunately, during installation, the connections to one of them were transposed and the starboard side of the ship was actively dissolved away, whilst corrosion on the piece of old iron used as the nominal anode was controlled instead. It was only when serious leaks occurred that the problem was realised.

In today's vessels, typical designs place anodes in symmetrical dispositions. In bulk carriers, however, the necessity for cable runs away from anodes and references, together with the need for internal access, usually precludes the siting of electrodes externally to storage tanks. In such cases, electrodes are placed either well forward or well aft, where the adjacent machinery spaces provide convenient access to the various pieces of equipment. Figure 16.10 shows this.[7]

Impressed current cathodic protection of ships is always used in conjunction with protective coatings, the latter being intended as the primary protection with the use of cathodic protection as a back-up in those areas where coating defects may be present. In the period immediately subsequent to the application of the coating, there is very little demand on the impressed current system. During the operational life of the ship, the

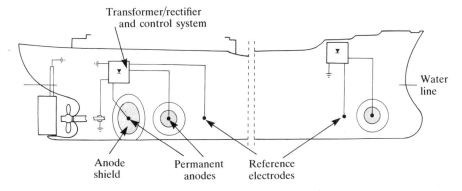

Transformer/rectifier and control system

Water line

Anode shield

Permanent anodes

Reference electrodes

Fig. 16.10 Typical impressed current cathodic protection system for a ship (after Jensen[7]).

coating deteriorates and the demand on the cathodic protection increases, in a manner described in section 16.1. Eventually, the demands placed upon the system may exceed the capabilities of the design, with high anode currents causing even more damage to the coating, particularly in the areas adjacent to the anodes where high concentrations of hydroxyl ion are present. In addition, the high local currents around the anodes may reduce the protection supplied to the rest of the structure. On a supertanker, the initial current of 10 A may rise to over 1000 A during the course of the ship's operational life. Thus, although it is possible to extend the period between dockings by use of impressed current rather than sacrificial anode cathodic protection, hull maintenance periods are still a part of any ship's programme. Note that although ICCP is used for hull protection, it is still common practice to use sacrificial zincs inside the numerous sea-water intakes where the effect of the impressed current may be inadequate to protect the steel.

Although it is commonly thought that the cathodic protection of ships is a well-understood subject, there is little published research to validate the principles which have been used in many designs. The fact that systems appear to have performed satisfactorily in the past seems to be the main criterion for current design. Areas of doubt remain:

(a) The standard of assessment of protection by current or current density level is always open to doubt when what is actually required is knowledge of the real potential of the whole of the structure. It is the difficulty of obtaining reliable potential data which leads to reliance on current density measurements as a means of assessment.

(b) Many ships' ICCP designs utilise no more than two or three reference electrodes, a fact which increases the uncertainty about the ability to protect a large area of structure, under very variable conditions. As will be discussed below, the siting of control electrodes is vital to the correct functioning of the system.

(c) There would appear to be very little understanding of the performance of impressed current systems under dynamic conditions. Much of the published research reports data obtained in static conditions, apparently because of the difficulties of obtaining consistent results in flowing environments. Thus little is known about the effect of the motion of a ship through the sea upon the distribution of protection potential over the hull. One trial for an Egyptian patrol boat indicated that although a current of about 4 A was sufficient to protect the hull while alongside a jetty, the requirement was about 35 A at 45 knots. Recent work[16,17] has shown that protection current demand on steel in the wave affected zone of an offshore platform can increase by up to 25 per cent, with concomitant change of potential to less negative values.

It has been shown[18] that the most significant part of a ship in determining the potential distribution over the hull is the stern area where non-ferrous alloys are used for the propeller. By the use of modelling techniques, results were obtained which accurately simulated data measured on a real warship under static conditions. In conditions which simulated cruising speed, the protection system was found to be unable to provide the required potential at the stern, Fig. 16.11(a). By adding an additional control electrode and repositioning the other control electrodes, good protection levels were obtained over the whole of the model, under both static and dynamic conditions, thus proving the critical nature of the reference position to the effective operation of the system, Fig. 16.11(b).

The large degree of uncertainty which, until recently, arose in the design of impressed current systems may account for the reluctance of designers to use such systems on offshore platforms. Very few of the enormously expensive structures in the North Sea are protected by impressed current installations, the designers preferring to opt for the more proven and dependable sacrificial anode method.

Tischuk[19] has reported the difficulties experienced with ICCP systems designed for the Piper and Claymore platforms, located about 120 miles north-east of Aberdeen.

Case 16.3: When the Piper drilling platform was installed in 1975, the current density for protection was 86 mA m^{-2}, but during the course of seven years of operation, current density was gradually increased to 130 mA m^{-2} on the Piper platform, and 160 mA m^{-2} on the modified Claymore system. Both systems utilised platinum–iridium anodes: 42 with 6 references on Piper, and 55 with 12 references on Claymore. Both systems suffered severely from the rigours of natural and man-made environments. On many occasions, cable runs to anodes were damaged, and anodes broken off by falling scaffolding. Many anodes were rendered inoperative because it was found that the pulsed current system was causing platinum consumption 30 times greater than had been expected.

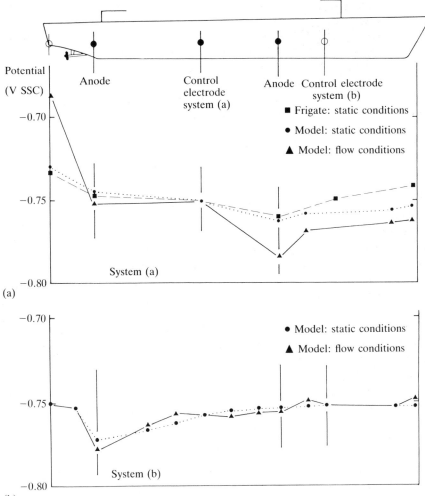

Fig. 16.11 The effect of reference electrode location on potential distribution on a ship's hull in both static and flow conditions.
(a) With one reference electrode amidships.
(b) With extra reference electrode aft and the original re-sited ahead of the forward anode. (After McGrath, Tighe-Ford and Hodgkiss[18]).

The most remarkable fact about Tischuk's catalogue of problems was that the platforms seemed to be resisting corrosion satisfactorily.

One of the most recent designs is that of CONOCO's Murchison platform, Fig. 16.12. Set in one of the most exposed parts of the northern North Sea its 80,000 m² surface area is protected by 100 anodes and 50 reference electrodes.

On land, ICCP of buried pipelines is, at present, the preferred method of protection. As with subsea pipelines, regular monitoring is an essential requirement, the following examples illustrating this point.

Fig. 16.12 CONOCO's Murchison platform. The field was ready for production by September 1980, only 5 years after a wild-cat well first struck oil on the location. (Photograph courtesy of Conoco (UK) Ltd.)

Case 16.4: After a survey on a length of pipe it was decided to investigate six possible defects, associated with inadequate potentials. In each case, extensive corrosion was revealed, with average pitting of 2.8 mm in just six years. Pipeline potential was found to vary between −1820 mV CSE and −560 mV CSE, depending on the resistivity of the soil.[20]

Case 16.5: The results of surveying 3560 miles of a gas pipeline over a period of five years, pinpointed 1567 possible problem areas: 625 were deemed to be serious enough to require investigation and repair. Without the survey there is no doubt that many leaks would have occurred, but more seriously, four defects were so serious that there was a real possibility of total rupture and a catastrophic accident.

The ICCP of buried pipelines is not without its difficulties. The method relies upon buried, localised arrays of anodes known as ground beds, which distribute current along the length of the line. The material in which the anodes are buried is not soil, but a specially formulated backfill which reduces the soil/anode resistance, allows the escape of anode gases, and increases the current capacity. Two common backfill materials are coal coke breeze and petroleum coke breeze which are both 95 per cent carbonaceous. Despite this, the great variation in water content and acidity of soils along the length of the pipeline can cause problems during the operation of the system, highlighting the need for regular monitoring. Anode materials vary from graphite, magnetite and silicon iron, to the more recently developed lead–platinum and platinised titanium. A comprehensive review of the subject has been published by Shreir and Hayfield.[21]

Impressed current methods have been used to protect steel cathodically in concrete in such applications as buried prestressed pipelines and tanks, bridge decks and marine structures. When embedded in a few centimetres of sound concrete, steel is passivated and corrosion resistant because of the highly alkaline conditions which exist (pH 12.5–13.5). Corrosion problems occur when the iron is depassivated by penetration of reactive species. The most common reaction occurs when carbon dioxide (acidic) penetrates the concrete and neutralises the alkaline constituents; pH falls and corrosion occurs. A much more dangerous reaction occurs when chloride ion, either from a marine environment or from de-icing salt, is able to reach the iron surface. This ion is able to depassivate iron, even at high pH, and corrosion problems occur similar to those suffered by the Pelham bridge (section 1.3). Not only is the concrete weakened, but the rusting steel sets up severe tensile stresses within the concrete and large pieces may become detached from the structure, a process known as *spalling*.

Case 16.6: In Ontario, Canada, 32 bridges have been given impressed current systems after a successful trial on Kingston's Division Street Bridge. Action became necessary after masonry falling onto Toronto's Gardiner Expressway necessitated the use of safety nets. In North America, in 1985, approximately 100 bridges had ICCP systems.[22]

When oxygen concentration is low iron is not passivated but seems to corrode at an insignificant rate. This is particularly true for concrete structures which are immersed in sea water. It is thought that the steel reinforcing bars are polarised to as low a potential as -800 to -1100 mV CSE in low oxygen, water-saturated conditions and steel in this environment is considered safe. The splash zone, where oxygen access is considerable, is the most vulnerable area of an offshore steel-reinforced concrete structure. There is good reason, therefore, to attempt to protect such concrete structures by impressed current techniques and it is reported that many such attempts have been successful.[23]

Figure 16.13 summarises data given by Arup[24] for the natural and impressed potentials for steel in concrete. It is currently believed that potentials more negative than -850 mV CSE should be used to provide adequate protection. The possibility of hydrogen evolution and consequent embrittlement of high strength steels has been considered a danger at

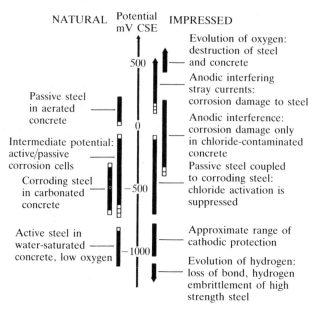

NATURAL   Potential   IMPRESSED
          mV CSE

Evolution of oxygen:
destruction of steel
and concrete

Anodic interfering
stray currents:
corrosion damage to steel

Passive steel
in aerated
concrete

Anodic interference:
corrosion damage only
in chloride-contaminated
concrete

Intermediate potential:
active/passive
corrosion cells

Passive steel coupled
to corroding steel:
chloride activation is
suppressed

Corroding steel
in carbonated
concrete

Active steel in
water-saturated
concrete, low oxygen

Approximate range of
cathodic protection

Evolution of hydrogen:
loss of bond, hydrogen
embrittlement of high
strength steel

Fig. 16.13   Natural and impressed potentials of steel in concrete (after Arup[24]).

potentials more negative than −1100 mV CSE, and overprotection at such potentials should be avoided.

One advantage of ICCP of steel in concrete is that current densities are lower than in many other applications, principally because the conductivity of the "electrolyte" is much lower. In a bridge deck, the worst case, values of 20 mA m^{-2} are typical, while concrete structures in soil require between 1 and 4 mA m^{-2}, depending upon the quality of the concrete. Values as low as <0.15 mA m^{-2} have been reported for steel in high quality concrete in sea water.

Such applications of ICCP are not without their difficulties. These can be:

(a) establishing effective electrical bonding of all the steelwork;
(b) distribution of the protective current through the concrete, a relatively high resistivity electrolyte;
(c) monitoring potentials and corrosion.

Probably the most serious is that of monitoring potential. Wilkins[23] has expressed doubts about the use of reference probes and it is essential that any system must be carefully designed and installed before the concrete is cast. Attempts to control corrosion by retrofit methods may be considered to be suspect.

16.6 ANODIC PROTECTION

In section 16.1 it was established that, in principle, the application of a potential to steel, such that the metal was anodically polarised from its free corrosion potential, could lead to the formation of a passive film and afford protection against corrosion. The criterion for protection must be a coherent, insulating film which is sufficiently robust to withstand mechanical damage.

In Fig. 16.14, a schematic plot of potential against lg i is shown for such a metal. Point O represents the freely corroding condition. The curve at potentials more negative than E_{corr} represents the metal in a cathodic regime of the type discussed in the preceding sections. As potential is made more positive than E_{corr}, the current density increases to a maximum, i_{max}, but then suddenly falls to an extremely low value, i_{pass}. In practical terms, the metal has achieved a passive condition in which it is very well protected by a film of corrosion product. Further increase of potential has little effect until, at sufficiently high potentials the free energy available for corrosion overcomes the protection, the film breaks down and corrosion accelerates once again. This domain is known as the transpassive state. The principle of anodic protection is now evident: if the potential of the metal can be maintained in the range which leads to passivity, then the corrosion current

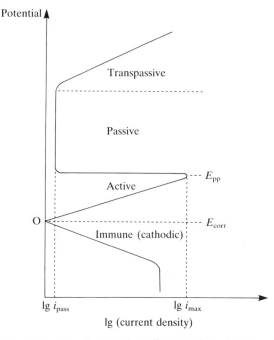

Fig. 16.14 The principle of anodic protection of steel: a schematic potentiodynamic scan for a material which exhibits the property of passivation.

density may be very low and very stable. A dangerous aspect is that if, for any reason, the metal potential wanders even slightly outside this range, corrosion may be worse than if anodic protection had not been used. Careful monitoring of the protection potentials and currents is therefore essential.

It is important to appreciate that such behaviour is not specific to the material alone, but to material/electrolyte combinations. Thus, for example, mild steel can be passivated in pure nitric acid, but not in acid which has been diluted with water, or in any concentration of hydrochloric acid. (See section 3.2, expts 3.12 and 3.13.) The determination of the values, E_{pp}, i_{pass} and i_{max} for those combinations which do possess this property is very important. Notice that two current densities are relevant: that which is necessary to passivate the metal; and that which is required to maintain the film once it has been formed. In many cases, i_{max} and i_{pass} may differ by factors of 100 to 1000. Thus, high current densities are needed in the first instance, but once passivated, large areas of metal can be efficiently protected by very small current densities.

The metals frequently protected by this means are iron, nickel, aluminium, titanium, molybdenum, zirconium, hafnium and niobium, together with alloys containing major amounts of these metals. Electrolytes can vary from very acid to very alkaline, and the method has a major advantage over comparable cathodic protection which is rarely applicable in such severe environments. Thus, anodically protected carbon steel can be used in the chemical and fertiliser industries to store a variety of oxidising acids, as well as caustic solutions. Similarly, anodically protected stainless steel is best employed for oxidising acid environments, but not for alkalis. Storage of non-oxidising acids such as hydrochloric and hydrofluoric acids is best done in protected chromium or titanium vessels. As in the case of cathodic protection, the method is only suitable in applications which involve continuous immersion.

The aggressive nature of the environments in which anodic protection is employed means that special reference electrodes are often necessary. All solid references made from the noble metals are usually suitable, for example, silver/silver chloride, or platinum/platinum oxide. Noble metals such as platinum are often used for the cathode too.

There are numerous advantages of using anodic protection. Industrial applications benefit from low operating costs and controlled, predictable conditions. Furthermore, it is often possible to replace an expensive alloy in unprotected plant with a cheaper material which is anodically protected, and the slow corrosion rates involved lead to less contamination of the products being stored or processed.

The disadvantages of the method are that failure of the electrical supply may be hazardous because of depassivation. In addition, the requirement for electrical current makes it useless for protection in organic liquid environments, or for components which are not continuously immersed. A comprehensive review of anodic protection has been published by Walker and Ward.[25]

16.7 REFERENCES

1. Pourbaix M 1966 *Atlas of electrochemical equilibria in solution*. Pergamon Press, pp 307–21
2. LaQue F L 1975 *Marine corrosion: causes and prevention*. John Wiley, p 203
3. CP1021:1973 *Code of practice for cathodic protection*. British Standards Institution, London
4. Davy Sir H 1824 *Philosophical transactions of the Royal Society* **114**(1): 197–204; also **114**(2): 242 and 1825, **115**(2): 328.
5. Schreiber C F 1986 Sacrificial anodes, in Ashworth V, Booker C J L (eds), *Cathodic protection theory and practice*. Chichester: Institution of Corrosion Science and Technology/Ellis Horwood, pp 78–93
6. Wyatt B S 1986 Cathodic protection of fixed offshore structures, in Ashworth V, Booker C J L (eds), *Cathodic protection theory and practice*. Chichester: Institution of Corrosion Science and Technology/Ellis Horwood, pp 143–71
7. Jensen J O 1986 Ships and semi-submersibles, in Ashworth V, Booker C J L (eds), *Cathodic protection theory and practice*. Chichester: Institution of Corrosion Science and Technology/Ellis Horwood, pp 128–42
8. International chart: North Sea, No. 4140 1985. Hydrographer of the Navy, Taunton
9. Fairhurst D 1984 Cathodic protection of the WB platform in the West Sole gas field – a review of 16 years' experience, in *Proceedings of UK corrosion 1984*. Institution of Corrosion Science and Technology
10. Jaske C E, Payer J H, Balint V S 1981 *Corrosion fatigue of metals in marine environments*. Columbus, Ohio: Batelle Press
11. Scott P M 1983 Design and inspection related applications of corrosion fatigue data, *Memoires et études scientifiques revue de metallurgie* Nov: 651–60
12. Martin B A 1986 Potential measurements on buried pipelines, in Ashworth V, Booker C J L (eds), *Cathodic protection theory and practice*. Chichester: Institution of Corrosion Science and Technology/Ellis Horwood, pp 276–92
13. Backhouse G, Holt R J 1984 Fundamental errors in subsea pipeline cathodic protection survey techniques, in *Proceedings of UK corrosion 1984*. Institution of Corrosion Science and Technology
14. Brand J W L F 1976 Impressed current anodes, in Shreir L L. (ed.), *Corrosion*. Newnes-Butterworths, section 11.3
15. Burbridge E 1986 Process plant and cooling water systems, in Ashworth V, Booker C J L (eds), *Cathodic protection theory and practice*. Chichester: Institution of Corrosion Science and Technology/Ellis Horwood, pp 191–213
16. Eliassen S, Steensland O 1977 *Journal of Metals* **29**: 12–18
17. Richter B 1980 *Germanischer Lloyds* **32**(10) and **32**(11)
18. McGrath J N, Tighe-Ford D J, Hodgkiss L 1985 Scale modelling of a ship's impressed current cathodic protection system, *Corrosion prevention and control* 32(2): 36–38, and 32(5): 89–91
19. Tischuk J L 1984 Operation and maintenance of impressed current cathodic protection systems, in Lewis J R, Mercer A D (eds), *Corrosion and marine growth on offshore structures*. Ellis Horwood, pp 61–68
20. von Baeckmann W G 1973. *Proceedings of 12th IGU conference*; paper C37.
21. Shreir L L, Hayfield P C S 1986 Impressed current anodes, in Ashworth V, Booker C J L (eds), *Cathodic protection theory and practice*. Chichester: Institution of Corrosion Science and Technology/Ellis Horwood, pp 94–127

22. *New Scientist* 1985 No. 1448, p 18
23. Wilkins N J M 1986 Cathodic protection of concrete structures, in Ashworth V, Booker C J L (eds), *Cathodic protection theory and practice*. Chichester: Institution of Corrosion Science and Technology/Ellis Horwood, pp 172–182
24. Arup H 1979 *Steel in concrete – electrochemistry and corrosion*. Glostrop, Copenhagen: Kerrosioncentralen; Newsletter No. 2
25. Walker R, Ward A 1969 The theory and practice of anodic protection, *Metallurgical Reviews* No. 137: 143–51

16.8 BIBLIOGRAPHY

Ashworth V, Booker C J L (eds) 1986 *Cathodic protection theory and practice*. Chichester: Institution of Corrosion Science and Technology/Ellis Horwood
Morgan J H 1959 *Cathodic protection*. Leonard Hill (Books)
Shreir L L (ed.) 1979 *Corrosion* (Vol. 2), Newnes-Butterworths, Ch. 11

17 CORROSION AT ELEVATED TEMPERATURES

Let all the greatest minds in the world be fused into one mind and let this
great mind strain nerve beyond its power; let it seek diligently on the earth
and in the heavens; let it search every nook and every cranny of nature;
it will only find the cause of the increased weight of the calcined metal in
air.

(Jean Rey: 1630)

The previous chapters have defined corrosion as the degradation of a metal
by an electrochemical reaction with its environment, but the environments
considered have been almost entirely aqueous. Corrosion on a metal surface
can occur even though a liquid electrolyte is not present, and, not surprisingly, the process is often referred to as dry corrosion. The definition of
corrosion which we have used so far is unchanged in dry corrosion
processes, as also is the description of a corrosion process by eqn [2.7].

Perhaps the most obvious dry corrosion process is the reaction of a metal
with the oxygen of the air. (Although nitrogen is the major constituent of
air, its role is unimportant when metals are heated in air because of the
dominating influence of oxygen. At high temperatures, nitrogen does react
with chromium, aluminium, titanium, molybdenum and tungsten, but reactions such as these are outside the scope of this text.) Though the reaction
with oxygen is, in principle, very simple, early scientists experienced great
difficulty in understanding the weight changes which accompanied the
calcination (oxidation) of a metal in air. Even today, studies of the oxidation
and other high temperature reactions of modern alloys have shown that the
processes involved are very complex. Therefore, the treatment in this text
is limited to an overview of this large and expanding subject.

Oxygen reacts readily with most metals, though the thermal energy
required to produce an oxidation rate of engineering significance may vary
considerably for different metals at the same temperature. At ambient
temperatures, most engineering materials are either already oxidised such
that the oxide layer protects the underlying metal, or react sufficiently
slowly in dry air for oxidation not to be a problem. At elevated temperatures, however, the rate of oxidation of metals increases. Thus, if an engineering component is exposed for a prolonged period in a high temperature
environment it may be rendered useless. For example, in pure, dry air at
temperatures below about 480 °C, a thin protective oxide film forms on the
surface of polished mild steel, but at a rate which is considered to be
negligible for engineering purposes. (A threshold rate has been defined[1] as
10^{-3} kg m^{-2} h.$^{-1}$) However, during hot rolling and forging of mild steel

(processes which take place at about 900 °C) the rate of oxidation is sufficiently great to produce a layer of oxide called mill scale, which is non-protective. We have already seen (section 15.1) that mill scale may have an important effect upon the corrosion rate of mild steel in aqueous environments (see section 15.1). On the other hand, the usefulness of metals such as aluminium and titanium depends upon their ability to form protective oxide films at room temperature (see sections 15.4 and 15.6).

In Chapter 1 we saw that not all corrosion processes are undesirable. The controlled oxidation of iron and steel in the manufacture of arms and armour was a well-established craft, intended to make the materials both decorative and long lasting. Decoration was achieved by the creation of colours on the metal surface. Titanium is oxidised electrochemically to produce beautifully coloured jewellery and works of art. In both cases, the effects are caused by oxide films. You may be more familiar with these effects if you have seen the spectral colours at the engine end of stainless steel motor bike exhausts.

Experiment 17.1

Take a piece of clean, brightly polished steel strip, approximately $300 \times 20 \times 3$ mm. Heat one end of the strip in a bunsen flame so that a steep temperature gradient is formed along its length (see Fig. 17.1). After a short time a series of coloured bands will be seen along the length of the strip. The colours are produced in transparent oxide films in the same way as in oil films on water. Interference takes place between the light reflected from a point such as A in the metal/oxide interface, and that reflected from point B, eliminating certain wavelengths from the spectrum. Hence the actual colour of the reflection will depend upon the thickness of the layer AB, and changes as AB increases. The temperature gradient causes the film to grow in the shape of a wedge, gradually thickening as heat flows along the strip. This in turn, causes the movement of the coloured bands with time. Only thin oxide films are transparent and show interference tints; eventually the oxide becomes sufficiently thick to be opaque and the colours disappear. At this point, the film is normally renamed a scale.

Before the days of sophisticated temperature control in heat treatment processes, the temperature of steel strip and bars was often judged by the colours developed on the metal surface during tempering heat treatments. The method is surprisingly accurate: for each step of 10 °C between 230 °C and 280 °C, the colour changes through the sequence, pale straw, dark straw, brown, brownish purple, purple and dark purple. The metal appears blue at 300 °C.

Until the development of the gas turbine engine for modern aircraft began with the Whittle engine in 1937, the uses of metals and alloys for

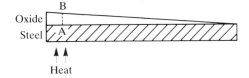

Fig. 17.1 The development of interference colours on a steel strip heated at one end. In a transparent oxide layer, the colour of the reflected light depends upon the thickness AB of the oxide film at that point. (AB is greatly exaggerated for clarity.) The movement of the colours reveals the gradual thickening of the oxide as heat travels along the strip.

engineering in high temperature environments were rarely severe enough to cause materials selection problems. Although the steam turbine had been developed in the late 1800s and used by Parsons in 1897 for marine propulsion, the operating temperatures required were low enough for existing materials to be used. The development of gas turbine aero-engines in the Second World War changed this situation dramatically.

The operating conditions were severe: materials were required which were capable of withstanding temperatures of 800 to 1000 °C, combined with large stress levels created by high speeds of rotation. This led to the development of a large class of alloys known as the **superalloys**. These are mostly nickel-based alloys, though there are also groups of iron- and cobalt-based materials. Today, superalloys are used in marine, aircraft, industrial and vehicular gas turbines, as well as in spacecraft, rocket engines, nuclear reactors, steam power plant, petrochemical plant and many other applications.

Steels still represent the major material for use in gas turbines, though their percentage share has declined in favour of the superalloys and titanium alloys. The contribution made by aluminium alloys to gas turbine development is small, but as we shall see, aluminium as an alloying addition is very important.

This chapter is a brief study of the behaviour of metals in elevated temperature, non-aqueous environments. The discussion will concentrate, in the main, upon oxide formation, although some other types of film will be introduced.

17.1 METAL OXIDES

Metal oxides (together with other compounds such as sulphides and halides) can be divided into two classes: those that are stable over the range of temperatures likely to be met in engineering structures, and those which are not. We will consider the unstable oxides first.

When an unstable metal oxide is heated, it decomposes to release the metal which may be deposited on to the substrate surface. For example, silver oxide decomposed above 100 °C, mercury(ii) oxide above 500 °C and cadmium oxide in the temperature range 900–1000 °C. Today, unstable oxides are of little use to the engineer, but they were of paramount importance to scientists in determining the fundamental mechanism of oxidation.

Early chemists, notably Stahl, had postulated the erroneous theory that a metal lost a substance which they called phlogiston when it was heated to form a calx (metal oxide):

$$\text{metal} - \text{phlogiston} \rightarrow \text{calx} \tag{17.1}$$

Stahl said that[2]:

> phlogiston is lighter than air, and, in combining with substances, strives to lift them, and so decrease their weight; consequently a substance which has lost phlogiston must be heavier than before.

In the 1780s, Lavoisier used the decomposition of mercury oxide to demonstrate that the phlogiston theory of oxidation was untenable. He heated mercury at just below its boiling point (357 °C) in a sealed apparatus and showed that approximately 20 per cent of the air was absorbed by the mercury. By collecting the red mercury oxide and heating it at about 500 °C he dissociated the unstable oxide to obtain a volume of gas equal to that lost in the first part of the experiment. He demonstrated that the recovered gas could support combustion, while that remaining in the apparatus after the first stage of the experiment could not. He further showed that the weights of mercury and recovered gas (oxygen) obtained by heating mercury calx, exactly equalled the weight of calx. Also, the increase in weight of the mercury in the formation of the calx is equal to the weight of the oxygen taken from the air. By this method it was established beyond doubt that the mechanism of oxidation was

$$\text{metal} + \text{oxygen} \rightarrow \text{metal oxide} \tag{17.2}$$

While this seems so obvious today, the problem had been vexing scientists for many years, as the quotation at the start of the chapter shows. The solution was a great advance for scientific thinking in the late 1700s.

The much broader class of stable oxides can again be divided into two groups: a minor group whose members volatilise at relatively low temperatures, and those which normally remain on the metal surface, unless they are physically or chemically removed.

Volatile oxides form on the metal surface, but immediately change to the gaseous state, leaving a fresh, reactive surface to continue the oxidation process until the metal is completely lost. The rate of reaction is constant and normally increases as the temperature rises. Molybdenum is the classic such example, oxidising at significant rates in air above 300 °C. Two oxide layers develop on the metal surface: an inner layer of MoO_2 and an outer layer of MoO_3. Above 500 °C, MoO_3 begins to volatilise, and at about 770 °C the rate of volatilisation equals the rate of oxidation. Further increases in temperature lead to extremely rapid metal loss. The effect becomes catastrophic when the MoO_3 begins to form a molten phase at temperatures above 815 °C.[3]

Case 17.1: Molybdenum plates of thickness 3 mm were employed as heat reflecting shields around a furnace used to determine the thermal conductivity of ceramics at temperatures up to 1400 °C. The whole unit was placed in a vacuum chamber and kept at a very low pressure during elevated temperature runs to prevent damage to the ceramics and the heat shields. One evening, while the chamber was operating at 900 °C, a small leak developed and air bled into the system overnight. In the morning, holes extended over 70 per cent of the shields and the metal which remained was severely thinned.

Stable, non-volatile oxides would be expected to remain on a metal surface and it may be thought that all such oxides would protect the substrate metal. Such is not the case. The rate of continued oxidation depends upon several factors, three of which are:

(a) The rate of diffusion of the reactants through the oxide film.

(b) The rate of supply of oxygen to the outer surface of the oxide.

(c) The molar volume ratio of oxide to metal.

The slowest rate process at any given temperature controls the rate of corrosion. In general, the rate will fall as the oxide thickens.

The molar ratio of the volume of oxide formed to the volume of metal consumed in producing the oxide, is a most important factor in determining the corrosion rate over an extended period of time. If M is the molecular mass of an oxide of density D, the volume occupied by 1 mol will be M/D. If m is the mass of metal in the mass M of oxide, and its density is d, a volume of metal m/d will have been converted into oxide. Table 17.1 lists the ratios $(M/D) \div (m/d)$ for a selection of metals. When the volume of oxide is smaller than that of the metal, i.e. $Md/mD < 1$, as in lithium, calcium and magnesium, the oxide is stretched over the metal surface to produce a porous, non-protective oxide. The oxidation process proceeds at a linear rate with time.

If the oxide volume is larger than the metal from which it formed, i.e. $Md/mD > 1$, then we would predict that the oxide is continuous and protective. In the case of aluminium, for example, this is so; however, other

Table 17.1 Values for the ratio of volume of oxide produced to volume of metal consumed in producing the oxide

Metal	Li	Ca	Mg	Al	Ni	Zr	Cu
Md/mD	0.57	0.64	0.81	1.28	1.52	1.56	1.68

Metal	Ti	Fe	U	Cr	Mo	W	
Md/mD	1.73	1.77	1.94	1.99	3.24	3.35	

complications may arise. Often compressive internal stresses are developed in the oxide as it thickens. Should a small stress develop it will tend to force together the sides of any cracks or defects, thus slowing down the rate of oxidation, but larger stresses tend to disrupt the bond between the oxide and the metal, the oxide then blisters and cracks. The disruption occurs because the fracturing of the interface between the metal and the oxide relieves the compressive internal stress. Obviously, the magnitude of the stress in the oxide will increase as it thickens. Thus, the oxidation rate may be small for long periods of time, there being sufficient compressive stress in a thin oxide film to maintain a tightly adherent compact barrier layer. As the film very slowly thickens, a stress level is generated where spontaneous rupture of the interface occurs and the oxidation rate suddenly increases. This is one type of **breakaway corrosion** and is dealt with in the next section.

The oxidation of metals which form stable, non-volatile oxides, then, occurs with an increase in weight of the sample which is relatively simple to measure in the laboratory. Indeed, much of the current knowledge of the mechanisms of oxidation has been obtained from studies of time-dependent weight increases. The rates of thickening have been found to fall mostly into three categories, examples of which are shown in Fig. 17.2. (Also shown is the weight loss with time which occurs when volatile oxides are formed.) In the equations which follow, y = oxide thickness, t = time, and c_1 to c_5 are constants.

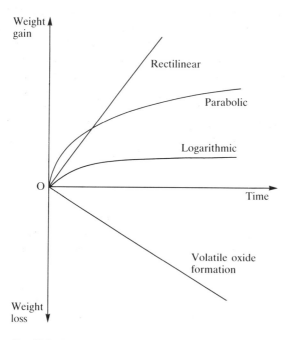

Fig. 17.2 Four oxidation rate laws: stable, non-volatile oxide formation leads to weight gain with linear, parabolic or logarithmic kinetics; the formation of volatile oxides leads to linear weight loss with time.

Parabolic growth

When the oxide film remains intact on the metal surface and offers a uniform barrier to the diffusion of metal or oxide ions through the film (section 17.3), the rate of growth of the oxide is inversely proportional to its instantaneous thickness:

$$dy/dt = c_1/y \qquad\qquad [17.3]$$

The integration of eqn [17.3] yields

$$y^2 = c_1 t \qquad\qquad [17.4]$$

(When $t = 0$, $y = 0$: hence no constant of integration is necessary.) Metals which oxidise according to the parabolic rate law are common and associated with thick, coherent oxides. Examples are cobalt, nickel, copper and tungsten, though these, as with the other examples, may follow a different rate law, depending upon the experimental conditions.

Rectilinear growth

The rate of oxidation is constant with time:

$$dy/dt = c_2 \qquad\qquad [17.5]$$

which, when integrated, gives:

$$y = c_2 t \qquad\qquad [17.6]$$

Rectilinear growth occurs whenever the oxide is unable to hinder the access of oxygen to the metal surface, as occurs when the oxide formed from a given volume of metal is too small to cover completely the surface of the metal. If the oxide cracks or spalls owing to large internal stresses, a series of short, parabolic-type weight increases will be observed which will appear linear overall. Such behaviour is termed **paralinear**. It may occur when the temperature cycles sufficiently for differential contraction and expansion between the metal and the oxide to spall the oxide from the metal.

Rectilinear growth is typically a high temperature process for the metal involved; two examples are iron above 1000 °C and magnesium above 500 °C.

Logarithmic growth

At low temperatures, a thin oxide film covers the surface. The rate of diffusion through the film is very low and after an initial period of rapid growth, the rate of thickening becomes virtually zero. The rate law is written:

$$y = c_3 \lg (c_4 t + c_5) \qquad\qquad [17.7]$$

Examples of metals which oxidise in such a manner are magnesium below 200 °C and aluminium below 50 °C.

17.2 BREAKAWAY CORROSION

We have already mentioned breakaway corrosion in connection with the compressive stresses developed in oxide scales (see section 17.1). The mechanism of breakaway can be very complicated and involves the interaction of many factors, including temperature, gas composition (particularly minor constituents), gas pressure, metal composition, component geometry and surface finish. It is an insidious form of attack and is frequently catastrophic in its outcome.

Two typical oxidation curves showing breakaway behaviour appear in Fig. 17.3. For long periods of time the oxidation rate appears to fall; low rates of weight increase are observed which may represent an acceptable and predictable rate of metal loss. Suddenly, the rate of oxidation increases. Two types of behaviour then ensue: either it mimics the parabolic type of kinetics at the start of the oxidation, as shown in curve A, or else the rate of oxidation continues at a high linear rate of loss to failure, as in curve B.

There have been many cases of breakaway corrosion. Zirconium, for example, undergoes breakaway in the conditions found in pressurised water atmospheres. Before the breakaway point is reached, the oxide is a shiny black, tightly adherent film, but after the transition, a loose powdery white oxide is formed. Thus, zirconium is used in pressurised water reactors in the form of 'zircaloy 2' for cladding fuel rods. This alloy contains tin to reduce the likelihood of breakaway corrosion.

A classic case of breakaway corrosion was found in 1969.[4] It illustrates not only the complex nature of the interaction of many variables in causing

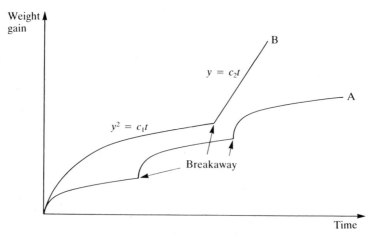

Fig. 17.3 Typical breakaway corrosion curves. Initially, the rate of oxidation falls with time and obeys a parabolic growth law. At the breakaway point, the existing oxide no longer protects the metal and linear growth ensues. In A, the new oxide is similar to the initial scale and the growth curve again follows a parabolic law until breakaway is repeated. In B, the new oxide no longer protects the substrate which corrodes according to the rectilinear growth law.

the phenomenon, but also the large financial risks which may be incurred when breakaway occurs. We shall discuss it in some detail.

Case 17.2: The first two generations of nuclear power station used in the United Kingdom, the Magnox and the Advanced Gas Cooled Reactor (AGR) used carbon dioxide as the cooling medium in the reactor. In 1983, the operating conditions were 370 °C and 28 atmospheres for the Magnox, and 650 °C and 41 atmospheres for the AGR. Maintenance and replacement of the components in a reactor are very difficult and many components are designed for a life of 30 years. Of necessity, an oxidation allowance is designed into the structure to accommodate the effects of metal wastage and deformation of bolted and welded components as a result of interfacial oxide growth. These allowances were derived by pre-testing the materials in simulated environments, many of the tests lasting for over a year. The materials selected formed protective oxides and showed an oxidation rate which decreased with time. These tests demonstrated that mild steel would be satisfactory up to 400 °C, meeting the requirements of the Magnox station, while that of the AGR could be met by using mild steel up to 400 °C, a steel containing 9 per cent Cr and 1 per cent Mo for components operating between 400 °C and 550 °C, and an austenitic stainless steel for those above this temperature.

In 1969, during a biennial inspection of a Magnox station, some fractured mild steel bolts were found. Subsequent examination showed that they had failed as a result of bolt straining caused by excessive oxidation of the interfaces between the bolts, washers and nuts. Such a rate of oxide growth was not predicted by extrapolation of the laboratory test data: breakaway oxidation had occurred. The porous oxide which formed after the breakaway occupied twice the volume of the metal consumed in producing it, and had the ability to continue to form even under the large compressive stresses generated at the interfaces by its formation. Immediately, the maximum operating temperature of all the Magnox stations had to be reduced with a consequent loss of generating capacity.

As a result of the problems with the mild steel, all the other steels were re-examined and it was shown in accelerated corrosion tests that the 9 per cent Cr steel had also suffered breakaway oxidation, although over a much longer time. Even so, on the basis of the then available data, it was predicted that there would be an unacceptable number of failures in boiler tubes. Only after further work, which resulted in a more complete understanding of oxidation mechanisms and an enlargement of the data base, could operating conditions be defined to achieve design life.*

---

* Case history reproduced with permission of the CEGB.

The mechanism of the breakaway oxidation in the nuclear reactors was found to be very complex. The protective oxide on a ferritic steel is in two layers, both of which are porous to the carbon dioxide coolant gas. The inner layer consists of small crystallites containing chromium and silicon if these elements are present in the steel. The outer layer has a columnar structure and consists of magnetite, Fe_3O_4. An equilibrium is established between the ingress of carbon dioxide and the outward, rate-controlling solid state diffusion of the iron. The carbon dioxide oxidises the iron, a reaction which produces carbon monoxide:

$$3Fe + 4CO_2 \rightarrow Fe_3O_4 + 4CO \qquad [17.8]$$

followed by deposition of carbon:

$$2CO \rightarrow CO_2 + C \qquad [17.9]$$

This carbon dissolves partly in the metal and partly in the oxide. When the carbon level of the inner oxide reaches about 10 per cent by weight, the individual crystallites become separated from each other by a grain boundary carbon film. This makes the oxide porous and it loses its ability to protect the substrate. Breakaway oxidation follows. Owing to the low solubility (<0.01 per cent) of carbon in ferrite, most of the carbon enters the oxide on mild steel components and breakaway occurs after about 1 to 5 years exposure. In the case of the 9 per cent chromium steel, carbon entering the metal is precipitated as chromium carbide. This allows a much greater proportion of the carbon to be absorbed by the metal and it takes a longer time to reach the critical carbon concentration for breakaway in the oxide film. The oxidation of mild steel follows curve B in Fig. 17.3, but there is evidence to show that 9 per cent Cr steel exhibits an oxidation pattern similar to curve A in some cases.

Many other factors influence the time at which breakaway occurs: temperature, the amount of water vapour in the carbon dioxide, and the silicon content of the mild steel (up to 0.2 per cent Si is very beneficial) are but three examples. Breakaway occurs preferentially at sharp corners on the 9 per cent Cr steel where three oxidising surfaces contribute carbon to a small volume element of metal, compared with two at an edge and one at a plane surface.

In the Magnox stations, the chance of component failure by steel oxidation has been limited by careful control of the gas composition and temperature to such an extent that all stations can be safely operated up to and beyond their original planned life.

However, carbon dioxide environments which are beneficial to steel oxidation (i.e. low moisture and hydrogen levels) can be harmful to the graphite used to moderate the reactor. A balance has to be drawn between the requirements for long steel life and that for graphite. The situation in the AGRs is more complicated and 'windows' have been defined for gas composition and temperature such that the needs of the graphite moderator, carbon deposition, coolant performance and 9 per cent Cr steel are optimised.

The above example clearly shows the complex interaction between a large number of variables and the different requirements of various parts of a plant. Such problems often confront engineers running large industrial processing and production units. It also emphasises once more the care needed when extrapolating short-term tests in simulated environments to real engineering structures.

17.3 THE MECHANISM OF OXIDE FILM GROWTH

We must now consider the mechanism by which oxygen and the metal are brought together through the film so that oxidation can continue. If this diffusion through the film did not occur, the oxidation would cease once a mono-molecular layer of oxide had formed all over the surface.

At one time it was assumed that oxidation always involved the movement of oxygen ions inwards through the film towards the metal. However, Pfeil was able to demonstrate that this was not the mechanism for oxide growth on iron. He took a clean, abraded piece of iron on which he coated a layer of chromium oxide. After the iron had been heated in air for some time, it was covered by a thick layer of iron oxide, but the chromium oxide was either on or just above the iron surface below the iron oxide. The position of the chromium oxide on the iron surface clearly indicated that the iron ions had diffused outwards to form the oxide, passing through the chromium oxide layer, not that oxide ions had diffused inwards. It has since been shown that copper ions also diffuse outwards to form an oxide film, whereas in zirconium and titanium, oxide ions move inwards to react at the metal/oxide interface.

Metal oxides are largely ionic compounds in which the metal and oxide ions are arranged in regular arrays in their respective crystal lattices. Some oxides contain excess metal ions which are located at interstitial positions (section 2.6); these are called n-type, or negative carrier type materials. Others are deficient in metal ions and vacant sites exist in the cation (metal ion) lattice; these are called p-type, or positive carrier type materials. Typical n-type oxides are ZnO, CdO and Al_2O_3, while some p-type oxides are Cu_2O, NiO, FeO and Cr_2O_3.

Let us consider how the diffusion of different species can occur through a layer such as Cu_2O, which might be expected to behave as an insulator. Figure 17.4(a) shows how the diffusion of copper ions occurs. Careful chemical analysis of the copper(I) oxide shows that there is a slight deficiency of copper from that expected from the chemical formula, Cu_2O. Such an oxide is said to be non-stoichiometric. Vacancies exist in the singly charged cuprous sub-lattice of the crystal structure, but to retain overall electrical neutrality there are sufficient doubly charged cupric ions to compensate. In particular, the number of vacancies is greater at the air/oxide interface than at the metal/oxide interface: the existence of a concentration gradient of vacancies causes the cuprous ions to migrate outwards to the air/oxide interface by a stepwise movement shown in Fig. 17.4(a). Conversely, the vacancy diffuses inwards to the metal/oxide interface, where a spare electron becomes available.

If this were the only process, it would lead to a surplus of electrons in the metal. However, oxygen atoms attach themselves to the surface layer, as shown in Fig. 17.4(b), where they become oxide ions:

$$O_2 + 4e^- \rightarrow 2O^{2-} \qquad [17.10]$$

The reaction in eqn [17.10] is achieved by oxidising positive $Cu(I)$ ions at the surface to $Cu(II)$ ions:

$$Cu^+ \rightarrow Cu^{2+} + e^- \qquad [17.11]$$

The electrons left behind when the Cu^+ ions diffused outwards can now diffuse to restore the $Cu(II)$ ions to the $Cu(I)$ state. This is a sequential process, an electron from an adjacent cuprous ion diffusing into the cupric ion and restoring it to the cuprous state:

$$Cu^{2+} + e^- \rightarrow Cu^+ \qquad [17.12]$$

leaving the adjacent cuprous ion in the cupric state. This process continues between adjacent ions until the electron left in the metal crosses the metal/oxide interface. This is equivalent to a flow of positive charges in the opposite direction, as shown in Fig. 17.4(c). The mechanism for the growth of the film is complete.

This type of mechanism can only occur for oxides of metals with variable valency, such as copper and iron. Aluminium oxide has only one valency and an oxide of fixed stoichiometry. In this case, oxidation would be expected to be very slow, as is indeed the case.

In n-type oxides, such as ZnO, where excess metal ions are located at interstitial positions, there must be an excess of electrons to maintain electrical neutrality. The simplest picture of oxide growth is then the simultaneous diffusion of positive zinc ions and negative electrons in opposite directions. The zinc ions move through the defect sites to the outer surface where they react with oxygen and build additional layers of oxide.

The situation occurring in oxidation processes is analogous to the basic wet corrosion cell, having four components, for the oxide acts as:

(a) an electrode for the oxidation of a metal (an anode in the wet corrosion cell);
(b) an electrode for the reduction of oxygen (a cathode);
(c) an ionic conductor (the electrolyte);
(d) an electron conductor (the external circuit).

The outward diffusion of metal ions can produce some peculiar effects. If iron wire is heated at about 800 °C, an oxide film forms on the outer surface. Iron ions diffuse outwards through this film, while vacancies diffuse inwards. Gradually, a tube of oxide is formed as the iron migrates to the outer surface to react with oxygen. Frequently, the oxide spalls during the

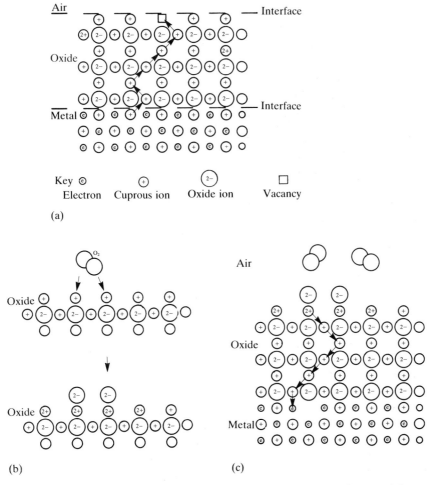

Fig. 17.4 Schematic diagram of the mechanism of oxidation of copper (after McGrath[5]).

(a) Diffusion of Cu⁺ from metal to air/oxide interface via cation vacancies. (Note how the presence of cation vacancies causes an equivalent number of copper ions to be in the divalent oxidation state.)

(b) Reaction of oxygen molecule with cuprous ions at the air/oxide interface. Reaction of one molecule leads to the attachment to the lattice of two oxide ions and the oxidation of four cuprous ions to cupric ions.

(c) Diffusion of positive charge inwards (electron outwards) to neutralise the excess of electrons in the metal.

process and a misshapen but still hollow section may be produced. The same process occurs with nickel heated in air at 1250 °C. However, in the oxidation of cobalt, cobalt oxide at this temperature is more plastic than nickel oxide and slowly collapses into the void, giving a solid oxide with only a few small holes at its centre.

17.4   THE OXIDATION OF ALLOYS

The influence of lattice defects on diffusion through an oxide film led both Hauffe and Wagner to propose a series of rules for the effects of metallic alloying additions on the oxidation rate of alloys. These rules, summarised in Table 17.2, are not inviolable, but they give general guidance in cases where the alloying metal is present in the oxide film of the parent metal. In some cases they predict unexpected, but nevertheless observable, effects.

For example, the addition of 0.1 per cent aluminium to zinc, which forms an n-type oxide, causes a reduction in the oxidation rate by a factor of about 100. Only two Al^{3+} ions instead of three Zn^{2+} ions are associated with three O^{2-} ions. This leaves an extra hole in the metal lattice which is occupied by one of the interstitial Zn^{2+} ions. This ion is trapped in the hole and is restricted in the part it can play in the diffusion of metal ions through the lattice; the oxidation rate is consequently reduced.

On the other hand, adding small quantities of chromium (which has a higher valency than that of the nickel ion) to nickel oxide (a p-type oxide) also increases the number of vacancies, but as the oxide is already deficient

Table 17.2   The effect of alloying upon the rate of oxidation

Oxide type	Valency of alloying element compared to parent metal	Effect	Diffusion controlled oxidation rate
p-type: e.g. Cu_2O NiO FeO Cr_2O_3	Higher valency	Increases number of vacancies; decreases number of parent metal ions with a higher valency	Increases
CoO Ag_2O MnO SnO	Lower valency	Decreases number of vacancies; increases number of parent metal ions with a higher valency	Decreases
n-type: e.g. ZnO CdO Al_2O_3	Higher valency	Decreases conc. of interstitial metal ions; increases number of free electrons	Decreases
TiO_2 V_2O_5	Lower valency	Increases conc. of interstitial metal ions; decreases number of free electrons	Increases

in metal ions, the additional holes make it easier for the nickel to diffuse; the oxidation rate rises.

If lithium, which forms a monovalent ion, is added to nickel oxide, two Li^+ ions are required to replace one Ni^{2+} ion. The number of vacant sites in the lattice is reduced in maintaining electrical neutrality; the nickel diffusion is hindered and the diffusion rate falls.

These examples illustrate two rather peculiar situations. Lithium, an active metal with a very low oxidation resistance, reduces the oxidation rate of nickel in oxygen, while chromium, an alloying addition noted for its oxidation resistance, increases the oxidation rate of nickel. (The latter effect occurs only for concentrations of below 5 per cent. Above 5 per cent chromium, the solubility of Cr^{3+} ions in the NiO lattice is exceeded: a separate protective layer of Cr_2O_3 forms on the metal surface and keeps the oxidation rate at very low levels. Thus, for example, nichrome electric fire elements last much longer than would nickel wire.)

The most effective way to control the oxidation of iron and steel is to form a stable surface layer of protective oxide from an alloying element. This hinders the diffusion of iron ions and electrons and so reduces the rate. A complex triple layer scale forms on the surface of iron heated in air at temperatures above about 500 °C. The inner layer, FeO, is by far the thickest. The outer layers are Fe_3O_4 and Fe_2O_3, the latter being outermost. Chromium and aluminium are the most effective alloying elements added to form stable scales on steels. Additional protection is conferred by adding nickel, silicon and some rare earth elements such as yttrium to iron/chromium alloys.

An alloy of Fe and 9 per cent Al compares favourably in oxidation resistance to 20 per cent Cr and 80 per cent Ni, but the iron/aluminium alloys have poor mechanical properties and are unsuitable for use as general engineering materials. This can be overcome to some extent by forming aluminium-rich outer layers on iron and steel in a process called **calorizing**. Small components are heated in a mixture of aluminium powder, aluminium oxide and ammonium chloride. The aluminium-rich outer layer resists oxidation while the inner core of aluminium-free metal provides the desired mechanical properties of the component.

Chromium additions confer good oxidation resistance upon iron and steel, and on many other alloy types too. The chromium enriches the innermost layers of the iron oxide coating and often generates a layer of chromium oxide on the metal surface below the iron oxides. These layers are more resistant to ion or electron diffusion than iron oxides alone, and the oxidation rate is decreased. Iron/chromium alloys containing 4 to 9 per cent chromium are used as oxidation-resistant metals in many areas, including oil processing plant. The 12 per cent chromium alloys make good blade materials for steam turbines, while those with up to 30 per cent chromium are used in the chemical industry and for heat treatment plant and burners. Further additions of silicon, nickel and yttrium make the high chromium alloys a suitable choice for petrol engine valves and other components which operate at high temperatures in aggressive environments. When 5 per cent aluminium is also added, excellent oxidation resistance is obtained, but the alloy has poor mechanical properties and suffers embrittlement at room temperature. Iron/aluminium alloys are used as windings in electrical furnaces where they give a long life provided they are protected from mechanical shock.

17.5 HOT CORROSION

Since the earliest gas turbine engines, their development has been great, yet the principle of their operation has remained the same. In brief, a gas turbine engine ingests air from the atmosphere, mixes it with fuel and compresses and ignites the mixture. This produces gases at temperatures in the region of 730 to 1370 °C. A fraction of the hot gas drives the turbine, which keeps the compressor running. The remainder of the hot gas is used to provide thrust, in the case of a turbojet engine, or shaft horsepower in a turboshaft engine.

Although the operating principle of the gas turbine engine has not changed, the improvements in performance have been remarkable. Specific fuel consumption has been reduced to a third of that in the early jet engines, while thrust/weight ratios have been tripled and time between overhauls increased 100-fold. This has been made possible only by advances in materials technology which have resulted in great weight reductions and allowed far higher operating temperatures to be used. In 1987, the good safety record of the aircraft industry is beyond doubt, an achievement made possible by meticulous inspection and maintenance procedures. There have been very few major disasters directly attributable to high temperature corrosion failure within aero-engines. Hence, the significant effect of high temperature corrosion is upon the life of an engine and the periods between major overhauls.

At present, there remains one particular problem associated with the operation of gas turbines in marine environments. This problem is known as **hot corrosion**. *It should not be confused with general high temperature oxidation.* Hot corrosion is a combination of oxidation and reactions with sulphur, sodium, vanadium and other contaminants which are present, either in the inlet air or in the fuel. It produces a non-protective oxide on the blade surface in place of the normal protective oxide of chromium or aluminium. Hot corrosion can severely reduce the life of turbine blades, and could lead to engine failure, although careful and regular inspection procedures can usually reduce considerably the likelihood of the latter.

Case 17.3: The nozzle guide segments of one engine corroded so badly in 800 operating hours that the engine failed completely.

Case 17.4: After only a few months of operation carrying out low level flights among islands in the Caribbean the turbine sections of Royal Air Force Harriers suffered a drastic loss of efficiency. The turbines of Royal Navy Sea Harriers in similar conditions were found to perform satisfactorily. The problem was caused because the turbine materials used in the RAF aircraft were not resistant to hot corrosion, in contrast to the Royal Navy aircraft which had been 'marinised' by employing different materials and coatings for increased corrosion protection.

Civil aviation gas turbines operating at high altitude or inland airports do not normally suffer from hot corrosion, and are designed primarily for good mechanical/creep strength and oxidation resistance. This is best achieved with alloys of low chromium but high aluminium content, the aluminium forming an oxide which provides an efficient barrier to further oxidation.

However, if the inlet air is laden with sea salt as well as having a high moisture content, conditions for hot corrosion are ideal and the protection offered by the aluminium oxide is much reduced.

Hot corrosion presents the design engineer with a materials selection problem. To obtain high creep strength (a pre-requisite for rapidly rotating turbine blades and discs) a high nickel content is required, while good resistance to hot corrosion is best conferred when large amounts of chromium are present in the alloy. Thus guide vanes, which are stationary and therefore operate at lower stress levels, can be made from high chromium alloys such as the cobalt-based X40. Rotating blades are made from high nickel alloys typified by the nimonic series.

Case 17.5: Figure 17.5 shows a rotor blade from a ship gas turbine engine. The blade is suffering from hot corrosion characterised by large green/black blisters with accompanying loss of dimensions, especially on concave surfaces and trailing edges.

Fig. 17.5 Hot corrosion of a marine gas turbine blade.

The mechanism of attack by hot corrosion is complex and not fully understood. Only a cursory discussion occurs here.

In marine environments, sodium chloride will either be ingested into the engine from the atmosphere, or be present as a contaminant in the fuel. In the hottest part of the engine, the sodium chloride reacts with the sulphur and other components of the gas stream to produce sodium sulphate:

$$2NaCl + S + \tfrac{3}{2}O_2 + H_2O \rightarrow Na_2SO_4 + 2HCl \qquad [17.13]$$

Sodium sulphate and sodium chloride react together to form a slag which melts on the surface of the component. The slag melts at approximately 620 °C, fluxes the layer of chromium and aluminium oxides which would otherwise protect the metal, and leaves it vulnerable to attack by the aggressive atmosphere in the engine.

Sulphur from the slag then diffuses into the alloy and reacts with the chromium to form internal sulphides. The substrate becomes depleted in chromium and the metal oxidises to yield a mixed nickel/chromium oxide with the spinel crystal structure. This oxide is much less protective than the chromium oxide alone. The chromium sulphide subsequently oxidises and releases sulphur which diffuses further into the metal where it reacts with more chromium to allow the oxidation process to continue in the newly chromium-depleted matrix metal. Thus metals which have a high chromium content are better able to resist hot corrosion as they are capable of maintaining the protective chromium oxide for longer periods.

Research is still in progress to determine the best method to reduce the incidence of hot corrosion. Clearly, improving the quality of the fuel/air mixture ingested by the engine will be beneficial; reducing the sulphur content of the fuel or filtering out the sodium chloride will minimise the formation of low melting point slag. Alloy coatings have also been applied to blades to provide a barrier between the metal and the aggressive atmosphere. Typical of these is 'Cocraly', an alloy of cobalt, chromium, aluminium and yttrium, although zirconium is now being substituted for yttrium.

17.6 REFERENCES

1. Fontana M G, Greene N D 1978 *Corrosion engineering*. New York: McGraw-Hill, p 369
2. Mellor J W 1946 *A comprehensive treatise on inorganic and theoretical chemistry*. Longmans, Green and Co, 63ff
3. Bentley J 1979 Molybdenum, in Shreir L L (ed.), *Corrosion*, Newnes-Butterworths, p 5:14
4. Rowlands P C, Garrett J C, Whittaker A 1983 *The corrosion of ferritic steels in gas-cooled nuclear reactors*. CEGB Research, Nov, pp 3–12
5. McGrath J N 1981 *Corrosion – a programmed text*. Institution of Metallurgists

17.7 BIBLIOGRAPHY

ASTM 1971 *Oxidation of metals and alloys*

ASTM 1967 *Hot corrosion problems associated with gas turbines*. ASTM-STP 421

ASTM 1980 *Superalloys: proceedings of the fourth international symposium on superalloys*

Coutsouradis D, Felix P, Fischmeister H, Habraken L, Lindblom Y, Speidel M O 1978 *High temperature alloys for gas turbines*. Applied Science

Evans U R 1960 *The corrosion and oxidation of metals*. Edward Arnold

Guttmann V, Merz M 1981 *Corrosion and mechanical stress at high temperatures*. Applied Science

Hart A B, Cutler A J B (eds) 1973 *Deposition and corrosion in gas turbines*. Applied Science

Hauffe K 1965 *Oxidation of metals*. New York: Plenum Press

Kofstad P 1966 *High temperature oxidation of metals*. John Wiley

Meetham G W (ed) 1981 *The development of gas turbine materials*. Applied Science

Michels H T, Friend W Z 1980 Nickel-base superalloys, in Friend W Z (ed.), *The corrosion of nickel and nickel-base alloys*. John Wiley

Sahm P R, Speidel M O 1974 *High temperature materials in gas turbines*. Elsevier

Shreir L L (ed) 1979 *Corrosion* (Vol. 1). Newnes-Butterworths, section 7

Sims C T, Hagel W C 1972 *The superalloys*. John Wiley

18 WORKED EXAMPLES AND PROBLEMS

Examinations are formidable even to the best prepared, for the greatest fool may ask more than the wisest man can answer.

(Charles Caleb Colton, 1780–1832)

18.1 WORKED EXAMPLES

E18.1 Express the standard electrode potential, E^o, of a metal in terms of the standard Gibbs free energy change, ΔG^o. Hence calculate the value of ΔG^o at standard temperature and pressure for the corrosion of iron, assuming a divalent reaction.

Ans:
$$\Delta G^o = -zE^oF$$
$$= -2(+0.44) \times 96{,}494$$
$$= -84.9 \text{ kJ mol}^{-1}$$

Note: 1. Negative ΔG^o indicates that corrosion occurs spontaneously.
2. The reduction potential (Table 4.1) is -0.44 V, so for the oxidation process, we use $+0.44$ V. See Section 4.5 for details of the sign convention.

E18.2 Iron is connected to copper and then immersed in a solution containing both Fe^{2+} and Cu^{2+} ions.
(a) Which metal corrodes?
(b) Write equations to describe the reactions which occur at each electrode, assuming each metal has a valency of 2.
(c) Calculate the maximum possible potential of the resulting corrosion cell.

Ans: The metal with the most negative reduction potential will be the anode. From Table 4.1, E^o for iron is -0.44 V, while for copper it is $+0.34$ V. The iron will thus be the anode. Convention requires that we write the cell as

$$Fe \mid Fe^{2+} \parallel Cu^{2+} \mid Cu$$

and the cell potential as the reduction potential of the electrode on the right *minus* the reduction potential of the electrode on the left.

354

Thus:

$$E_{(cell)} = E_{(Cu\ redn)} - E_{(Fe\ redn)}$$
$$= (+0.34) - (-0.44)$$
$$= +0.78\ V$$

The electrode on the left is the anode, and thus the electrode reactions are:

$$Fe \rightarrow Fe^{2+} + 2e^-$$
$$Cu^{2+} + 2e^- \rightarrow Cu$$

Note: A positive cell potential leads to a negative ΔG^o and tells us that the cell, as described, will corrode spontaneously.

E18.3 Calculate the rest potential, versus the saturated calomel electrode, of a piece of copper in equilibrium with a solution containing 10^{-6} M copper ions.

Ans: Using the Nernst equation:

$$E = E^o + \frac{0.059}{2}\ \lg\ (10^{-6})$$
$$= +0.34 + (-0.177)$$
$$= +0.163\ V\ SHE$$

To convert from SHE to SCE we subtract 0.242 V, thus:

$$E = 0.163 - 0.242$$
$$= -0.079\ V\ SCE$$

E18.4 A metal, M, of valency, z, atomic mass, W, and density, D kg m^{-3}, is corroding uniformly over its exposed surface area with a current density of i_{corr} A m^{-2}. Derive an expression for the number of millimetres of metal which will be lost during one year, assuming that the build-up of corrosion product does not stop the corrosion reaction.

Ans: Over 1 m^2 of exposed metal, the number of coulombs passed in one year will be:

$$i_{corr} \times 60 \times 60 \times 24 \times 365 = 3.154 \times 10^7 \times i_{corr}$$

1 mol of metal of valency z converted into ions gives:

$$z \times 96{,}494\ coulombs$$

(Remember that a mole is expressed in grams.)
Thus the number of mol per square metre lost in a year is

$$= \frac{3.154 \times 10^7 \times i_{corr}}{z \times 96{,}494}$$
$$= \frac{326.8 \times i_{corr}}{z}$$

Converting mol into kg, the number of kg lost per square metre per year is

$$= \frac{326.8\ W \times i_{corr}}{1000 \times z}$$

The metal is of density D kg m^{-3}, thus if the mass lost were D kg m^{-2} then the depth of penetration of the corrosion would be 1 metre (1000 mm). The actual penetration (in mm) is therefore:

$$= \frac{326.8 \; W \times 1000 \times i_{corr}}{1000 \times z \times D}$$

$$= \frac{326.8 \; W \times i_{corr}}{z \times D}$$

Note: The expression derived above is commonly used as a measure of corrosion rate. It is known as the wastage rate, is measured in mmpy, and is popular because it gives engineers a better 'feel' for the rate of corrosion. In the case of copper, a corrosion current density of 0.01 A m^{-2} is commonly observed. Thus, with $z = 2$, $W = 63.5$, and $D = 8960$ kg m^{-3}, we calculate a wastage rate of:

$$\frac{326.8 \times 63.5 \times 0.01}{2 \times 8960}$$

$$= 1.16 \times 0.01 \; \text{mmpy}.$$

Thus we see that the wastage rate for copper is almost the same as the corrosion current density of copper in A m^{-2}. The units of i_{corr} are important. It is both surprising and fortuitous that for many metals the quantity, $W/(zD)$, is approximately constant. Hence the observation for copper is applicable to many other metals too:

$$i_{corr} \simeq \text{mmpy}$$

E18.5 A univalent metal, M, has $\beta_a = +0.2$ V, $\beta_c = -0.2$ V, and $i_o = 20$ mA m^{-2}. Calculate values for i_a and i_c when the metal is anodically polarised to $+0.20$ V. Hence determine the experimentally measured current density for the same polarisation.

Ans: The figure plots the anodic and cathodic current densities for M. At $+0.20$ V:

$$i_a = 200 \; \text{mA m}^{-2}$$
$$i_c = 2 \; \text{mA m}^{-2}$$
and $$i_{meas} = i_a - i_c$$
$$= 198 \; \text{mA m}^{-2}$$

Alternatively, we can use the Tafel equation, thus:

$$\eta = \beta \; \lg (i/i_o)$$

Taking antilogs we get for the anodic reaction:

$$10^{\eta/\beta} = i_a/i_o$$

But $\eta/\beta = 1$, thus:

$$i_a = 10 i_o$$
$$= 200 \; \text{mA m}^{-2}$$

Similarly for the cathodic current,

$$\eta/\beta = -1, \text{ thus:}$$
$$i_c = 0.1 i_o$$
$$= 2 \; \text{mA m}^{-2}$$

Therefore $i_{meas} = 198$ mA m^{-2}, as before.

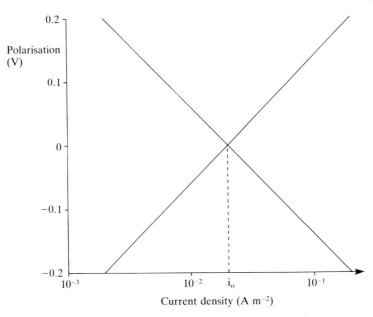

Current density (A m^{-2})

Note: The slope of the Tafel plot – the beta constant (+0.2 or −0.2 in the example) – is obtained by measuring the voltage change per logarithmic decade of current, e.g. the voltage change between 10^{-1} and 10^{-2} A m^{-2}.

E18.6 (a) Calculate the theoretical capacity of zinc sacrificial anode material.
(b) The measured capacity is 780 Ah kg^{-1}. Calculate the efficiency of the material.

Ans: (a) Atomic weight of zinc — — — — — — — — = 65.4
1 kg of zinc contains 1000/65.4 — — — — — — = 15.29 mol
But 1 mol of zinc converted into divalent
zinc ions yields 2 × 96,494 — — — — — — — — = 192,988 C
Therefore, 15.29 mol yield — — — — — — — — = 2.95 × 10^6 C
But 1 coulomb = 1 ampere second
Therefore charge created — — — — — — — — — = 2.95 × 10^6 As
— = 819 Ah

Thus, the maximum capacity of zinc is 819 Ah kg^{-1}.
(b) Allowing for the presence of impurities and other irregularities in the dissolution process, the practical quoted capacity of 780 Ah kg^{-1} leads to a dissolution efficiency of 780/819 × 100 per cent = 95 per cent.

E18.7 Two gas pipelines, each 75 km long and 1 m diameter, are laid in different parts of a Middle Eastern oil state at the same time. Pipeline A receives a very poorly applied coal tar epoxy coating, while pipeline B is given a high quality protective coat. Each line is further protected by a standard ICCP installation which uses a 100 V, 100 A (10 kW) rectifier/groundbed system to cover the whole length of the line.

After several years it is found that pipeline A required a current density of 1.72 mA m^{-2} to achieve the protection potential, while line B required 86 μA m^{-2}. The recommended protection current density is 130 μA m^{-2}. For each pipeline:

(a) Comment upon any excess or shortfall in the capacity of the protection system.

(b) Describe the effect of the corrosion protection system.

Ans: Total area of each pipeline: $\pi \times 1 \times$ 75,000 = 239,000 m^2
Total current: line A 1.72 $\times$ 10^{-3} $\times$ 239,000 = 411 A
Total current: line B 86 $\times$ 10^{-6} $\times$ 239,000 = 21 A
Line A requires more current than the equipment is able to provide. Much of the line is thus unprotected by the ICCP and with its poor coating is therefore liable to leak or rupture. (This question is based upon a real case history. In fact only 7 km of line A was protected by the ICCP.) Line B requires only 21 A of the 100 A available and is seriously overdesigned. (Only 120 W of the 10 kW capacity was required.) In the case of line A which required such a high current, the best solution is to use more current drain points, each with a lower current.

E18.8 It is planned to place an uncoated steel drilling platform in the sea. The continuously immersed parts of the structure will be protected with a sacrificial anode cathodic protection system which must have a life of 10 years. The anodes available are semi-cylindrical, 100 mm diameter, 400 mm long, with a steel core suitable for welding the plane side to the structure. (The ends and backs may be ignored for calculation purposes.)

(a) Select a suitable anode material.

(b) Calculate the minimum number of anodes necessary to meet the requirement at the start of the period.

(c) Show by calculation whether 2510 anodes will be sufficient to provide protection throughout the whole of the 10 years.

Data: Wetted surface area of legs = 2000 m^2
Wetted surface area of cross members = 500 m^2
Current requirement for uncoated steel = 110 mA m^{-2}

Ans: (a) The alloy should be either zinc or aluminium, not magnesium.
(b) Assume all anodes corrode at a constant rate.
Output of both zinc and aluminium alloy (Table 16.1) is 6.5 A m^{-2}
Total current requirement = 2500 m^2 $\times$ 110 mA m^{-2} = 275 A
Anode area required to supply 275 A is 275/6.5 = 42.3 m^2

Approx. area of one anode = $3.142 \times 0.4 \times 0.05$ = $0.063 \ m^2$
Minimum number of anodes = $42.3/0.063$ = 671
(c) 275 A are required for 10 years   = 2750 Ay
This output is required from 2510 anodes,
therefore output required from each anode =
2750/2510   = 1.10 Ay
Approx. volume of one anode
$$= 3.142 \times 0.05 \times 0.05 \times 0.4/2 = 1.57 \times 10^{-3} \ m^3$$
From Table 16.1, volume wastage rate for zinc is $1.518 \times 10^{-3} \ m^3 \ Ay^{-1}$; thus total wastage by volume of one zinc anode = $1.518 \times 10^{-3} \times 1.1 = 1.67 \times 10^{-3} \ m^3$
Thus all zinc anodes will be consumed after 10-year period.
Similarly, for anode of aluminium, wastage by volume (Table 16.1)   = $1.18 \times 10^{-3} \times 1.1 = 1.3 \times 10^{-3} \ m^3$
Volume left after 10 years
$$= (1.57 \times 10^{-3}) - (1.3 \times 10^{-3}) = 0.27 \times 10^{-3} \ m^3$$
Assuming regular dissolution, the radius of anode remaining is given by the following expression:
$$0.27 \times 10^{-3} = 3.142 \times r^2 \times 0.4/2; \qquad r = 21 \ mm$$
Area left after 10 years:
$$= 3.142 \times 0.021 \times 0.4 \times 2500 = 66.2 \ m^2$$
Since $42.3 \ m^2$ is required to supply the current, then aluminium anodes are still effective after 10 years.

18.2 PROBLEMS

P18.1 For each of the following give *one* possible reason why:
(a) So few metals exist in the uncombined state in nature.
(b) Gold exists in the uncombined state in nature.
(c) Well-preserved iron artefacts have been uncovered after centuries of immersion in peat bogs.

P18.2 Briefly define *each one* of the following:
(a) Corrosion
(b) Standard electrochemical potential
(c) Exchange current
(d) Polarisation
(e) Double layer

P18.3 Consider the following statements:
(a) All natural corrosion reactions are spontaneous.
(b) Measurements of electrical potential form an important part of corrosion monitoring.
Show how it is possible to relate the two statements.

P18.4 Draw and label a diagram to describe *each* of the following:
(a) The use of a polarisation curve to calculate a cathodic Tafel constant.
(b) An anodic polarisation curve exhibiting passivation.

P18.5 Illustrate *each of the following* with a simple, labelled diagram:
(a) The measurement of a standard electrochemical potential.
(b) The measurement of a free corrosion potential in 3.5 per cent sodium chloride solution.

P18.6 For the electrochemical cell having nickel and cadmium electrodes in equilibrium with solutions of their ions:
(a) Write *two* equations to describe the reactions which occur.
(b) Determine the greatest potential which may be obtained from such a cell under standard conditions.
(c) State *one* other condition necessary to achieve this potential.

P18.7 Calculate the rest potential, versus the saturated calomel electrode, of a piece of nickel in equilibrium with a solution containing 10^{-6} M nickel ions.

P18.8 (a) Give *two* useful functions of E/pH diagrams.
(b) Give *two* limitations to their use.

P18.9 (a) Explain what is meant by the term 'passivity' in the context of corrosion.
(b) Give *two* examples of natural passivity of metals.

P18.10 The following couples are immersed in fresh water:
$$Fe/Cd \quad Fe/Ti \quad Fe/Zn \quad Fe/Cu$$
(a) In which *one* of the four will the iron corrode the fastest?
(b) Which combination offers the best protection to the iron?

P18.11 (a) Give the general name for the form of corrosion which can occur when turbulent sea water flows through brass pipes.
(b) Give *two* likely causes of this turbulence.

P18.12 (a) What precipitates are formed when weld decay occurs in an unstabilised stainless steel?
(b) Where are the precipitates formed?
(c) Explain the mechanism of weld decay.
(d) What elements are added to stainless steels to reduce weld decay?
(e) State *two* other measures which may be taken to alleviate the problem.

P18.13 (a) Why is the combination of pitting corrosion and oscillating stress a serious problem in a metal component?
(b) Suggest two ways of assessing the severity of pitting corrosion.

P18.14 Explain how the addition of an anodic inhibitor can aggravate a corrosion problem.

P18.15 Outline the role of sulphur in the hot corrosion of a Ni/Cr alloy turbine blade.

P18.16 (a) Sketch *two* curves which illustrate the weight gained by materials exhibiting rectilinear and parabolic oxidation kinetics.
(b) What is the role played by the oxide in *each* case?

P18.17 Explain why, in the oxidation of metals, the ratio:
$$\frac{\text{molar volume of oxide produced}}{\text{molar volume of metal oxidised}}$$
is important.

P18.18 (a) Explain clearly the meaning of the following terms when used in the context of E/pH diagrams:
immunity   corrosion   passivation   pH
(b) (i) Use Fig. 4.17 to obtain a value for the free corrosion potential of zinc in water at pH 8.
(ii) Calculate the same value by means of the Nernst equation and thus show how the equation is used in constructing part of the E/pH diagram.
(iii) Describe, in principle, how it is possible to complete the remainder of the chart.
(c) (i) Explain how inspection of the chart suggests *two* possible methods for controlling the corrosion of zinc in water.
(ii) What factors, if any, might make these suggestions impractical?
(iii) What other factors limit the application of E/pH diagrams to real situations?
(d) State the effect of an increase in temperature upon the potential you found in parts (b)(i) and (ii). Give the reason for your answer.

P18.19 Imagine that you are asked to investigate the rate of corrosion of Inconel 625 in the estuarine water of a dockyard.
(a) Describe the basic equipment you would need to carry out a laboratory investigation.
(b) Explain the role of *each* piece of equipment mentioned in (a).
(c) Describe *one* experiment which you would perform, clearly stating the conditions for the experiment.

(d) Explain how you would interpret data obtained from your experiment.

(e) List the other parameters which might usefully be varied and explain how further information relating to the corrosion performance in estuarine water might be obtained as a result.

(f) Discuss the limitations of the experiment you have described in the context of other forms of corrosion to which the material may be susceptible.

P18.20 In a programme of experiments to determine the resistance of a series of alloys to crevice corrosion, suitable specimens were subjected to two different tests which consisted of:
1. potentiodynamic scanning measurements in sea water at 25 °C;
2. sea water exposure trials for 4 years, followed by weight loss measurements.

A typical plot of potential versus current density obtained from potentiodynamic scans is shown in the figure. Values of $E_c - E_p$ obtained from the scans for the range of alloys, together with the results of the exposure trials are summarised in the table.

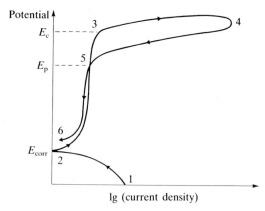

lg (current density)

Material type	% composition			$E_c - E_p$ (mV)	Weight loss (mg ml⁻¹)
	Cr	Ni	Mo		
316	18	8	3	80	9.2
304	18	8	—	110	10.8
410	12	—	—	220	26.7
430	18	—	—	160	18.9
446	25	—	—	140	17.4
Hastelloy C	16	59	16	0	0.2

(a) Describe the sequence of events taking place at the metal/electrolyte interface at each of the stages represented by 1–6 in the figure, and interpret the significance of the parameters denoted by E_{corr}, E_c and E_p.

(b) By means of a suitable graph compare the laboratory-derived data with those from the exposure trials.

(c) Hence comment on the usefulness of the experiments in determining the resistance to crevice corosion and the relative performance of the alloys listed.

(d) Show how the data can be used to make a useful prediction of the crevice corrosion resistance of an alloy containing 15 per cent chromium and 4 per cent nickel.

P18.21 A number of notched specimens were used in an experimental programme in which they were subjected to constant tensile load equal to 0.5 times the yield stress, while immersed in an electrolyte of pH 6 at 25 °C. A different potential was applied to each specimen such that both anodic and cathodic current densities (i) were obtained, and the times to failure (t_f) recorded. The data obtained from the experiments are summarised in the table.

$i(mA\ mm^{-2})$	$t_f(min)$	i	t_f
+8.0	12	−0.33	415
+5.1	13	−2.4	92
+3.0	15.5	−4.1	39
+1.2	31	−7.5	21
0	90	−8.8	19

(a) Plot a graph of i versus lg (t_f).

(b) Interpret the significance of the graph in relation to possible corrosion failure modes.

(c) Describe the mechanisms of the corrosion processes involved.

P18.22 (a) Discuss the contribution which the 'stress-corrosion spectrum', as devised by Parkins, makes to the understanding of stress-corrosion cracking.

(b) Corrosion fatigue behaviour has traditionally been examined using endurance testing and the creation of $S-N$ curves.

(i) Explain how this would be done for a structural steel in a moist atmosphere.

(ii) Draw a diagram of the sort of results you would expect to obtain, compared to the performance of the same material in dry air.

(iii) Highlight the significant features of the diagram you have drawn.

(c) Before constructing an offshore platform, corrosion fatigue tests were carried out on the structural steel which was to be used. Testing methods used the principles of linear elastic fracture

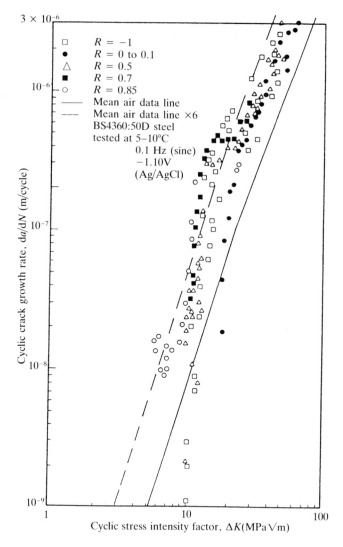

mechanics to quantify susceptibility to corrosion fatigue. The figure shows the experimental data obtained from the tests which used steel of the following composition: Fe, 0.17 per cent C, 0.35 per cent Si, 1.35 per cent Mn, in sea water. Specimens were of the single-edge-notched type. Explain the significance of:

(i) the axes;
(ii) the values of R;
(iii) the frequency;
(iv) the applied potential.

P18.23 In a programme of experiments to investigate the corrosion fatigue behaviour of a ferrous alloy in sea water, a series of 17 compact tension test specimens were manufactured to conform with the specifications required for a valid plane strain corrosion fatigue test. The test pieces, of width $W = 50$ mm and thickness $B = 25$ mm were pre-cracked such that the ratio of crack length to width, $a/W = 0.50$. Next they were exposed to the following environments:

1. Specimens 1–8, dry air at 298 K.
2. Specimens 9–17, sea water at 298 K.

Cyclic loading patterns were applied such that in every case, frequency $= 0.1$ Hz and the ratio of minimum to maximum load, $R = P_{min}/P_{max} = 0$. The table lists for each specimen the maximum load and the calculated crack growth rate in the appropriate environment. If the plane strain intensity factor $K_I = PY_2/BW^{\frac{1}{2}}$ and Y_2, the stress intensity factor coefficient $= 9.60$ for $a/W = 0.50$:

(a) Plot a graph which can be used to predict the susceptibility of the alloy to corrosion fatigue in sea water.

(b) Identify *three* important points on your graph and discuss the significance of each in relation to the mechanism of crack growth under the conditions used for the test.

(c) Given that such a material/environment combination is unavoidable, discuss the operating conditions which you consider would lead to acceptable corrosion fatigue resistance.

Specimen number	P_{max} (kN)	da/dN (mm cycle^{-1})	Specimen number	P_{max} (kN)	da/dN (mm cycle^{-1})
1	2.10	1.2×10^{-7}	9	1.43	1.3×10^{-7}
2	2.39	5.5×10^{-7}	10	1.63	5.0×10^{-7}
3	3.49	1.8×10^{-6}	11	2.10	1.3×10^{-6}
4	5.82	7.25×10^{-6}	12	3.49	4.8×10^{-6}
5	12.52	6.5×10^{-5}	13	5.82	1.7×10^{-5}
6	21.00	2.7×10^{-4}	14	12.52	1.1×10^{-4}
7	45.10	2.5×10^{-3}	15	21.00	4.0×10^{-4}
8	58.23	1.0×10^{-2}	16	45.10	2.6×10^{-3}
			17	58.23	1.0×10^{-2}

P18.24 (a) List *three* factors which control the rate of atmospheric corrosion on steel, and give brief details of the effect each has on the rate.

(b) A number of precision components made from mild steel are to be transported by sea. They will travel and be stored together in a single closed wooden case which has a hinged lid. On arrival in the islands the case will be placed in an unheated warehouse. Individual components will be removed from the case at approximately three-weekly intervals over a period of two years. It has been suggested that perforated containers of vapour-phase inhibitors (VPI) placed in the case will give adequate protection to the components during transit and storage.

(i) What type(s) of VPI would you recommend for this purpose? Mention any special characteristics which make them suitable for the task

(ii) What additional measures would be necessary to make the VPIs effective?

(iii) How would the decision to use VPIs be affected if the components were cadmium plated?

(iv) Explain the roles of the cation and anion in a VPI.

(v) Describe the mechanism of the particular corrosion hazard which wood presents to such components.

P18.25 The figure shows details of the cooling system for an engine in an open, wooden hulled launch operating in estuarine waters which are polluted by ammonium and sulphide ions.

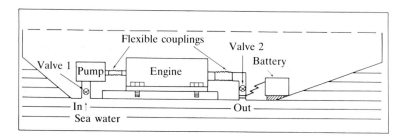

The pipe system is made from 70/30 brass tube with cast 60/40 brass elbow and tee joints. There are two flexible, stainless steel bellows adjacent to the engine which has a grey cast iron cylinder block. The engine is bolted to wooden bearers using mild steel fasteners.

The pump has a cast iron body, a stainless steel shaft and a nickel aluminium bronze impeller, while the valves are made with leaded bronze bodies and brass trims.

A copper clip is used to connect the earth from the battery to the outlet pipe just before the pipe passes through the hull.

If the engine block must be made from cast iron, identify four areas in which corrosion is likely to occur and suggest remedial actions which will minimise corrosion damage.

P18.26 The figure shows a section from the machinery space in a bulk carrier. A cold, sea-water pipe passes through the bulkhead.

(a) Identify the corrosion problems which are likely to occur.

(b) Suggest modifications which will minimise corrosion damage.

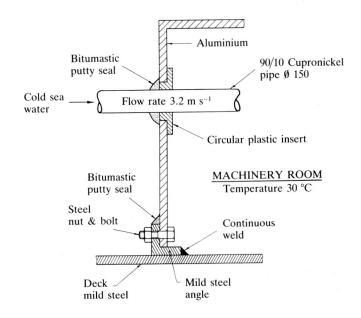

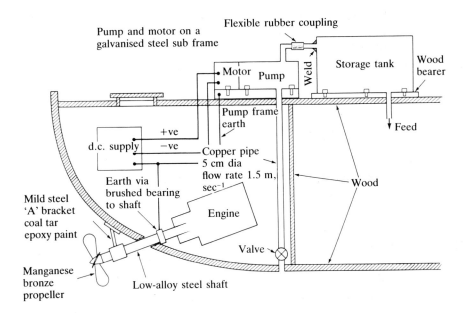

P18.27 The figure shows the arrangement of the stern fittings of a sea-going wooden hulled vessel. The electrically driven pump fills the water storage tank at two-day intervals, working for about two hours on each occasion. The system has a projected life of 10 years. The table lists some components and their materials.

Component		Material
Pump	Body	Cast iron
	Impeller	Nickel aluminium bronze
	Shaft	Mild steel
	Motor case	Aluminium
Valve	Body	Nickel aluminium bronze
	Trim	Cupronickel (Alloy 400)
Storage tank		Mild steel, galvanised inside and out to a thickness of 130 μm
Fasteners	Pump and tank	Cadmium plated mild steel

(a) Identify the corrosion problems which are likely to occur.
(b) Comment on the protection measures already taken.
(c) Suggest modifications which will further minimise corrosion.

P18.28 The figure shows the layout of a subsidiary heat exchanger unit for a large land-based power station sited on an estuary. The unit has a projected life of 20 years and is closed for major overhaul at two-yearly intervals. The pipe sizes, flow rates and the order of the valves and pumps were formulated in a computer aided design study, these cannot be changed. The vapour side of the condenser is compatible with all conceivable tube/tube plate materials. The table lists some of the components and their materials. The system is in continuous use.

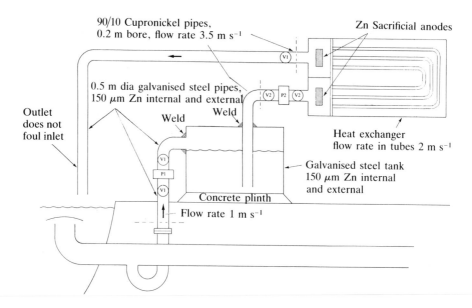

Component		
Estuarine water	Oxygen level high; pH 8.0; NH_4^+ and S^{2-} present as pollutants; inlet temperature 4–10 °C, outlet 40 °C	
Valve V1	Body	Cast iron
	Trim	Brass
Valve V2	Body	Leaded bronze (Cu + Sn 6%, Zn 4%, Pb 2%)
	Trim	Alloy 400 (Ni + Cu 32%, Fe 1%, Mn 1%)
Pump P1	Body	Cast iron
	Impeller	Nickel aluminium bronze (Cu + Al 8%, Ni 4.5%)
	Shaft	316 Stainless steel
Pump P2	Body	Nickel aluminium bronze (Cu + Al 8%, Ni 4.6%)
	Impeller and shaft	Alloy 400 (Ni + Cu 32%, Fe 1%, Mn 1%)
Heat exchanger	Water box	Low-alloy steel with Zn sacrificial anodes
	Tube plate	Naval brass (Cu + Zn 39%, Sn 1%)
	Tubes	Aluminium brass (Cu + Zn 29%, Al 2%, As 0.05%)

Identify the corrosion problems which are likely to occur. Comment on the protection measures already taken and suggest modifications to further minimise corrosion.

P18.29 It is proposed to use unstabilised austenitic stainless steel to construct a temporary staging for use in a tidal basin. Tubes will be used for the general platform, held together by clamps which are galvanically compatible with the steel in this environment. The deck plates will be welded to the tubes. It is expected that the staging will be in place for 9 to 12 months.

(a) Describe the development of *two* major corrosion phenomena you would expect to find in the structure.

(b) Give details of the methods normally used to control the corrosion you have described.

(c) What materials and corrosion control methods would you recommend for such a structure?

P18.30 The figure shows part of the superstructure of an offshore platform and a sea-water cooling system.
(a) List the locations in which you would expect to find corrosion.
(b) Describe the measures you would take to control the corrosion you have identified in (a).

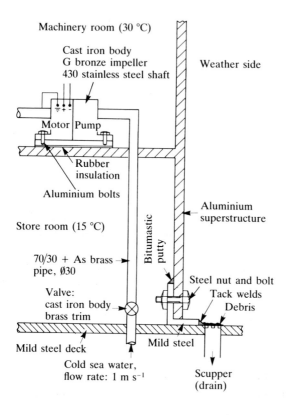

P18.31 (a) Describe the mechanisms whereby the semi-permeable nature of paints may give rise to breakdown of paint films on mild steel plates in a marine environment.
(b) The corrosion problems caused by diffusion through paint films on mild steel plates may be alleviated by the incorporation of sacrificial or inhibitive pigments in the primer. Describe the mechanisms of the corrosion protection afforded by these two classes of pigments giving examples of each.
(c) The design for a low-alloy steel shaft called for a micro-cracked chromium coating in preference to conventional chromium plating. Use diagrams to compare and contrast the corrosion behaviour of each coating, stating clearly the advantages that the designer hoped to gain.

P18.32 One method of protecting the hull of a ship is to use impressed current cathodic protection. For a system which uses lead/silver anodes rated at 20 A.

(a) Describe the nature, function and location of *all* the components.

(b) Outline the theoretical and practical principles of the design.

(c) Using the data given below, calculate the minimum number of anodes necessary to provide the protective current.

(d) The lead anodes do not dissolve. Suggest what the electrode reactions might be.

Data: Wetted surface area: Hull and appendages = 1390 m²
Propellers = 18 m²

Current requirements for protection:
Hull and appendages = 32 mA m⁻²
Propellers = 540 mA m⁻²

P18.33 It is required to protect the hull and appendages of a ship with aluminium alloy sacrificial anodes in longitudinal arrays. The designer has specified the use of 85 sacrificial anodes, each semi-cylindrical of diameter 100 mm and length 400 mm. If the designer's specifications are carried out:

(a) Estimate the maximum length of time which may elapse before the spent anodes must be replaced, assuming that current requirements are just satisfied.

(b) Calculate the current output of the anodes and hence comment on the designer's specifications.

(c) Give details of any approximations or assumptions implicit in your method of calculation. Discuss the importance of any errors which may have arisen from them.

(d) Briefly describe where you would expect the sacrificial anodes to be located on the hull.

Assume that all anodes corrode at the same rate.

Data: Total wetted surface area: Hull and appendages = 972 m²
Propellers = 14 m²

Current requirement for protection:
Hull and appendages = 32 mA m⁻²
Propellers = 540 mA m⁻²

Other useful data may be found in Table 16.1.

P18.34 (a) Sketch a plan of a modern ship on which is shown the layout of an impressed current cathodic protection system. Label all the components and on the circuit diagram indicate the direction of the electron flow.

(b) Describe in detail the nature and role of *each* component.

(c) Over a period of nine months during the operation of the ship, problems were experienced in obtaining stability of the protection potential. After several months of high current readings it was decided to switch the system off. Upon subsequent docking and inspection, damage was found in local areas on the hull. Describe where you would expect to find this damage and suggest reasons why it might have occurred.

P18.35 Two methods of predicting the current requirement for the ICCP of a coated, buried pipeline have been proposed. Method 1 assumes that the contractor is likely to apply the coating with an efficiency of 95 per cent and that a current density of 12 mA m^{-2} is required for protection. Method 2 uses the assumption that the coating will completely cover the pipe and that a current density of less than 130 μA m^{-2} will be sufficient for a pipe of diameter >760 mm. Assume that you must design a protection system for a pipeline similar to those in E18.7.
(a) Calculate the protection current density for a line 75 km long and 1 m diameter.
(b) With the case history described in E18.7 in mind, comment upon the suitability of each method for use in the design.

P18.36 Imagine that you have been tasked to decide upon a cathodic protection system for a new type of small ship. The choice is between:
1. A commercially supplied impressed current system costing £20,000 per ship, all inclusive.
2. A sacrificial system, fitted by labour in your own dockyard for £10,000 per ship, excluding the cost of the anodes.
  Rectangular zinc anodes measuring 400 × 100 × 50 mm can be purchased for £45 each. They have an output of 6.5 A/m^2 and a consumption rate of 1.518 × 10^{-3} m^3 Ay^{-1}. Current density requirements for protection are 32 mA m^{-2} for hull and fittings, and 540 mA m^{-2} for propellers. The wetted surface areas are respectively 1000 m^2 and 15 m^2. Anodes are fixed with the largest face in contact with the hull. This area can be ignored in calculations.
(a) Calculate the current needed to protect the ship.
(b) Calculate the minimum number of anodes necessary to protect the ship. Hence compare the cost of the two cathodic protection systems.
(c) How long will the zinc anodes last before replacement is necessary?
(d) Discuss the other factors which would have to be taken into consideration before you would be able to state which system should be adopted.

Index

Index